BMA

Management of High Altitude Pathophysiology

Management of High Altitude Pathophysiology

Edited by

Kshipra Misra

Priyanka Sharma

Anuja Bhardwaj

ELSEVIER

ACADEMIC PRESS

An imprint of Elsevier

Academic Press is an imprint of Elsevier
125 London Wall, London EC2Y 5AS, United Kingdom
525 B Street, Suite 1650, San Diego, CA 92101, United States
50 Hampshire Street, 5th Floor, Cambridge, MA 02139, United States
The Boulevard, Langford Lane, Kidlington, Oxford OX5 1GB, United Kingdom

Notices
Knowledge and best practice in this field are constantly changing. As new research and experience broaden
our understanding, changes in research methods, professional practices, or medical treatment may become
necessary.

Practitioners and researchers must always rely on their own experience and knowledge in evaluating
and using any information, methods, compounds, or experiments described herein. In using such
information or methods they should be mindful of their own safety and the safety of others, including
parties for whom they have a professional responsibility.
To the fullest extent of the law, neither the Publisher nor the authors, contributors, or editors, assume
any liability for any injury and/or damage to persons or property as a matter of products liability,
negligence or otherwise, or from any use or operation of any methods, products, instructions, or ideas
contained in the material herein.

Library of Congress Cataloging-in-Publication Data
A catalog record for this book is available from the Library of Congress

British Library Cataloguing-in-Publication Data
A catalogue record for this book is available from the British Library

ISBN 978-0-12-813999-8

For information on all Academic Press publications
visit our website at https://www.elsevier.com/books-and-journals

Working together
to grow libraries in
developing countries

www.elsevier.com • www.bookaid.org

Publisher: John Fedor
Acquisition Editor: Mary Preap
Editorial Project Manager: Timothy Bennett
Production Project Manager: Punithavathy Govindaradjane
Cover Designer: Greg Harris

Typeset by SPi Global, India

Dedication

Dedicated
To
Dr. B. B. Sarkar

for his altruistic and inspirational services to Indian soldiers.

Contents

SECTION II   Herbal Therapeutics for High Altitude Problems

CHAPTER 7   *Ganoderma* sp.: The Royal Mushroom for
High-Altitude Ailments 115
Anuja Bhardwaj, Kshipra Misra

CHAPTER 8   *Curcuma* sp.: The Nature's Souvenir for
High-Altitude Illness...153
Jigni Mishra, Anuja Bhardwaj, Kshipra Misra

SECTION III Nonherbal Therapeutics for High-Altitude Illness

SECTION IV   Nonmedical Therapies for High Altitude Ailments

Contributors

Rajesh Arora   Department of Biochemical Sciences (DBCS), Defence Institute of Physiology and Allied Sciences (DIPAS), Delhi, India

Murugesan Balamurugan   Department of Chemistry, Biomedical Research Lab, VHNSN College (Autonomous), Virudhunagar, India

Anuja Bhardwaj   Chemistry Division, Department of Biochemical Sciences (DBCS), Defence Institute of Physiology and Allied Sciences (DIPAS), Delhi, India

Jonathan C. Claussen   Mechanical Engineering, Iowa State University, Ames, IA, United States

Arul J. Duraisamy   Kresge Eye Institute, Wayne State University School of Medicine, Detroit, MI, United States

Shefali Gola   Department of Basic and Applied Sciences, School of Engineering, GD Goenka University, Haryana, India

Chandran Karunakaran   Department of Chemistry, Biomedical Research Lab, VHNSN College (Autonomous), Virudhunagar, India

Deepti Majumdar   Ergonomics Division, Defence Institute of Physiology and Allied Sciences (DIPAS), Delhi, India

Preenon Majumdar   Kalinga Institute of Medical Sciences, Bhubaneshwar, India

Dhurjati Majumdar   Defence Research and Development Organization, New Delhi, India

Manimaran Manickam   Research and Development, PathGene Healthcare Private Limited, Tirupathi, India

Jigni Mishra   Chemistry Division, Department of Biochemical Sciences (DBCS), Defence Institute of Physiology and Allied Sciences (DIPAS), Delhi, India

Kshipra Misra   Department of Biochemical Sciences (DBCS), Defence Institute of Physiology and Allied Sciences (DIPAS), Delhi, India

Mamta Pal   Division of Forensic Science, School of Basic and Applied Sciences, Galgotias University, Greater Noida, India

Syed Rahamathulla   Research and Development, PathGene Healthcare Private Limited, Tirupathi, India

Rakhee   Department of Chemistry, University of Delhi, Delhi, India

Paulraj Santharaman   Department of Chemistry, Biomedical Research Lab, VHNSN College (Autonomous), Virudhunagar, India

Priyanka Sharma   Cardiorespiratory Department, Defence Institute of Physiology and Allied Sciences (DIPAS), Delhi, India

Raj K. Sharma   Department of Chemistry, University of Delhi, Delhi, India

Sushil Kumar Singh   Functional Materials Group, Solid State Physics Lab, Defence Research and Development Organization, Timarpur, India

Preface

Towering, majestic mountains, with their snow-capped peaks and rarefied atmosphere, have always fascinated, inspired, and attracted people. Many people travel to and stay at high altitudes for varying lengths of time, either for recreation (such as tourists and mountaineers) or for work (such as armed forces), exposing them to hypoxia and the entailed risk of acute mountain sickness. Such an exposure, at times, can lead to high-altitude pulmonary edema (HAPE) and high-altitude cerebral edema (HACE) that could cause serious health complications. Every person, regardless of age, gender, or physical characteristics, exposed to high altitudes is vulnerable to high-altitude-induced hypobaric hypoxia, which is an irreversible, lifelong condition, imposing severe physiological distress. Hypoxia is caused by rarefied atmosphere at high altitudes, reflected in low barometric pressure, and the consequent low oxygen density in inhaled air compared to the oxygen density at sea level. Oxygen density progressively declines with the increasing altitude, causing correspondingly higher stress to the human physiological systems, because a steady, uninterrupted supply of oxygen is essential for normal functioning of mitochondrial metabolism.

Given the almost inevitable physiological consequences of hypoxia, it is a major challenge to maintain the normal level of functional performance of people who live and work at high altitude. Though the instinctive adaptation of the physical and mental capabilities of human beings in the face of such a harsh working environment could be viewed as mitigating the risk to an extent, that is not optimal, and people in that environment do become the victims of the deleterious impact of hypoxia to varying degrees. With a view to combating the multiple health complications arising from hypobaric hypoxia, several therapies have been developed over the years, and some of them are being practiced worldwide.

The main objective of writing this book is to provide the readers a holistic view of prevalent environmental conditions at high altitudes, including hypoxia and low temperatures, their effects on human physiology, and the various measures adopted for the effective management of the pathophysiological impact of that environment. We have tried to present our views on the current and emerging scenario in the field of management of high-altitude pathophysiology. The scheme of presentation in the book is summarized below.

Section I presents an overview of the unique environmental features observed at the high altitudes, and their physiological effects on human beings.

Sections II and III highlight herbal and nonherbal therapeutics, respectively, being practiced and researched for the management of high-altitude pathophysiology. Each chapter of these sections seeks to present comprehensively the current state of research in this area.

The role of exercise and yogic interventions has been found to be invaluable in maintaining physical wellbeing at the high altitudes. This aspect forms the subject matter of Section IV and is discussed in detail in its two chapters.

We would like to thank all those, too numerous to name here, who have been a source of inspiration for initiating this project, and those who helped us in completing this book. We are especially grateful to Dr. A. K. Datta, former Chief Controller, R& D, Defence Research & Development Organisation, India for his enduring support and encouragement all along. The authors are thankful to the Director of Defence Institute of Physiology and Allied Sciences (DIPAS) for granting permission to write the book. We would like to thank all the authors who readily agreed to contribute their research findings to this book and without whose support this book could not have materialized. We are also grateful to Elsevier publishers for providing us an opportunity to publish this book. Last but not the least, we are grateful to the Indian soldiers whose lives inspired us to do something, however small and indirect, toward improving their wellbeing in the midst of their harsh working conditions at high altitudes.

This small effort in compiling current research in the area of high-altitude pathophysiology, we hope, also will throw some light on the direction of further research needed in this field.

Kshipra Misra, Priyanka Sharma, Anuja Bhardwaj

Human Performance at High Altitude

CHAPTER 1

High Altitude and Hypoxia

Priyanka Sharma*, Kshipra Misra†

*Cardiorespiratory Department, Defence Institute of Physiology and Allied Sciences (DIPAS), Delhi, India, †Department of Biochemical Sciences (DBCS), Defence Institute of Physiology and Allied Sciences (DIPAS), Delhi, India

List of Abbreviations

mmHg millimeters of mercury
PO_2 partial pressure of oxygen in ambient air

1.1 INTRODUCTION

Altitude has different meaning based on the context in which it is used. In general, altitude is vertical distance between a reference datum and a point or object. The term altitude, also called true altitude, commonly is used to mean the height of a location above sea level. The distance of a point from the ground below is sometimes called absolute altitude, as described in *ISO 6709:2008(en) preview* (ISO, 2008). Altitude is related to air pressure and can be determined by measuring air pressure with an altimeter. As altitude rises, air pressure drops because of two reasons: First, Earth's gravitational pull keeps air as close as possible to its surface. Second, the number of molecules in the air decreases as the altitude increases, making the air less dense. Less dense air also is cooler because fewer molecules have much less chance to collide with each other. Altitude could be further divided based on the distance from the sea level (Fig. 1.1). Regions on the Earth's surface (or in its atmosphere) that are high above mean sea level are referred to as "high altitudes." High altitudes sometimes are defined to begin at 8000 ft above sea level (Chamberlin, 2015).

Generally, prolonged stay at high altitudes as a part of adventure sport or country's defense means living in stressful environment leading to a need to study the effects of such an alien locale. The most significant and connatural factor of high altitudes is hypoxia. Hypoxia or rarefied atmosphere with low oxygen availability affects both physical and mental performances of people because oxygen is even more essential to life than food and water.

3

Management of High Altitude Pathophysiology. https://doi.org/10.1016/B978-0-12-813999-8.00001-X

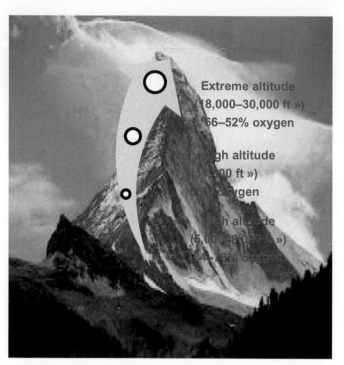

FIG. 1.1
Percentage of oxygen at different altitudes.

This chapter introduces readers to the concepts of high altitudes, the worldwide scenario of high altitudes, and the related hypoxia (i.e., hypobaric hypoxia). This basic knowledge is essential for better understanding of high-altitude pathophysiologies and their management.

1.2   HIGH ALTITUDES: WORLDWIDE SCENARIO

High altitudes of mountains have always been a source of great fascination and inspiration for people. The serenity and calmness of mountains allure thinkers, spiritualists, and philosophers; the exigent environment challenges mountaineers, and unexplored resources lure people to inhabit areas of even higher reaches of mountains. Establishment of a major civilization in the Himalayan region (3500–4500 m) is a well-known example of successful adaptation to high altitude (Aldenderfer, 2003; Shi et al., 2008; Wu, 2001; Yuan et al., 2007; Zhao et al., 2009). A large population also lives at 4500 m in the Peruvian Andes, and caretakers of a mine in Chile have lived at nearly 6000 m. Lhasa, in Tibet, altitude 3658 m, has more than 130,000 inhabitants. The highest point on Earth is the summit of Mt. Everest (8848 m), and well-acclimatized climbers can just reach that altitude without oxygen (Encyclopedia.com, 2001).

1.3 HIGH ALTITUDES ENVIRONMENT AND HYPOXIA

The mountain environment is composed of several factors that are alien to plain dwellers and therefore evokes modifications in the physiology and homeostasis. The most crucial and inherent factors of such an environment are hypoxia (lack of oxygen because of thinning of air), cold (for every 150 m rise, the temperature drops 1°C), wind (adds to the effect to cold leading to chill), increased solar radiation, ozone concentration, absence of flora and fauna, isolation from civilization. Individually, these factors are potent psycho-physiological stress-producing elements. When they exist together in variable proportions in different geographical locations and seasons they evoke an alarming challenge to human adaptability and capacity to survive. Hypoxia, however, remains the primary factor defining all other stress factors. At high altitude, it arises because atmospheric pressure is lower than that at sea level. This situation exists because of two physical effects: gravity, which causes the air to be as close to the ground as possible; and the heat content of the air, which causes the molecules to bounce off each other and expand (Hackett and Roach, 2001). High-altitude hypoxia persistently affects people living in this environment because oxygen is even more essential to life than food and water (Julian et al., 2009).

1.3.1 Defining the Term "Hypoxia"

Van Liere and Stickney (1963) wrote a book about hypoxia in which they described anoxemia as deficiency of oxygen in the blood (Van Liere and Stickney, 1963). Other scientists, however, proposed that body suffers deficiency in oxygen even though there is no deficiency in the blood. For such conditions, Barcroft (1920) suggested the term "anoxia," a condition of oxygen deprivation in the body regardless of the cause. In his studies, he identified three types of anoxia: anoxic, anemic, and stagnant (Barcroft, 1920). The term "anoxia," however, also was objected to because it actually meant "without oxygen/or total lack of oxygen," a condition that does not exist. Sometimes, hypoxia is confused with another term, "asphyxia." It should be clearly understood that in asphyxia, carbon dioxide accumulates in the body's lungs and tissues. The most acceptable term was determined to be "hypoxia" [*hypo* = a Greek word meaning under/sub and oxygen] that means less than a normal amount of oxygen. This is summarized in Fig. 1.2. Hypoxia can be combined with the terms normobaric, hypobaric, stagnant, anoxic, anemic, and histotoxic to express the exact type of oxygen deficiency.

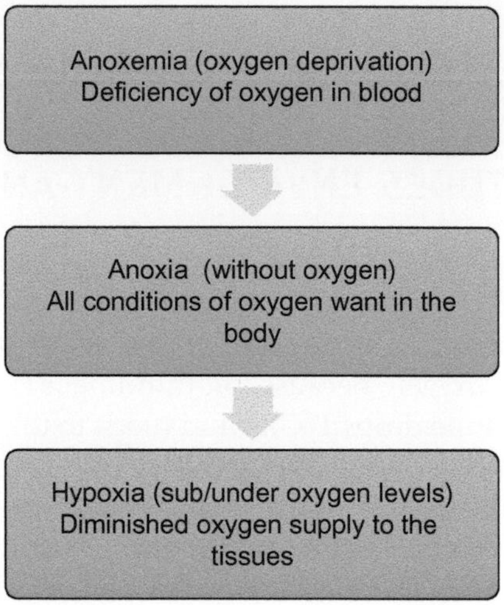

FIG. 1.2

Origin of the term "hypoxia."

1.3.2 Types of Hypoxia

Various researchers describe the different types of hypoxia depending on the type of oxygen deficiency (Van Liere and Stickney, 1963; Millet et al., 2012).

Normobaric hypoxia: When there is reduction in inspired fraction of oxygen below 20% although the barometric pressure remains at 760 mmHg.

Hypobaric hypoxia: When the body is deprived of a sufficient supply of oxygen from the air to supply the body tissues whether in quantity or molecular concentration.

Anoxic hypoxia: When there is lack of oxygen in arterial blood that leads to unsaturation of hemoglobin to its normal extent; sometimes called hypoxemia or arterial hypoxia.

Anemic hypoxia: When oxygen tension (partial pressure of oxygen) in arterial blood (normally 75–100 mmHg) is sufficient but it cannot be carried because of lack of functional hemoglobin.

Stagnant hypoxia: When arterial blood holds a normal amount of oxygen under normal tension but it is not given off to the tissues; arises because of insufficient supply of blood in capillaries because of some heart problem.

Histotoxic hypoxia: When the tissues are unable to use the oxygen that is supplied to it as happens in cyanide poisoning.

1.4 HYPOBARIC HYPOXIA AND PHYSIOLOGY

One of the major stresses faced at high altitudes is hypobaric hypoxia, which occurs as the direct result of the nearly exponential fall in barometric pressure as one ascends from sea level. Because the relative concentration of oxygen in the troposphere (lowest atmospheric layer) is constant at 20.93%, the partial pressure of oxygen in ambient air (PO_2) for a given elevation is obtained by multiplying 0.2093 by the corresponding barometric pressure. At sea level, PO_2 is 0.2093 multiplied by 760 mmHg, or 159.1 mmHg. At the summit of Mt. Everest, the ambient PO_2 is only 52.9 mmHg. Such low pressures of oxygen (hypoxia) result in stress-related ailments (Hackett and Roach, 2001; Schoene, 2001). In the next chapter, we will discuss these ailments, their causes, and effects.

1.5 CONCLUSION

This chapter introduced the readers to the concepts of high altitudes, the worldwide scenario of high altitudes, and the related hypoxia (i.e., hypobaric hypoxia). This chapter is crucial for better comprehension of high-altitude associated pathophysiologies and the various approaches applied or applicable for their management.

References

Aldenderfer, M.S., 2003. Moving up in the world: archaeologists seek to understand how and when people came to occupy the Andean and Tibetan plateaus. Am. Sci. 91 (6), 542–550.

Barcroft, J., 1920. On anoxaemia. Lancet 2 (4), 485.

Chamberlin, R., 2015. Survival medicine tips techniques & secrets. The Prepper Pages.com. February 4th 2015. Available from: https://www.prepperwebsite.com/. A survival medicine & medical preparedness blog sharing tips, techniques and secrets for building the perfect first aid kit and using it to treat injuries and illnesses preppers encounter during disasters (accessed 23 October 2017).

Encyclopedia.com, 2001. Altitude facts, information, pictures. Encyclopedia.com articles about altitude. Available from: http://www.encyclopedia.com/medicine/encyclopedias-almanacs-transcripts-and-maps/altitude (accessed 23 December 2017).

Hackett, P.H., Roach, R.C., 2001. High-altitude illness. N. Engl. J. Med. 345 (2), 107–114.

ISO 6709:2008(en) preview. www.iso.org. ISO. 2008. Retrieved 8 June 2016.

Julian, C.G., Wilson, M.J., Moore, L.G., 2009. Evolutionary adaptation to high altitude: a view from in utero. Am. J. Hum. Biol. 21 (5), 614–622.

Millet, G.P., Faiss, R., Pialoux, V., 2012. Point/counterpoint: hypobaric hypoxia induces/does not induce different responses from normobaric hypoxia. J. Appl. Physiol. 112 (10), 1783–1784.

Schoene, R.B., 2001. Limits of human lung function at high altitude. J. Exp. Biol. 204 (18), 3121–3127.

Shi, H., Zhong, H., Peng, Y., Dong, Y.L., Qi, X.B., Zhang, F., Liu, L.F., Tan, S.J., Ma, R.Z., Xiao, C.J., Wells, R.S., 2008. Y chromosome evidence of earliest modern human settlement in East Asia and multiple origins of Tibetan and Japanese populations. BMC Biol. 6 (1), 45.

Van Liere, E.J., Stickney, J.C., 1963. Hypoxia. Library of Congress Catalog Card Number: 63-16722, The University of Chicago Press, Chicago and London.

Wu, T., 2001. The Qinghai–Tibetan plateau: how high do Tibetans live? High Alt. Med. Biol. 2 (4), 489–499.

Yuan, B., Huang, W., Zhang, D., 2007. New evidence for human occupation of the northern Tibetan plateau, China, during the late Pleistocene. Chin. Sci. Bull. 52 (19), 2675–2679.

Zhao, M., Kong, Q.P., Wang, H.W., Peng, M.S., Xie, X.D., Wang, W.Z., Jiayang, Duan, J.G., Cai, M.C., Zhao, S.N., Cidanpingcuo, Tu, Y.Q., Wu, S.F., Yao, Y.G., Bandelt, H.J., Zhang, Y.P., 2009. Mitochondrial genome evidence reveals successful late Paleolithic settlement on the Tibetan plateau. Proc. Natl. Acad. Sci. U. S. A. 106, 21230–21235.

CHAPTER 2

High Altitude Ailments: Causes and Effects

Shefali Gola*, Kshipra Misra†

*Department of Basic and Applied Sciences, School of Engineering, GD Goenka University, Haryana, India, †Department of Biochemical Sciences (DBCS), Defence Institute of Physiology and Allied Sciences (DIPAS), Delhi, India

Abbreviations

AMS	acute mountain sickness
ENaCs	epithelial sodium channels
EPO	erythropoietin
Erg-1	ETS-related gene-1
GAPDH	glyceraldehyde phosphate dehydrogenase
HA	high altitudes
HACE	high-altitude cerebral edema
HAPE	high-altitude pulmonary edema
HIF	hypoxia-inducible factor
HPV	hypoxic pulmonary vasoconstriction
NF-κB	nuclear Factor Kappa B
NSAIDs	nonsteroidal antiinflammatory drugs
PaO_2	arterial partial pressure of oxygen
PiO_2	partial pressure of inspired O_2
PO_2	pressure of oxygen
p53	tumor protein p53
RNFL	retinal nerve fiber layer
VEGF	vascular endothelial growth factor

2.1 INTRODUCTION

The association of humans with the mountains is almost as old as mankind. Mountain ranges comprise approximately one-fifth of Earth's land surface and all are inhabited to some extent. About one-tenth of the world's people derive their life support from mountains. In spite of long strides on scientific and technological fronts, the challenges offered by high-altitude (HA) conditions remain to be overcome. The misty weather and distant high peaks attract people and promise great adventure. Millions of people sojourn to HA each

9

Management of High Altitude Pathophysiology. https://doi.org/10.1016/B978-0-12-813999-8.00002-1

year for recreation and adventure purposes and test the limits of human endurance. The numbers of mountaineers, rescue teams, defense personnel, and natives who travel to extreme HA regions, especially Alps in Europe, Himalayas in Asia, Andes in South America, and Rockies in the United States, have increased enormously in the past decades (Peacock, 1998). Understanding the physiological effects of HA exposure and discovering how to ameliorate these effects is of chief importance for securing the health and livelihood of the large number of humans subjected to these stressful environmental conditions.

HA ranges from intermediate altitude (1500–2500 m), high altitude (2500–3500 m), very high altitude (3500–5800 m), to extreme altitude (>5800 m). It is deemed that human life is not possible above 5500 m permanently, although moderate altitudes sometimes can be tolerated without supplementary oxygen (Hackett, 1999a; Hackett and Roach, 2001a). HA is chiefly characterized by several adverse environmental conditions, including low barometric pressure, low air humidity, low atmospheric temperature, high aridity, high ultraviolet radiation, and, most prominently, hypobaric hypoxia. At HA, most of the undesirable physiological effects result from the decrease in atmospheric pressure and low availability of oxygen, additional deterioration results from extreme cold, exposure to ionizing radiation, and high wind velocity. The unenviable pathophysiologies encountered at HA are discussed in subsequent sections.

2.1.1   Hypobaric Hypoxia at High Altitudes

The main environmental stress at HA is hypobaric hypoxia, which is caused by the fall in barometric pressure with increasing altitude and the fewer oxygen molecules in a breath of air as compared with sea level. Although oxygen availability in air is 21% at any altitude, the partial pressure and thus the bioavailability of oxygen decreases with altitude, causing a condition called hypobaric hypoxia. The decreased barometric pressure of the ambient atmosphere at HA results in diminished alveolar oxygen tension and, as a result, arterial partial pressure of oxygen (PaO_2) falls significantly (Fedele et al., 2002). Reduced oxygen partial pressure causes the arterial hemoglobin to be inadequately saturated with oxygen (Peacock and Jones, 1997), resulting in hypoxia that not only limits human physical performance (Pugh, 1964; West, 1988) but also brings out many physiological changes and putting the body in further jeopardy. For example, at the altitude of La Paz, Bolivia (4000 m), partial pressure of inspired O_2 (PiO_2) is 86.4 mm Hg, which is equivalent to breathing 12% oxygen at sea level. The decrease in PaO_2 is the most significant environmental change caused by HA; therefore, HA environment commonly is referred to as one of hypobaric hypoxia. The ascent rate and period of stay at HA greatly influence the effects of hypobaric hypoxia. These effects progressively become more intense with increasing altitude and period of stay, thus stressing the biological systems (Fedele et al., 2002).

2.1.2 Physiological Adaptations at High Altitudes: The Acclimatization Process

An organism's survival often depends on its ability to acclimatize to environmental stresses. A typical example of adaptation to a stressful environment is acclimatization to HA. The process of acclimatization allows the gradual adjustment of individuals to hypoxia and enhance their survival and performance under hypoxia (Paralikar and Paralikar, 2010). On ascent to HA, the oxygen cascade in human body get disrupted. Acclimatization to hypoxic condition involves several organ systems and tissues of the human body. These physiological responses commence after ascent to an altitude of about 1500 m, generally within the first few minutes to hours. The onset of HA-induced hypoxia triggers an immediate rise in alveolar ventilation, which is regulated by carotid body, causing an increase in arterial oxygen content. Physiological effects observed at HA include increased production of hemoglobin along with increase in hematocrit and oxygen-carrying capacity of blood; elevated 2,3-bisphosphoglycerate production; pulmonary vasoconstriction; raised mass of lung and liver; increased mass of left ventricle; elevated tidal volume and rate of ventilation; increased capillary density; anorexia and subsequent weight loss (Aaron and Powell, 1993; Benjamini and Hochberg, 1995; Branco et al., 2006; Gopfert et al., 1996; Mortola, 1999; Shukla et al., 2005). Furthermore, while the above physiological changes are proceeding, body alters its metabolic capacity to reduce the oxygen requirement. This involves elevated anaerobic glycolysis and glucose consumption, decreased rate of metabolism, and decrease in body temperature (Gautier, 1996; Hochachka et al., 1996; Semenza et al., 1994; Steiner and Branco, 2002).

Substantial information is available about the physiological acclimatization to hypoxic exposure, however, some studies have focused primarily on the changes involving principal molecular genetics. Several studies about the effects of hypoxia on genes and protein expressions have focused on the responses of cells to hypoxia exposure. Alterations in transcription factors, such as nuclear factor kappa B (NF-κB), tumor protein p53 (p53), AP-1, Myc family of proteins and ETS-related gene-1 (Erg-1), take place in response to cellular hypoxia. These transcription factors play a crucial role in the processes of cell development regulation, cell proliferation and differentiation, inflammation and apoptosis, and respond to various stressors (Cummins and Taylor, 2005; Kenneth and Rocha, 2008). One of the transcription factors that respond to hypoxic exposure is hypoxia-inducible factor (HIF). Under hypoxic conditions, the HIF-

1α subunit accumulates rapidly and dimerizes with the HIF-1β subunit, leading to the transcription of more than 70 hypoxia responsive genes (Hammond et al., 2002; Haouzi et al., 2008; Jungermann and Kietzmann, 2000). Many of these genes are linked with various physiological changes caused by hypoxia (Sarkar et al., 2003). For example, under hypoxic conditions, HIF triggers upregulation of erythropoietin (EPO), transferrin, genes related with angiogenesis, such as vascular endothelial growth factor (VEGF), and genes linked with various metabolic substrates, such as glyceraldehyde phosphate dehydrogenase (GAPDH) and pyruvate kinase, which enhance the uptake of glucose and anaerobic glycolysis (Bracken et al., 2003; Semenza, 2000). An insufficiency of HIF-1α, however, hinders long-term physiological responses to hypoxia. Therefore, both responses and physiological acclimatization to hypoxia are controlled immensely by the HIF pathway (Aimee et al., 1999).

2.1.3   Physical Factors at High Altitudes

In addition to hypoxia, HA also is characterized by cold ambient temperatures, increased UV radiation, and high wind velocity. The physical and mental performance at HA is highly affected by these physical and physiological stresses as well as such as dehydration and lack of antioxidant nutrients in the diet (Askew, 1995; Cymerman, 1996; Huey and Eguskitza, 2001).

2.2   HIGH-ALTITUDE AILMENTS

Altitude exposure and acclimatization long have been areas of research. The immediate (acute) effects of lowered ambient pressure of oxygen (PO_2) pertaining to the human response and the adaptations to prolonged exposure (chronic) are complex. Several body systems (i.e., muscular, cardiovascular, pulmonary, endocrine, hepatic) are affected by HA-induced hypoxia. High-altitude ailments represent physiological alterations associated with environmental challenges imposed by hypobaric hypoxia.

The term HA ailments or sickness is used mainly for a class of pathophysiological conditions caused by rapid ascent to HA of above 2500 m (Litch, 2007). The three main types of altitude sickness are acute mountain sickness (AMS), high-altitude cerebral edema (HACE), and high-altitude pulmonary edema (HAPE) (Basnyat, 2005; Chao et al., 1999; Fiore et al., 2010; Hartman-Ksycińska et al., 2016; Khodaee et al., 2016; Luks, 2015; Luks et al., 2017; Parise, 2017; Singh and Selvamurthy, 1993; Smedley and Grocott, 2013). Unacclimatized sojourners are at high risk from HA conditions. Cerebral and pulmonary anomalies are not subtle, and might lead to death if unrecognized or ignored. The various risk factors that are general to all altitude sicknesses are the absolute altitude, the speed of ascent, individual predisposition, and lack of acclimatization (Schneider et al., 2002) (Table 2.1).

Table 2.1 Pathophysiological Conditions at High Altitudes

Condition	Clinical Sign and Symptoms	Etiology
Acute mountain sickness (AMS)	Headache, anorexia, malaise, nausea, vomiting, fatigue, dizziness, weakness, difficulty in sleeping	Develops on rapid ascent to altitude of 2000–3000 m within 6–36 h
High-altitude cerebral edema (HACE)	Ataxia, altered consciousness and changes in mental status (confusion, impaired mentation, drowsiness, stupor), which can progress to coma	Occurs later than AMS at higher altitudes >3000–4000 m within 24–36 h
High-altitude pulmonary edema (HAPE)	Dyspnea even at rest, cough, decreased exercise performance, chest tightness/congestion, tachypnea, tachycardia, pulse oximetry reveals marked hypoxemia, wheezing	Commences after arrival at higher altitudes >3000 m in 2–4 days
High-altitude retinopathy	Optic disc swelling, vitreous hemorrhage, retinal and choroidal blood flow alteration and retinal hemorrhages, generally regresses within 2–3 weeks	Occurs at altitudes above 4000 m
Sleep disturbance	Poor sleep quality, increased arousals, restless sleep	Develops at altitude greater than 2500 m

Adapted from Fiore, D.C., Hall, S., Shoja, P., 2010. Altitude illness: risk factors, prevention, presentation, and treatment. Am. Family Phys. 82(9), 1103–1110; McFadden, D.M., Houston, C.S., Sutton, J.R., Powles, A.P., Gray, G.W., Roberts, R.S., 1981. High-altitude retinopathy. JAMA 245(6), 581–586; Paralikar, S.J., Paralikar, J.H., 2010. High-altitude medicine. Ind. J. Occupat. Environ. Med. 14(1), 6–12; Parise, I., 2017. Travelling safely to places at high altitude—understanding and preventing altitude illness. Aust. Family Phys. 46(6), 380–384; Rennie, D., Morrissey, J., 1975. Retinal changes in Himalayan climbers. Arch. Ophthalmol. 93(6), 395–400; Schoene, R.B., 2008. Illnesses at high altitude. Chest 134(2), 402–416; Weil, J.V., 2004. Sleep at high altitude. High Alt. Med. Biol. 5(2), 180–189.

2.2.1 Acute Mountain Sickness

Altitude sickness, also known as acute mountain sickness (AMS), altitude illness, hypobaropathy, or soroche, is a pathological effect of HA on humans. AMS might progress to HAPE or HACE, which are potentially fatal (Cymerman and Rock, 2009). Typical symptoms of AMS are anorexia, dizziness, fatigue, headache, nausea, sleep disturbance, and vomiting (Roach et al., 1993).

AMS occurs after at least 4–6 hours of ascent to an altitude above 2000–2500 m. The reported prevalence of AMS ranges from 8% to 25% at 2500–3000 m and from 40% to 60% at 4500 m (Honigman et al., 1993; Maggiorini et al., 1990; Montgomery et al., 1989). Although the pathophysiological mechanisms that cause AMS are unclear, AMS generally is most critical after the first night of ascent to a new and higher altitude. Its severity is increased and aggravated by strenuous physical exercise (Roach et al., 2000). The symptoms of AMS usually subside in 24–48 hours if further ascent and intense exercise is avoided. Further ascent to new altitude is not recommended if symptoms of AMS are still present, as HACE might ensue.

The physiological processes of severe AMS range from mild to moderate AMS, which is described by impaired gas exchange, hypoventilation, fluid retention and redistribution, and augmented sympathetic drive. An early finding in AMS is absence of normal high-altitude diuresis; however, this symptom is not always present (Paralikar and Paralikar, 2010). The mechanism of fluid retention by the kidneys might result from elevated levels of an anti-diurectic hormone, activation of rennin-angiotensin-aldosterone system, and an enhanced sympathetic drive (Swenson, 1997). Moderate to severe AMS is associated with edema of white matter in the brain. The study of Hackett et al. (1998) provided the most convincing evidence in this matter, and suggested that edema is vasogenic in origin along with an elevated endothelium permeability. The increased intravascular pressure or hypoxemia might be the cause of the leak (Hackett et al., 1998).

2.2.2   High Altitude-Induced Headache

Headache has been considered the most common and most prominent symptom at HA with the incidence of 25% at less than 2956 m and 47%–62% at 4928 m (Basnyat and Murdoch, 2003; Basnyat et al., 2006; Gertsch et al., 2004; Gertsch et al., 2010; Ziaee et al., 2003). The supply of oxygen to the brain decreases at HA and results in increased blood flow in the brain up to 26% (Severinghaus et al., 1966). The brain swells, leading to the traction of the meninges and headache because of local traction-irritable mechanoreceptors (Cymerman and Rock, 1994; Sanchez and Moskowitz, 1999). Furthermore, hypoxemia is one of the main causes of releasing inflammatory mediators, causing vasodilatation (Hartmann et al., 2000; Maggiorini et al., 2009; Strassman and Levy, 2006). The possible mechanism involved in HA-induced headache was proposed by Sanchez and Moskowitz (1999), who said that HA-induced headache is multifactorial, with various chemical and mechanical factors activating the trigeminovascular system. Triggering factors might include nitric oxide, arachidonic acid metabolites, serotonin, and histamine, which sensitize less in myelinated fibers conveying pain and accumulation proximity to trigger aminovascular fibers, causing headache. Response to nonsteroidal antiinflammatory drugs (NSAIDs) and steroids provide indirect evidence for involvement of the arachidonic-acid pathway and inflammation in the genesis of HA-induced headache (Broome et al., 1994; Burtscher et al., 1998; Ferrazzini et al., 1987).

2.2.3   High-Altitude Cerebral Edema

HACE usually is considered to be the end stage of AMS and is characterized by ataxia, altered consciousness, and changes in mental status, which might progress to coma. Clinically and pathophysiologically, HACE is purportedly a protraction of AMS. A change in the mental status of an individual in HACE is characterized by drowsiness, impaired mentation, confusion, stupor, and coma. Clinical findings might reveal papilloedema, ataxia, retinal hemorrhages, and focal neurological deficits (Hackett and Roach, 2001a).

In cases of HACE, neuroimaging shows vasogenic edema (Hackett et al., 1998; Matsuzawa et al., 1992). Hemodynamic factors such as sustained vasodilatation (Jensen et al., 1996), impaired cerebral autoregulation (Levine et al., 1999) and elevated cerebral capillary pressure (Krasney, 1994) most likely contribute to the formation of edema but cannot explain the process entirely (Hackett, 1999b). Moreover, hypoxia-induced biochemical alteration of the blood-brain barrier also might be an important factor. Possible mediators of these biochemical alterations, some triggered by endothelial activation, include VEGF, inducible nitric oxide synthase, and bradykinin (Clark et al., 1999; Schilling and Wahl, 1999; Severinghaus, 1995).

2.2.4 High-Altitude Pulmonary Edema

High altitude pulmonary edema (HAPE) is responsible for most deaths related to HA (Hackett and Roach, 2001a). It is a noncardiogenic form of edema that is linked with elevated capillary pressure and pulmonary hypertension. HAPE archetypally commences at altitudes above 3000 m. It occurs because of rapid ascent from sea level and also might affect healthy individuals who had not suffered HAPE earlier, even with repetitive altitude exposure. Generally, it develops within the first 2–4 days of ascent and usually on the second night at HA.

HAPE is characterized by nonproductive cough and dyspnea (i.e., shortness of breath), especially with intense exercise. HAPE advances to a devitalizing grade of dyspnea even at rest and a cough produces pink frothy sputum. Other characteristics of HAPE include tachypnea, tachycardia, and marked hypoxemia. Thorax imaging shows patchy opacities with inconsistent predominance of location; however, infiltrates often are seen initially in the region of right middle lobe (Schoene, 2008). Paralikar and Paralikar, 2010, discussed the key pathophysiological mechanisms of HAPE: pulmonary hypertension, stress failure of pulmonary capillaries, and disturbed alveolar fluid clearance.

Pulmonary hypertension: Hypoxia leads to hypoxic pulmonary vasoconstriction (HPV). This vasoconstriction is uneven because smooth muscles in different parts of the lung react differently to hypoxia. These differences cause elevated pressure and flow in the perfused areas, resulting in pulmonary hypertension and subsequent edema (Hackett and Roach, 2007; Barrett et al., 2009).

Stress failure of the pulmonary capillaries: In HAPE cases, high-permeability type of pulmonary edema occurs with proteins and white blood cells leakage. Therefore, as suggested by West et al., 1991, stress failure of the pulmonary capillaries is the main cause of edema, which occurs because of the mechanical failure of the thin walls of pulmonary capillaries when pressure inside them rises to very high values (40–60 mm Hg) (West et al., 1991).

Disturbed alveolar fluid clearance: Hypoxia causes inhibition of apical epithelial sodium channels (ENaCs) and basolateral sodium-potassium ATPase pumps activity. Water reabsorption is associated with sodium, which causes fluid accumulation in the alveoli (Höschele and Mairbäurl, 2003).

2.2.5   High-Altitude Retinopathy

On ascent to high altitude, exposure to hypobaric hypoxia leads to compensatory mechanism in retinal vasculature to maintain retinal oxygenation. Some individuals cannot adapt to the environmental stress and are unable to generate adequate autoregulatory response. This leads to morphological and functional changes in the retina and therefore, development of one of HA-related clinical pathologies known as high-altitude retinopathy. HA retinopathy generally occurs at altitudes above 4000 m from sea level. Its symptoms can include vitreous hemorrhage, optic disc swelling, dilated vessels, peripapillary hyperemia, retinal and choroidal blood flow alteration, and retinal hemorrhages (Bhende et al., 2013; McFadden et al., 1981; Rennie and Morrissey, 1975). The symptoms of high-altitude retinopathy might regress within 2–3 weeks without causing any permanent damage to the retina (Wilczyński et al., 2014). The pathophysiology of high-altitude retinopathy is not well understood; however, increase in intracranial pressure under hypobaric hypoxia is considered to be the main cause. Many cases have been reported, and human and animal studies have been conducted to understand the prevalence and cause of high-altitude retinopathy. One case was reported by a 35-year-old woman, who suffered from decreased vision in her right eye, scotomas, and high-altitude retinopathy after ascending to an altitude more than 7000 m above sea level (Wilczyński et al., 2014). In another case of maladaptive auto-regulatory response of three patients aged 27–52 years at HA, scotoma and/or visual acuity deficit was reported after their return from HA expedition above 6000 m (Russo et al., 2014).

Ho et al. (2011) conducted study on a group of six experienced climbers who suffered retinal vascular engorgement, retinal hemorrhage, and tortuosity in varying degrees in both eyes after ascent to Mt. Aconcagua to an altitude of 6962 m. The laboratory studies on these subjects revealed a high level of antiphospholipid antibodies and a significant reduction of the left ocular blood flow in one subject's eye by Doppler examination (Ho et al., 2011).

To understand the impact of rapid ascent to HA on retinal health, a study was conducted recently on 91 participants of Chinese military assigned to the Tibetan plateau. The results showed a significant increase in retinal nerve fiber layer (RNFL) thickness in the temporal and nasal quadrants of the optic disc and significant decrease in RNFL thickness in the inferior optic disc. A significant increase in RNFL thickness in the superior and inferior macula also was observed, with a significant increase in the ganglion cell layer thickness in the superior macula (Tian et al., 2018). Another case study showed that a 31-year-old male doctor developed sudden unilateral blurring of vision without any other symptoms even after 6 weeks of volunteering at HA in Nepal. The fundoscopic examination in his left eye revealed macular retinal hemorrhage. Later, he recovered his vision after few weeks in Kathmandu and his retinal hemorrhages regressed (Bhandari et al., 2017).

An animal study with rats was conducted to explore the effects of hypobaric hypoxia on the retina and assess the protective effect of resveratrol. The results showed that resveratrol had an antioxidative role by modulating hypoxia stress-associated genes and an antiapoptosis role by regulating apoptosis-related cytokines (Xin et al., 2017).

2.2.6 Sleep Disturbance

People arriving at HA from lowlands often complain of insomnia, frequent awakening, and restless sleep (Bloch et al., 2015; Tseng et al., 2015; Weil, 2004). Unacclimatized people are more prone to poor sleep quality in the first few nights after arriving at HA; however, this situation improves with acclimatization (Nussbaumer-Ochsner and Bloch, 2010). Sleep disturbance develops at altitude greater than 2500 m. The important feature of sleep disturbance is the increased incidence of central sleep apnea. The central sleep apnea develops due to interaction between the ventilator response to hypoxia and the feedback-control system. The feedback-control system governs breathing during sleep when cortical influences on breathing are absent. Periodic breathing might be more common in individuals with stronger ventilatory responses to hypoxia (Masuyama et al., 1989) and could persist or worsen during long stays at HA as ventilatory responses continue to increase (Bloch et al., 2010).

Sleep disturbances at HA were studied earlier in individuals ascending to about 3048 m and more than 4300 m. At about 3048 m, people suffered poor sleep; however, individuals sleeping at more than 4300 m had marked sleep disturbance (Coote, 1994; Nicholson et al., 1988). A recent comparative study has been reported on 1689 participants (aged 18–84 years) of Himalayan and sub-Himalayan regions of India at three different altitudes (400 m, 1900–2000 m, and 3200 m) to assess the effects of altitude on subjective sleep quality. The results showed that poor quality of sleep was approximately twice as prevalent at high altitudes as at low altitudes (Gupta et al., 2018). In another study, 50 subjects aged between 18 and 25 years were assessed to investigate the relationship between acclimatization levels and sleep architecture changes in migrants at HA. The results suggested that the acclimatization ability of migrants to plateau varied as per the arrived altitude and the length of stay (Ha et al., 2017).

2.2.7 Existing Medical Conditions as Risk Factors at High Altitudes

The susceptibility of individuals to HA might vary with respect to age, gender, health condition, and presence of medical complications (Honigman et al., 1993; Hackett et al., 1976; Hackett and Roach, 2001b; Yaron et al., 1998). State of physical endurance might not ensure protection against HA ailments (Milledge et al., 1991). Changes in barometric pressure at HA pose a significant effect on medical conditions such as chronic obstructive arrhythmias, congenital heart disease, coronary artery disease, pulmonary disease, pulmonary hypertension, and sickle cell disease. Consequently, exposure to HA means greater risk for individuals suffering from these health problems (Alexander, 1995; Basnyat and Murdoch, 2003; Erdmann et al., 1998; Green et al., 1971; Hackett and Roach, 2001a; Hultgren, 1992; Lane and Githens, 1985; Roach et al., 1995).

For example, patients with chronic obstructive pulmonary disease are at risk of developing symptoms of HA illnesses because they might show hypoxemia symptoms even at low altitude (Hackett and Roach, 2007; Luks and Swenson, 2007; Rodway et al., 2004). This occurs because chronic hypercapnia diminishes the response of the carotid body, which in turn might decrease the production of hypoxic ventilatory response (Cogo et al., 2004). Furthermore, cold air inhalation increases pulmonary artery pressure and pulmonary vasoconstriction (Luks and Swenson, 2007).

In people sojourning to HA, the degree of pressure change depends on level of hypoxic stress, exercise, diet, and cold conditions. Thus, significant individual variation is observed in blood pressure at HA (Hackett, 2001). Also, the periodic breathing during sleep at HA enhances the risk of aggravated hypertension, thereby causing cardiac arrhythmias (Hultgren, 1992), but no cases of coronary artery disease associated with HA illness have been evidenced (Erdmann et al., 1998; Pollard and Murdoch, 2003). Patients with congenital heart disease, however, show increased pulmonary arteriolar vasoconstrictor response to hypoxic stress, making them vulnerable to pulmonary hypertension and HAPE development (Hackett and Roach, 2007; Hultgren, 1992).

Sickle cell disease patients are at high risk of sickle cell crisis and splenic infarct at altitude above 2000 m so their travel is contraindicated (Claster et al., 1981; Hackett, 2001; Pollard and Murdoch, 2003). Evidence also has shown an association of splenic crisis with exposure to HA hypobaric hypoxia (Franklin and Compeggie, 1999; Sheikha, 2005), and it has been reported that severe exertion is linked with sickle cell crisis and sudden death of patients with sickle cell disease (Kark et al., 1987; Le Gallais et al., 1996). In one study, Thiriet et al. (1994) suggested that sickle cell disease patients are capable of severe exercise at HA, but their performance reduces significantly.

2.3 CONCLUSIONS

We have reviewed in this chapter various physiological HA ailments and their pathophysiology. Complete information about HA ailments could aid in better prevention and care of the sojourners before and after their ascent to HA. The preventive and treatment strategies of these ailments also could assist medical practitioners in better services. Understanding the mechanism of pathophysiology and genetics of the ailments could lead to exploration of new line of managing HA pathophysiology.

Acknowledgments

The authors are thankful to the vice chancellor of GD Goenka University, Gurgaon, Haryana, India, and the director of Defense Institute of Physiology and Allied Sciences, Delhi, India, for their constant support and encouragement.

References

Aaron, E.A., Powell, F.L., 1993. Effect of chronic hypoxia on hypoxic ventilatory response in awake rats. J. Appl. Physiol. 74 (4), 1635–1640.

Aimee, Y.Y., Shimoda, L.A., Iyer, N.V., Huso, D.L., Sun, X., McWilliams, R., Beaty, T., Sham, J.S., Wiener, C.M., Sylvester, J.T., Semenza, G.L., 1999. Impaired physiological responses to chronic hypoxia in mice partially deficient for hypoxia-inducible factor 1α. J. Clin. Investig. 103 (5), 691–696.

Alexander, J.K., 1995. Coronary problems associated with altitude and air travel. Cardiol. Clin. 13 (2), 271–278.

Askew, E.W., 1995. Environmental and physical stress and nutrient requirements. Am. J. Clin. Nutr. 61 (3), 631S–637S.

Barrett, K.E., Barman, S.M., Boitano, S., Brooks, H.L., 2009. Ganong's Review of Medical Physiology, 23rd ed. Tata-McGraw-Hill, New Delhi, pp. 619–620.

Basnyat, B., 2005. High altitude cerebral and pulmonary edema. Travel Med. Infect. Dis. 3 (4), 199–211.

Basnyat, B., Gertsch, J.H., Holck, P.S., Johnson, E.W., Luks, A.M., Donham, B.P., Fleischman, R.J., Gowder, D.W., Hawksworth, J.S., Jensen, B.T., Kleiman, R.J., 2006. Acetazolamide 125 mg BD is not significantly different from 375 mg BD in the prevention of acute mountain sickness: the prophylactic acetazolamide dosage comparison for efficacy (PACE) trial. High Altitude Med. Biol. 7 (1), 17–27.

Basnyat, B., Murdoch, D.R., 2003. High-altitude illness. Lancet 361 (9373), 1967–1974.

Benjamini, Y., Hochberg, Y., 1995. Controlling the false discovery rate: a practical and powerful approach to multiple testing. J. R. Stat. Soc. Ser. B (Methodological) 57 (1), 289–300.

Bhandari, S.S., Koirala, P., Regmi, N., Pant, S., 2017. Retinal hemorrhage in a high-altitude aid post volunteer doctor: a case report. High Alt. Med. Biol. 18 (3), 285–287.

Bhende, M., Karpe, A., Pal, B., 2013. High altitude retinopathy. Ind. J. Ophthalmol. 61 (4), 176–177.

Bloch, K.E., Buenzli, J.C., Latshang, T.D., Ulrich, S., 2015. Sleep at high altitude: guesses and facts. J. Appl. Physiol. 119 (12), 1466–1480.

Bloch, K.E., Latshang, T.D., Turk, A.J., Hess, T., Hefti, U., Merz, T.M., Bosch, M.M., Barthelmes, D., Hefti, J.P., Maggiorini, M., Schoch, O.D., 2010. Nocturnal periodic breathing during acclimatization at very high altitude at Mount Muztagh Ata (7546 m). Am. J. Resp. and Crit. Care Med. 182 (4), 562–568.

Bracken, C.P., Whitelaw, M.L., Peet, D.J., 2003. The hypoxia-inducible factors: key transcriptional regulators of hypoxic responses. Cell. Mol. Life Sci. CMLS 60 (7), 1376–1393.

Branco, L.G., Gargaglioni, L.H., Barros, R.C., 2006. Anapyrexia during hypoxia. J. Therm. Biol. 31 (1), 82–89.

Broome, J.R., Stoneham, M.D., Beeley, J.M., Milledge, J.S., Hughes, A.S., 1994. High altitude headache: treatment with ibuprofen. Aviat. Space Environ. Med. 65 (1), 19–20.

Burtscher, M., Likar, R., Nachbauer, W., Philadelphy, M., 1998. Aspirin for prophylaxis against headache at high altitudes: randomized, double-blind, placebo-controlled trial. Br. Med. J. 316 (7137), 1057–1058.

Chao, W.H., Askew, E.W., Roberts, D.E., Wood, S.M., Perkins, J.B., 1999. Oxidative stress in humans during work at moderate altitude. J. Nutr. 129 (11), 2009–2012.

Clark, I.A., Awburn, M.M., Cowden, W.B., Rockett, K.A., 1999. Can excessive iNOS induction explain much of the illness of acute mountain sickness? (abstract). In: Roach, R.C., Wagner, P.D., Hackett, P.H. (Eds.), Hypoxia: Into the Next Millennium. Vol. 474 of Advances in Experimental Medicine and Biology. Kluwer Academic/Plenum, New York, p. 373.

Claster, S., Godwin, M.J., Embury, S.H., 1981. Risk of altitude exposure in sickle cell disease. West. J. Med. 135 (5), 364–367.

Cogo, A., Fischer, R., Schoene, R., 2004. Respiratory diseases and high altitude. High Alt. Med. Biol. 5 (4), 435–444.

Coote, J., 1994. Sleep at high altitude. In: Cooper, R. (Ed.), Sleep. Chapman Hall, London, pp. 243–264.

Cummins, E.P., Taylor, C.T., 2005. Hypoxia-responsive transcription factors. Pflüg. Arch. Eur. J. Physiol. 450 (6), 363–371.

Cymerman, A., 1996. The physiology of high altitude exposure. In: Marriott, B.M., Carlson, S.J. (Eds.), Nutritional Needs in Cold and High-Altitude Environments. National Academy Press, Washington, DC, pp. 295–317.

Cymerman, A., Rock, P.B., 1994. Medical problems in high mountain environments. A handbook for medical officers. US army research institute of environmental medicine, Natick, MA, Feb. Report date: Feb 1994.

Cymerman, A., Rock, P.B., 2009. Medical problems in high mountain environments. A handbook for medical officers. USARIEM-TN94-2. US army research institute of environmental medicine thermal and mountain medicine division technical report. Retrieved 2009-03-05.

Erdmann, J., Sun, K.T., Masar, P., Niederhauser, H., 1998. Effects of exposure to altitude on men with coronary artery disease and impaired left ventricular function. Am. J. Cardiol. 81 (3), 266–270.

Fedele, A.O., Whitelaw, M.L., Peet, D.J., 2002. Regulation of gene expression by the hypoxia-inducible factor. Mol. interv. 2 (4), 229–243.

Ferrazzini, G., Maggiorini, M., Kriemler, S., Bärtsch, P., Oelz, O., 1987. Successful treatment of acute mountain sickness with dexamethasone. Br. Med. J. (Clin. Res. Ed.) 294 (6584), 1380–1382.

Fiore, D.C., Hall, S., Shoja, P., 2010. Altitude illness: risk factors, prevention, presentation, and treatment. Am. Fam. Physician 82 (9), 1103–1110.

Franklin, Q.J., Compeggie, M., 1999. Splenic syndrome in sickle cell trait: four case presentations and a review of the literature. Mil. Med. 164 (3), 230–233.

Gautier, H., 1996. Interactions among metabolic rate, hypoxia, and control of breathing. J. Appl. Physiol. 81 (2), 521–527.

Gertsch, J.H., Basnyat, B., Johnson, E.W., Onopa, J., Holck, P.S., 2004. Randomized, double-blind, placebo-controlled comparison of ginkgo biloba and acetazolamide for prevention of acute mountain sickness among Himalayan trekkers: the prevention of high altitude illness trial (PHAIT). Br. Med. J. 328 (7443), 797.

Gertsch, J.H., Lipman, G.S., Holck, P.S., Merritt, A., Mulcahy, A., Fisher, R.S., Basnyat, B., Allison, E., Hanzelka, K., Hazan, A., Meyers, Z., 2010. Prospective, double-blind, randomized, placebo-controlled comparison of acetazolamide versus ibuprofen for prophylaxis against high-altitude headache: the headache evaluation at altitude trial (HEAT). Wilderness Environ. Med. 21 (3), 236–243.

Gopfert, T., Gess, B., Eckardt, K.U., Kurtz, A., 1996. Hypoxia signalling in the control of erythropoietin gene expression in rat hepatocytes. J. Cell. Physiol. 168 (2), 354–361.

Green, R.L., Huntsman, R.G., Serjeant, G.R., 1971. The sickle-cell and altitude. Br. Med. J. 4 (5787), 593–595.

Gupta, R., Ulfberg, J., Allen, R.P., Goel, D., 2018. Comparison of subjective sleep quality of long-term residents at low and high altitudes: SARAHA study. J. Clin. Sleep Med 14 (1), 15–21.

Ha, Z.D., Pan, K.L., Jian, X.L., Luo, J.P., Guan, S.Q., Guo, W.W., 2017. Changes to sleep patterns in young migrants at high altitude. Chin. J. Tubercul. Resp. Dis. 49 (9), 689–692.

Hackett, P., 2001. Altitude and common medical conditions. In: Wilkerson, J.A. (Ed.), Medicine for Mountaineering and Other Wilderness Activities, fifth ed. The Mountaineers Books, Seattle, pp. 240–258.

Hackett, n.dHackett, P.H., Roach, R.C., High-altitude medicine, In: Auerbach, P.S. (Ed.), Wilderness Medicine, fifth ed. Mosby, Elsevier, Philadelphia, pp. 2–36.

Hackett, P., Rennie, D., Levine, H., 1976. The incidence, importance, and prophylaxis of acute mountain sickness. Lancet 308 (7996), 1149–1155.

Hackett, P.H., 1999a. The cerebral etiology of high-altitude cerebral edema and acute mountain sickness. Wilderness Environ. Med. 10 (2), 97–109.

Hackett, P., 1999b. High-altitude cerebral edema and acute mountain sickness: a pathophysiology update. In: Roach, R.C., Wagner, P.D., Hackett, P.H. (Eds.), Hypoxia: Into the Next Millennium. Vol. 474 of Advances in Experimental Medicine and Biology. Boston, US: Springer, pp. 23–45.

Hackett, P.H., Roach, R.C., 2001a. High-altitude illness. N. Engl. J. Med. 345 (2), 107–114.

Hackett, P.H., Roach, R.C., 2001b. High-altitude medicine. In: Auerbach, P.A. (Ed.), Wilderness Medicine. Mosby, Elsevier, St. Louis, pp. 2–43.

Hackett, P.H., Yarnell, P.R., Hill, R., Reynard, K., Heit, J., McCormick, J., 1998. High-altitude cerebral edema evaluated with magnetic resonance imaging: clinical correlation and pathophysiology. JAMA 280 (22), 1920–1925.

Hammond, K.A., Chappell, M.A., Kristan, D.M., 2002. Developmental plasticity in aerobic performance in deer mice (Peromyscus maniculatus). Comp. Biochem. Physiol. Mol. Integr. Physiol. 133 (2), 213–224.

Haouzi, P., Notet, V., Chenuel, B., Chalon, B., Sponne, I., Ogier, V., Bihain, B., 2008. H 2 S induced hypometabolism in mice is missing in sedated sheep. Resp. Physiol. Neurobiol. 160 (1), 109–115.

Hartman-Ksycińska, A., Kluz-Zawadzka, J., Lewandowski, B., 2016. High-altitude illness. Przegl. Epidemiol. 70 (3), 490–499.

Hartmann, G., Tschöp, M., Fischer, R., Bidlingmaier, C., Riepl, R., Tschöp, K., Hautmann, H., Endres, S., Toepfer, M., 2000. High altitude increases circulating interleukin-6, interleukin-1 receptor antagonist and C-reactive protein. Cytokine 12 (3), 246–252.

Ho, T.Y., Kao, W.F., Lee, S.M., Lin, P.K., Chen, J.J., Liu, J.H., 2011. High-altitude retinopathy after climbing Mount Aconcagua in a group of experienced climbers. Retina 31 (8), 1650–1655.

Hochachka, P.W., Buck, L.T., Doll, C.J., Land, S.C., 1996. Unifying theory of hypoxia tolerance: molecular/metabolic defense and rescue mechanisms for surviving oxygen lack. Proc. Natl. Acad. Sci. 93 (18), 9493–9498.

Honigman, B., Theis, M.K., Koziol-McLain, J., Roach, R., Yip, R., Houston, C., Moore, L.G., Pearce, P., 1993. Acute mountain sickness in a general tourist population at moderate altitudes. Ann. Intern. Med. 118 (8), 587–592.

Höschele, S., Mairbäurl, H., 2003. Alveolar flooding at high altitude: failure of reabsorption? Physiology 18 (2), 55–59.

Huey, R.B., Eguskitza, X., 2001. Limits to human performance: elevated risks on high mountain. J. Exp. Biol. 204 (18), 3115–3119.

Hultgren, H.N., 1992. Effects of altitude upon cardiovascular diseases. J. Wilderness Med. 3 (3), 301–308.

Jensen, J.B., Sperling, B., Severinghaus, J.W., Lassen, N.A., 1996. Augmented hypoxic cerebral vasodilation in men during 5 days at 3,810 m altitude. J. Appl. Physiol. 80 (4), 1214–1218.

Jungermann, K., Kietzmann, T., 2000. Oxygen: modulator of metabolic zonation and disease of the liver. Hepatology 31 (2), 255–260.

Kark, J.A., Posey, D.M., Schumacher, H.R., Ruehle, C.J., 1987. Sickle-cell trait as a risk factor for sudden death in physical training. N. Engl. J. Med. 317 (13), 781–787.

Kenneth, N.S., Rocha, S., 2008. Regulation of gene expression by hypoxia. Biochem. J. 414 (1), 19–29.

Khodaee, M., Grothe, H.L., Seyfert, J.H., VanBaak, K., 2016. Athletes at high altitude. Sports Health 8 (2), 126–132.

Krasney, J.A., 1994. A neurogenic basis for acute altitude illness. Med. Sci. Sports Exerc. 26 (2), 195–208.

Lane, P.A., Githens, J.H., 1985. Splenic syndrome at mountain altitudes in sickle cell trait: its occurrence in nonblack persons. JAMA 253 (15), 2251–2254.

Le Gallais, D., Bile, A., Mercier, J., Paschel, M., Tonellot, J.L., Dauverchain, J., 1996. Exercise-induced death in sickle cell trait: role of aging, training, and deconditioning. Med. Sci. Sports Exerc. 28 (5), 541–544.

Levine, B.D., Zhang, R., Roach, R.C., 1999. Dynamic cerebral autoregulation at high altitude. In: Roach, R.C., Wagner, P.D., Hackett, P.H. (Eds.), Hypoxia: Into the Next Millennium. Vol. 474 of Advances in Experimental Medicine and Biology. Boston: Springer, pp. 319–322.

Litch, J.A., 2007. High altitude illnesses. In: Rakel, R.E., Bope, E.T. (Eds.), Conn's Current Therapy 2008. Elsevier Saunders, Philadelphia.

Luks, A.M., 2015. Physiology in medicine: a physiologic approach to prevention and treatment of acute high-altitude illnesses. J. Appl. Physiol. 118 (5), 509–519.

Luks, A.M., Swenson, E.R., 2007. Travel to high altitude with pre-existing lung disease. Eur. Respir. J. 29 (4), 770–792.

Luks, A.M., Swenson, E.R., Bärtsch, P., 2017. Acute high-altitude sickness. Eur. Respir. Rev. 26 (143).

Maggiorini, M., Bühler, B., Walter, M., Oelz, O., 1990. Prevalence of acute mountain sickness in the Swiss Alps. BMJ 301 (6756), 853–855.

Maggiorini, M., Streit, M., Siebenmann, C., 2009. Dexamethasone decreases systemic inflammatory and stress response and favors vasodilation in high altitude pulmonary edema susceptible at 4559m. In: 16th International Hypoxia Symposium. Chateau Lake Louise, Alberta, Canada, pp. 10–15.

Masuyama, S., Kohchiyama, S., Shinozaki, T., Okita, S., Kunitomo, F., Tojima, H., Kimura, H., Kuriyama, T., Honda, Y., 1989. Periodic breathing at high altitude and ventilatory responses to O_2 and CO_2. Japan. J. Physiol. 39 (4), 523–535.

Matsuzawa, Y., Kobayashi, T., Fujimoto, K., Shinozaki, S., Yoshikawa, S., 1992. Cerebral edema in acute mountain sickness. In: Ueda, G., Reeves, J.T., Sekiguchi, M. (Eds.), High-Altitude Medicine. Shinshu University, Matsumoto, Japan, pp. 300–304.

McFadden, D.M., Houston, C.S., Sutton, J.R., Powles, A.P., Gray, G.W., Roberts, R.S., 1981. High-altitude retinopathy. JAMA 245 (6), 581–586.

Milledge, J.S., Beeley, J.M., Broome, J., Luff, N., Pelling, M., Smith, D., 1991. Acute mountain sickness susceptibility, fitness, and hypoxic ventilatory response. Eur. Respir. J. 4 (8), 1000–1003.

Montgomery, A.B., Mills, J., Luce, J.M., 1989. Incidence of acute mountain sickness at intermediate altitude. JAMA 261 (5), 732–734.

Mortola, J.P., 1999. How newborn mammals cope with hypoxia. Resp. Physiol. 116 (2), 95–103.

Nicholson, A.N., Smith, P.A., Stone, B.M., Bradwell, A.R., Coote, J.H., 1988. Altitude insomnia: studies during an expedition to the Himalayas. Sleep 11 (4), 354–361.

Nussbaumer-Ochsner, Y., Bloch, K.E., 2010. Air travel and altitude. In: Ayres, J.G., Harrison, R.M., Nichols, G.L., Maynard, R.L. (Eds.), Environmental Medicine. Hodder Arnold, London, pp. 547–561.

Paralikar, S.J., Paralikar, J.H., 2010. High-altitude medicine. Ind. J. Occupat. Environ. Med. 14 (1), 6–12.

Parise, I., 2017. Traveling safely to places at high altitude: understanding and preventing altitude illness. Aust. Fam. Phys. 46 (6), 380–384.

Peacock, A.J., 1998. Oxygen at high altitude. Br. Med. J. 317 (7165), 1063–1067.

Peacock, A.J., Jones, P.L., 1997. Gas exchange at extreme altitude: results from the British 40th Anniversary Everest Expedition. Eur. Respir. J. 10 (7), 1439–1444.

Pollard, A.J., Murdoch, D.R., 2003. Chronic disease, pregnancy and contraception at altitude. In: Pollard, A.J., Murdoch, D.R. (Eds.), The High-Altitude Medicine Handbook. third ed. Radcliffe Medical Press, Abingdon, pp. 81–87.

Pugh, L.G., 1964. Man at high altitude: studies carried out in the Himalayas. Sci. Basis Med. Annu. Rev., 32–54.

Rennie, D., Morrissey, J., 1975. Retinal changes in Himalayan climbers. Arch. Ophthalmol. 93 (6), 395–400.

Roach, R.C., Bartsch, P., Oelz, O., Hackett, P.H., 1993. The Lake Louis acute mountain scoring system. In: Sutton, J.R., Houston, C.S., Coates, G. (Eds.), Hypoxia and Molecular Medicine. Queen City Printers, Burlington, VT, pp. 272–274.

Roach, R.C., Houston, C.S., Honigman, B., Nicholas, R.A., Yaron, M., Grissom, C.K., Alexander, J.K., Hultgren, H.N., 1995. How well do older persons tolerate moderate altitude? West. J. Med. 162 (1), 32–36.

Roach, R.C., Maes, D., Sandoval, D., Robergs, R.A., Icenogle, M., Hinghofer-Szalkay, H., Lium, D., Loeppky, J.A., 2000. Exercise exacerbates acute mountain sickness at simulated high altitude. J. Appl. Physiol. 88 (2), 581–585.

Rodway, G.W., Hoffman, L.A., Sanders, M.H., 2004. High-altitude-related disorders—Part II: prevention, special populations, and chronic medical conditions. Heart Lung J. Crit. Care 33 (1), 3–12.

Russo, A., Agard, E., Blein, J.P., Chehab, H.E., Lagenaite, C., Ract-Madoux, G., Dot, C., 2014. High-altitude retinopathy: report of 3 cases. J. Fr. Ophtalmol. 37 (8), 629–634.

Sanchez, D.R.M., Moskowitz, M.A., 1999. High-altitude headache. Lessons from headaches at sea level. Adv. Exp. Med. Biol. 474, 145–153.

Sarkar, S., Banerjee, P.K., Selvamurthy, W., 2003. High altitude hypoxia: an intricate interplay of oxygen responsive macroevents and micromolecules. Mol. Cell. Biochem. 253 (1–2), 287–305.

Schilling, L., Wahl, M., 1999. Mediators of cerebral edema. In: Roach, R.C., Wagner, P.D., Hackett, P.H., eds. Hypoxia: Into the Next Millennium. Vol. 474 of Advances in Experimental Medicine and Biology. Boston, US: Springer, pp. 123-41.

Schneider, M., Bernasch, D., Weymann, J., Holle, R., Bartsch, P., 2002. Acute mountain sickness: influence of susceptibility, pre-exposure, and ascent rate. Med. Sci. Sports Exerc. 34 (12), 1886–1891.

Schoene, R.B., 2008. Illnesses at high altitude. Chest 134 (2), 402–416.

Semenza, G.L., 2000. HIF-1: mediator of physiological and pathophysiological responses to hypoxia. J. Appl. Physiol. 88 (4), 1474–1480.

Semenza, G.L., Roth, P.H., Fang, H.M., Wang, G.L., 1994. Transcriptional regulation of genes encoding glycolytic enzymes by hypoxia-inducible factor 1. J. Biol. Chem. 269 (38), 23757–23763.

Severinghaus, J.W., 1995. Hypothetical roles of angiogenesis, osmotic swelling, and ischemia in high-altitude cerebral edema. J. Appl. Physiol. 79 (2), 375–379.

Severinghaus, J.W., Chiodi, H., EGER II, E.I., Brandstater, B., Hornbein, T.F., 1966. Cerebral blood flow in man at high altitude. Circ. Res. 19 (2), 274–282.

Sheikha, A., 2005. Splenic syndrome in patients at high altitude with unrecognized sickle cell trait: splenectomy is often unnecessary. Can. J. Surg. 48 (5), 377–381.

Shukla, V., Singh, S.N., Vats, P., Singh, V.K., Singh, S.B., Banerjee, P.K., 2005. Ghrelin and leptin levels of sojourners and acclimatized lowlanders at high altitude. Nutr. Neurosci. 8 (3), 161–165.

Singh, S.B., Selvamurthy, W., 1993. Effect of intermittent chronic exposure to hypoxia on feeding behaviour of rats. Int. J. Biometeorol. 37 (4), 200–202.

Smedley, T., Grocott, M.P., 2013. Acute high-altitude illness: a clinically orientated review. Br. J. Pain 7 (2), 85–94.

Steiner, A.A., Branco, L.G., 2002. Hypoxia-induced anapyrexia: implications and putative mediators. Annu. Rev. Physiol. 64 (1), 263–288.

Strassman, A.M., Levy, D., 2006. Response properties of dural nociceptors in relation to headache. J. Neurophysiol. 95 (3), 1298–1306.

Swenson, E.R., 1997. High-altitude dieresis: fact or fancy. In: Houston, C.S., Coates, G. (Eds.), Hypoxia: Women at Altitude. Queen City Printers, Burlington, VT, pp. 272–283.

Thiriet, P., Le Hesran, J.Y., Wouassi, D., Bitanga, E., Gozal, D., Louis, F.J., 1994. Sickle cell trait performance in a prolonged race at high altitude. Med. Sci. Sports Exerc. 26 (7), 914–918.

Tian, X., Zhang, B., Jia, Y., Wang, C., Li, Q., 2018. Retinal changes following rapid ascent to a high-altitude environment. Eye (Lond) 32 (2), 370–374.

Tseng, C.H., Lin, F.C., Chao, H.S., Tsai, H.C., Shiao, G.M., Chang, S.C., 2015. Impact of rapid ascent to high altitude on sleep. Sleep Breath. 19 (3), 819–826.

Weil, J.V., 2004. Sleep at high altitude. High Alt. Med. Biol. 5 (2), 180–189.

West, J.B., 1988. High points in the physiology of extreme altitude. Adv. Exp. Med. Biol. 227, 1–15.

West, J.B., Tsukimoto, K., Mathieu-Costello, O., Prediletto, R., 1991. Stress failure in pulmonary capillaries. J. Appl. Physiol. 70 (4), 1731–1742.

Wilczyński, M., Kucharczyk, M., Filatow, S., 2014. High-altitude retinopathy—case report. Klinika Oczna 116 (3), 180–183.

Xin, X., Dang, H., Zhao, X., Wang, H., 2017. Effects of hypobaric hypoxia on rat retina and protective response of resveratrol to the stress. Int. J. Med. Sci. 14 (10), 943–950.

Yaron, M., Waldman, N., Niermeyer, S., Nicholas, R., Honigman, B., 1998. The diagnosis of acute mountain sickness in preverbal children. Arch. Pediatr. Adolesc. Med. 152 (7), 683–687.

Ziaee, V., Yunesian, M., Ahmadinejad, Z., Halabchi, F., Kordi, R., Alizadeh, R., Afsharjoo, H.R., 2003. Acute mountain sickness in Iranian trekkers around Mount Damavand (5671m) in Iran. Wilderness Environ. Med. 14 (4), 214–219.

Further Reading

Fedele, A.O., Whitelaw, M.L., Peet, D.J., 2002. Regulation of gene expression by the hypoxia-inducible factors. Mol. Interv. 2 (4), 229–243.

Herbal Therapeutics for High Altitude Problems

CHAPTER 3

Hippophae sp.: A Boon for High-Altitude Maladies

Manimaran Manickam*, Anuja Bhardwaj[†],
Syed Rahamathulla*, Arul J. Duraisamy[‡]

*Research and Development, PathGene Healthcare Private Limited, Tirupathi, India, [†]Chemistry Division, Department of Biochemical Sciences (DBCS), Defence Institute of Physiology and Allied Sciences (DIPAS), Delhi, India, [‡]Kresge Eye Institute, Wayne State University School of Medicine, Detroit, MI, United States

Abbreviations

AAS	atomic absorption spectroscopy
AD	atopic dermatitis
AFB1	aflatoxin B1.
AIA	adjuvant-induced arthritis
ALP	alkaline phosphatase
ALT	alanine aminotransferase
AMPKα	AMP-activated protein kinase-α
AST	aspartate aminotransferase
CAT	catalase
CHOP	C/EBP homologous protein
CKMB	creatine kinase isoenzyme
COX	cyclooxygenase
CS	citrate synthase
CT	computed tomography
DFRL	Defence Food Research Laboratory
DIHAR	Defence Institute of High Altitude Research
DIPAS	Defence Institute of Physiology and Allied Sciences
DRDO	Defence Research Development Organization
EGFR	epidermal growth factor receptor
ER	endoplasmic reticulum
ERK	extracellular signal-regulated kinase
FFA	free fatty acids
G-6-PD	glucose-6-phosphate dehydrogenase
GC	gas chromatography
GGT	glutamyl transpeptidase
GLUT4	glucose transporter 4
GOT	glutamate oxaloacetate transferase
GPT	glutamate pyruvate transferase

Management of High Altitude Pathophysiology. https://doi.org/10.1016/B978-0-12-813999-8.00003-3

GR	glutathione reductase
GRP78	glucose-regulated protein
GSH	reduced glutathione
GSSG	glutathione disulfide
GST	glutathione-*S*-transferase
GT	gasping time
H. rhamnoides	*Hippophae rhamnoides*
HA	high altitudes
Hb	hemoglobin
HbO$_2$	oxygenated hemoglobin
HIF	hypoxia inducible factor
HK	hexokinase
HMP	hexose monophosphate
HSP	heat shock protein
ICPMS	inductively coupled plasma mass spectrometry
IL-1β	interleukin-1 beta
IL-6	interleukin-6
iNOS	inducible nitric oxide synthase
ISO	isoproterenol
JAK	Janus-activated kinase
JNK	Jun N-terminal kinase
LD	lactic acid
LDH	lactate dehydrogenase
LLC	Lewis lung carcinoma
LPS	lipopolysaccharide
LSM	liver stiffness measurement
MDA	malondialdehyde
MIRI	myocardial ischemia reperfusion injury
NAFLD	nonalcoholic fatty liver disease
NDOs	nondigestible oligosaccharides
NF-κB	nuclear factor kappa B
NO	nitric oxide
PCNA	proliferating cell nuclear antigen
PFK	phosphofructokinase
PKR	protein kinase R
PRF	phenolic-rich fraction
PUFA	polyunsaturated fatty acids
QTP	Qinghai-Tibet Plateau
ROS	reactive oxygen species
SBTLE	supplementation of its leaf extract
SOD	superoxide dismutase
SP	sea buckthorn polysaccharide
SSD	silver sulfadiazine
STAT	signal transducers and activators of transcription
TBARS	thiobarbituric acid reactive substance
TC	total cholesterol
TFH	total flavones of *H. rhamnoides*
TG	triglyceride

TLR4	toll-like receptor 4
TNF-α	tumor necrosis factor α
VEGF	vascular endothelial growth factor

3.1 INTRODUCTION

Sojourners to high-altitude (HA) venues scale heights of mountains for various reasons, such as trekking, mountaineering, mining, scientific activities, and military duties. The number of visitors to mountains has increased appreciably in recent years because of improvements in communications, logistics, and medical care available at HA locations. At such extreme conditions, however, the normal physiology of the body is altered, and consequently, the body tends to acclimatize to restore its normal functioning. Usually, the normal physiological processes of altitude acclimatization involve hyperventilation, cardio-acceleration, and increase in red blood cell count, hemoglobin (Hb) and 2,3-bisphosphoglyceric acid levels. These physiological manifestations help in the dissociation of oxygen from oxygenated hemoglobin (HbO_2) so that more oxygen is delivered to the tissues for use by mitochondrial enzymes (Selvamurthy and Basu, 1998).

The HA region of the trans-Himalayan cold desert has unfavorable climatic conditions for human survival. These adverse conditions can induce modifications in normal physiological functions, resulting in sustained energy deficit, malnutrition, and metabolic disorders (Sherpa et al., 2013). Several epidemiological studies confirm that consumption of botanical products is associated with a decreased risk of several chronic diseases (Korekar et al., 2011; Tayade et al., 2013). Herbal plants at such HA regions have been a part of traditional systems of medicine both as prophylactics and therapeutics for alleviating HA-associated disorders (Dhar et al., 2013).

The *Hippophae* species is considered to be the most enriched among other HA plants in terms of various medicinally important bioactive metabolites and therefore has attained great importance among researchers. This chapter illustrates the wide range of medicinal benefits offered by the *Hippophae* species, particularly *Hippophae rhamnoides*, and also their application to overcome HA maladies.

3.2 HISTORY AND TRADITIONAL USAGE

Hippophae is a Latin word derived from the words *"hippo"* meaning horse and *"phaos"* meaning shine (Krejcarová et al., 2015; Suryakumar and Gupta, 2011). In ancient times, the leaves and twigs of *Hippophae* were given as a feed to race-horses for weight gain and shining coat; hence, its name (Suryakumar and

Gupta, 2011; Zeb, 2004). Its common name, sea buckthorn, comes the fact that it often grows near the sea and has spines and thorns. In traditional medicine, it is called other regional names such as shaji, culiu, suanci, dhar-bu, and star-bu (Junzeng et al., 2015). The nutritional and medicinal uses of *Hippophae* have been referred in ancient Greek texts and in Chinese and Tibetan medicinal systems (Suryakumar and Gupta, 2011; Wani et al., 2016). It has been archived in the classical Tibetan medicinal text "the RGyud Bzi" (The Four pharmaco-poeia) compiled during Tang Dynasty (618--907 AD) (Suryakumar and Gupta, 2011; Zeb, 2004). It also has been included in regional standards: The List of Medicinal and Edible Plants of China and Ayurvedic Pharmacopeia of India (Liu et al., 2017).

Traditionally, *Hippophae* has been used to treat various human diseases, including cardiovascular diseases, HA diseases, lung disorders, mucous membrane injuries of the stomach, skin injuries, and for the enhancement of blood circulation (Liu et al., 2017; Suryakumar and Gupta, 2011; Wani et al., 2016).

3.3   BOTANICAL DESCRIPTION AND NATURAL DISTRIBUTION

Sea buckthorn is a deciduous, dioecious spinescent shrub that reaches a height of 2–4 m. Its leaves are alternate, lanceolate, narrow, and are dark green with a silver-gray tone at the bottom. The female *Hippophae* plant produces berry-like fruits that are pulpy and rich in oils. The ripened berries are orange to red in color and consist of a single seed that is dark brown, glossy, and ovoid or ellip-tical in shape (Li and Schroeder, 1996; Suryakumar and Gupta, 2011). All parts of *Hippophae* namely—bark, berries (fruits), leaves, and roots—are a rich source of various biologically active phytochemicals. The content, However, varies according to the species (or subspecies), origin, climatic conditions, time of harvesting, and procedures followed during their processing (Krejcarová et al., 2015).

This berry crop is regarded as a pioneer plant because of its wide ecological adaptability, fast growth, ability to provide protection against wind and soil drift, and efficient nitrogen fixation (Buzoianu et al., 2014; Wani et al., 2016). *Hippophae* also possesses a broad pectrum of bioactivities such as anti-oxidant, immunomodulatory, antiatherogenic, antistress, hepato-protective, radioprotective, and tissue repair (Liu et al., 2017; Wani et al., 2016). These pharmacological and medicinal effects are attributable to the presence of a wide variety of bioactive compounds such as carotenoids, fatty acids, phenolics, phy-tosterols, vitamins, and minerals. All these potentials of the *Hippophae* species has led to its importance in human health and nutrition in addition to its role in land rehabilitation (Buzoianu et al., 2014; Suryakumar and Gupta, 2011).

Hippophae is widely distributed throughout the temperate zone of Asia and northwestern Europe and throughout subtropical zones, especially at high altitudes (Krejcarová et al., 2015; Zeb, 2004). The species, *H. rhamnoides* L., grows widely at high altitudes of 7000–15,000 ft of the northwest Himalayan region and is endemic to Eurasia (Suryakumar and Gupta, 2011; Wani et al., 2016). The genus grows in low humidity (15%), alluvial gravel, wet landslips, various soil conditions, hills, ravine tops, and riverside with brown tarnished, scaly shoots (Banjade, 1999; Basistha, 2001). Belonging to the family Elaeagnaceae, *Hippophae* has seven species: *H. salicifolia, H. rhamnoides Linn, H. tibetana Schlecht, H. neurocarpa, H. gyantsensis Lian, H. goniocarpa,* and *H. litagenesis* (Yadav et al., 2006). Some of these are discussed in the following sections.

3.4 *HIPPOPHAE* SPECIES

3.4.1 *Hippophae rhamnoides*

H. rhamnoides L. (sea buckthorn) (Fig. 3.1) is a wild and cultivated berry plant of Ladakh, where the regional name is tsermang and its berries are called as tsestalullu (Singh et al., 2006). In India, this *Hippophae* species is found mainly in the cold arid zone of Arunachal Pradesh, Ladakh, Lahaul-Spiti, parts of Chamba and upper Kinnaur districts of Himachal Pradesh, and Sikkim (Kumar et al., 2011). *H. rhamnoides* is a multipurpose plant, with each part endowed with various pharmaceutical and therapeutic benefits. For example, the fruits (berries) are used in the food industry, in traditional medicine, as part of drugs, and in the cosmetic industry. The leaves are used as feed, particularly for ruminants (Kumar et al., 2011; Wani et al., 2016). The plant also is used for

FIG. 3.1

Hippophae rhamnoides seeds (A); the plant (B), and berries (C).

land retrieval or as shelterbelt because of its ability to tolerate erosion, nutrient deprivation, and soil salinity (Wani et al., 2016).

3.4.2 *Hippophae tibetana*

H. tibetana is native to Tibet, China, and India (Acharya et al., 2010) and is commonly called as Tibetan sea buckthorn (Raina et al., 2012). It is endemic to the Qinghai-Tibet Plateau (QTP), covering the Himalayas in the west and extending to the Northeast QTP (Xu et al., 2014a, b). In India, *H. tibetana* grows in the Trans-Himalayan cold deserts of the Spiti region at altitudes above 2500–4500 m and thrives well under extreme temperature variations (−40°C to 30°C), low precipitation, and low oxygen in air (Ranjith et al., 2006). It is a small dioecious shrub of height 0.2–0.6 m (Stobdan et al., 2011a, b) and can be propagated by seeds or horizontal roots (Xu et al., 2014a, b). The seed of this *Hippophae* species is elliptical and dull white in color; the seeds of other *Hippophae* species (such as *H. rhamonoides*) are light to dark brown in color (Stobdan et al., 2011a, b). It is rich in lipophilic carotenoids and tocols (Ranjith et al., 2006) and has the highest oil content (~16%–18%) in comparison to *H. rhamonoides* (~7%–15%) and *H. salicifolia* (~11%) (Stobdan et al., 2011a, b). Traditionally, its fruit, usually 15–21 g (Stobdan et al., 2011a, b), is made into an aqueous decoction after being dried and crushed for consuming as tea to cure cough, congestion, jaundice, and for blood purification (Singh, 2012). Its fruit juice, which contains high levels of carotenoids, flavonoids, and vitamin C, is used to cure gastrointestinal disorders (Uprety et al., 2010). It has also been used to cure irregular menstruation cycles and to help with weight reduction (Singh, 2012). All these health benefits of *H. tibetana* are described in traditional medicine only. Most of the wild resources are protected as grassland sources (Rongsen et al., 2013). Considering its promising medicinal properties as reported in traditional medicine systems and its ability to thrive well under environmental extremes, this species could be further taken up for evaluating its bioactivities against HA illness.

3.4.3 *Hippophae neurocarpa*

H. neurocarpa generally grows as flat-topped shrubs in valley bottoms (at an altitude of 3400–4400 m) or as dwarf trees in flood plains, river banks, and terraces, with a height ranging from 0.6 m to 3.5 m (Chiej, 1984; Kou et al., 2014). *H. neurocarpa* is found in Qinghai-Xizang plateau and Gansu regions of China and in Tibet (Kapoor, 2017). It is morphologically distinct from other members of *Hippophae* species with respect to its cylindrical, curved, brown fruit that has several ribs and is almost non-juicy (Kou et al., 2014; Sun et al., 2002). It includes the subspecies, *neurocarpa* and *stellatopilosa*, which are described mainly on the basis of variation in leaf trichomes (Kapoor, 2017).

In Tibetan medicine, *H. rhamnoides sinensis*; *H. gyantsensis*; *H. neurocarpa*, and *H. tibetana* typically are described under one name—*Shaji*—as a cure for cardio-vascular diseases and lung disorders (Liu et al., 2016). The tender branches and leaves of this *Hippophae* species also are used to generate oil, which is used as an ointment for treating burns. Its fruit can be crushed and applied to wounds as an emergency measure for stopping bleeding (Chiej, 1984; Kapoor, 2017). It also is used as a remedy for eczema and radiation injury (Kapoor, 2017). These health benefits of *H. neurocarpa* highlight its potential curative role in wound healing and against radiation injury, which often is encountered at high altitudes.

3.4.4 *Hippophae gyantsensis*

H. gyantsensis is found exclusively to the west of QTP (Jia et al., 2016; Kapoor, 2017; Rongsen et al., 2013). It grows on gravelly river terraces and dried river beds with an altitude range of 3500–5000 m above sea level (Jia et al., 2016). It shares some of its morphological features with *H. rhamnoides yunnanensis* and others with *H. neurocarpa*. The fruits of this species are elliptical and the plants have white bark and narrow leaves with a whitish tone at the bottom (Jia et al., 2016). It is supposed that the berries were used in Tibetan medicine (Kapoor, 2017; Rongsen et al., 2013). Most of it, however, is protected as wild forest species (Rongsen et al., 2013). No research has been conducted concerning the therapeutic health benefits of *H. gyantsensis*, but, like other *Hippophae* species, it could be explored for its efficacy against HA-induced maladies.

3.4.5 *Hippophae goniocarpa*

Hippophae goniocarpa is among the few *Hippophae* specimens that is native to Qinghai and Sichuan regions of China (Wang et al., 2008). The species grows in mountainous regions in Nepal and China, where it is found on flood lands, mountain slopes, river banks, and valley terraces. The growth altitude ranges between 2650 and 3700 m (Yongshan et al., 2003). It is a diploid hybrid species that originated from *H. rhamnoides*? ssp. *sinensis* and *H. neurocarpa* ssp. *neurocarpa* S.W. Liu & T.N. He and has morphological features intermediate to its parental species (Zhou et al., 2010). It includes two subspecies: subp. *litangensis* and subp. *goniocarpa* (Wang et al., 2008; Yongshan et al., 2003; Zhou et al., 2010). The fresh berries of *H. goniocarpa* are fleshy and yellow in color—a trait attributable to its parental species *H. rhamnoides* ssp. *sinensis*. The ripened berries are black–brown or deep green, without any bright yellow coloration, terete, and ridged; these features are similar to those of *H. neurocarpa* (Wang et al., 2008). The subp. *litangensis* differs from the typical subspecies by the young branchlets and the lower surface of leaves (Yongshan et al., 2003). Almost all the parts—berries, newly formed branches, and leaves—are enriched with bioactive compounds and used for oil production. However, it is reported that the fruit-based oil is of higher quality than oil from the branches and leaves (Matthews, 1994). The oil is used as an ointment for treating burns, eczema, and radiation injury and is consumed internally in the treatment of stomach and intestinal diseases, as well as being used in the treatment of cardiovascular disorders. The fruit is a rich source of vitamins (A, C, and E) and minerals and flavonoids. It is suggested that it can reduce the incidence of cancer and can inhibit or reverse the growth of some types of cancers (Matthews, 1994). Such medicinal and therapeutic benefits of *H. goniocarpa* accounts for its potential use as an alternative medicine for the HA disorders (Matthews, 1994).

3.4.6   *Hippophae salicifolia*

In India, two *Hippophae* species are common: *H. rhamnoides* and *H. salicifolia* (Rathor et al., 2015). Among these two species, *H. salicifolia* has not been studied much for its high value medicinal benefits (Gupta et al., 2011). *H. salicifolia* belongs to the family Eleagnaceae and is commonly known as willow leaved sea buckthorn (Ramu et al., 2014). It differs from *H. rhamnoides* in two major aspects: It is a shrub that could grow up to a tree (4–10 m) at 1500–3200 m above sea level and is limited in its biogeographical distribution to the Himalayas (central and northern) (Gupta et al., 2011; Kaushal et al., 2013). *H. rhamnoides*, on the other hand, is a bushy tree that is distributed widely in both Asia and Europe at higher altitudes (Chakraborty et al., 2016; Usha et al., 2014). It is found in Tibet of China, Bhutan, Nepal, and India of the Himalayas (Ramu et al., 2014). In India, it dwells at HA regions of Himachal Pradesh, Jammu and Kashmir, Sikkim, and Uttar Pradesh (Usha et al., 2014). Indigenously, the ethnobotanical uses of *H. salicifolia* by the regional people of Central Himalaya include animal feed, cosmetics, food, fuel, medicine, veterinary care, and biofencing (Thakur et al., 2015). In Northern Himalaya, its berries are used to treat malnutrition, liver disorders, jaundice, microbial infections, pain, skin diseases, ulcers, and tumors and to promote tissue regeneration (Thakur et al., 2015). It also is used as a nerve protector from toxicants and to strengthen the immune system (Ramu et al., 2014). This versatile plant is of great importance because of its cosmetic, pharmaceutical, nutritional, and therapeutic applications (Chakraborty et al., 2016; Kaushal et al., 2013). Its berries are edible and nutritious and are consumed in the form of jellies, juices, pickles, and squash (Thakur et al., 2015). The bark of this plant is reported to possess antioxidant qualities and traditionally was employed for its antidiarrheal and antitumor benefits, and its ash was used in cosmetics and for healing benefits. It has been reported that the plant has antibacterial and antifungal properties (Chakraborty et al., 2016; Ramu et al., 2014). The presence of numerous bioactive compounds such as vitamin A and C, tocopherols (vitamin E), dietary minerals, amino acids, carotenoids, flavonoids, isorhamnetin, quercetin, omega-3, omega-6 fatty acids, kaempferol, α-carotene, β-carotene, catechins, proanthocyanidins, and chlorogenic acids (Gupta et al., 2011; Kaushal et al., 2013; Thakur et al., 2015; Usha et al., 2014) supports its medicinal and nutritional benefits. The vitamin C content in *H. salicifolia* is highest among all *Hippophae* species (Gupta et al., 2011). Thus, the metabolic profile and its various health benefits highly demand research to evaluate its efficacy against health problems including HA diseases.

3.5   CHEMICAL CONSTITUENTS OF *H. RHAMNOIDES*

H. rhamnoides exhibit numerous medicinal and nutritive properties because each part (vegetative: bark, leaves, roots, and shoots or reproductive: fruits/berries; flower) is enriched with various bioactive constituents. As such, *H. rhamnoides* sometimes is called a wonder plant (Gupta and Deswal, 2012; Sharma et al., 2016). A significant variability, however, exists in the content of phytochemicals with respect to species (or subspecies), origin, climatic conditions, time of harvesting, and extraction procedures followed during their processing (Chirila et al., 2014; Krejcarová et al., 2015; Šnē et al., 2013; Zakynthinos and Varzakas, 2015). Phytochemicals such as amino acids, carotenoids (α-, β-, δ-carotene, lycopene), flavonoids, lipids, organic acids (malic acid, oxalic acid), phenolic compounds, phytosterols (amyrins, ergosterol, lansterol, stigmasterol), polysaccharides and sugars, polyunsaturated fatty acids, sterols, tannins, tocopherols, vitamins (A, C, E, K, riboflavin, folic acid), and minerals have been reported in *H. rhamnoides* (Chauhan and Varshneya, 2012; Chirila et al., 2014; Maheshwari et al., 2011; Olas, 2016; Šnē et al., 2013; Zakynthinos and Varzakas, 2015). The presence of such a vast array of bioactive compounds of *H. rhamnoides* contributes to its commercial status as a "super fruit" (Kalia et al., 2011) and subsequent use of its seeds, berries, leaves, extracts, and fractions as sources of bioactive ingredients for the production of a wide variety of cosmetics and health products (Yang et al., 2006). Some authors, however, suggest that most of the research has been carried out on *H. rhamnoides* berries, seeds, and oils only with major emphasis on phenolic compounds. Also, most of these studies are not detailed and have become obsolete (Šnē et al., 2013). It is important to have a prior knowledge of the metabolite composition of any herbal plant before its medicinal and therapeutic usage. Therefore, in the following sections, we present a detailed description of various bioactive phytochemicals in *H. rhamnoides*.

3.5.1   Carbohydrates

The sugars and fruit acids are important components that influence the sensory properties of *H. rhamnoides*. Both the absolute content and relative abundance of these components play a vital role in determining the flavor and the consumer acceptance of the berry and its products (Beveridge et al., 2002; Šnē et al., 2013; Stobdan et al., 2011a, b). The leaves and berries of *H. rhamnoides* are reported to contain sugars such as sucrose, glucose, and fructose (Beveridge et al., 2002; Ranjith and Arumughan, 2009; Šnē et al., 2013). Presence of sugar alcohols such as mannitol, sorbitol, and xylitol also have been found, but at low levels (Stobdan et al., 2011a, b). Lignans, namely secoisolariciresinol and matairesinol, in seeds, fruit pulp/peel, and whole berries of *H. rhamnoides* have been analyzed by isotope dilution gas chromatography-mass spectrometry (Yang et al., 2006).

HRWP, an important water-soluble polysaccharide, also has been isolated by hot-water extraction and purified by DEAE-Cellulose ion-exchange chromatography by Ni et al. (2013). It is a yellowish-brown powder mainly composed of HG-type pectin and some glucans. It exhibited a significant antifatigue effect as indicated by its ability to mobilize triglyceride (TG) during exercise and protection of corpuscular membrane by inhibiting lipid oxidation via regulation of several enzymatic activities (Ni et al., 2013). The same group, in 2015, isolated a high-methoxyl homogalacturonan pectin, designated as HRWP-A from the *H. rhamnoides* berries by using DEAE-Cellulose ion-exchange chromatography, Sephadex G-75 gel filtration chromatography and DEAE-Sepharose Fast Flow ion-exchange chromatography. The structural analysis revealed that HRWP-A is a pectic polysaccharide with repeating units of $(1 \rightarrow 4)$-β-d-galactopyranosyluronic. It demonstrated a remarkable antitumor effect on intragastric administration against Lewis lung carcinoma (LLC) growth in tumor-bearing mice model (Wang et al., 2015). Recently, this group further investigated that HRWP-A has immuno-modulatory activity enhanced through TLR4/MyD88 pathway mediated activation of macrophages. These successive studies indicated use of HRWP-A as a potential immunostimulant and anticancer adjuvant (Wang et al., 2018). Another water-soluble polysaccharide from *H. rhamnoides* leaf tea (WPHT), a commercially available product, has revealed potential antioxidant activity in addition to HRWP-A. The results revealed that WPHT administration at 50 mg/(kg/d) and 100 mg/(kg/d) elevated the activities of plasma and liver homogenate superoxide dismutase (SOD) and liver catalase and decreased malondialdehyde (MDA) content in both plasma and liver homogenate, with a significant difference from the model group ($P < 0.05$) (Li et al., 2011).

An interesting investigation has revealed presence of nondigestible oligosaccharides (NDOs) such as prebiotics in *H. rhamnoides* berries that have shown the ability to enhance the activity of the health-promoting bacteria of human gut flora: *Lactobacillus acidophilus* and *Bifidobacterium lactis* (Gunenc et al., 2016).

3.5.2 Lipids

H. rhamnoides is a source of simple, complex, and derived lipids. Numerous lipids have been identified and isolated from *H. rhamnoides*, and we have tabulated some of the major lipids in Table 3.1. The seeds of *H. rhamnoides* are rich in polyunsaturated fatty acids (PUFA), and the skin and the pulp have prolific content of saturated and monounsaturated fatty acids (Pavlović et al., 2016). Linoleic acid (omega-3) and linolenic acid (omega-6), in a ratio of 1:1, are the two PUFAs present in the seed oil from *H. rhamnoides*. This characteristic quality, along with the high oleic acid content (56.4%) in seed oil, makes it a medicinally valuable oil (Zeb and Ullah, 2015). A ratio of 1:1 between the two omegas is considered to be an important factor in regulation of several metabolic processes through prostaglandin pathways (Pavlović et al., 2016; Stobdan et al., 2013). These essential fatty acids are asserted to cure dermatitis, relieve chronic eczema, and maintain healthy skin (Stobdan et al., 2013).

The proportion of oil content in the pulp oil is comparatively lower to that in seeds and usually varies from 2% to 11% in the whole fresh berries, including the seeds (Ranjith and Arumughan, 2009; Zeb and Khan, 2008). The oil from

Table 3.1 Major Lipids Found in Various Parts of *Hippophae rhamnoides*

Lipids	References
Complex lipids: Galactolipids [rich in 18:1 ($n-9$) and 18:3 ($n-9, n-6, n-15$) fatty acids]; phospholipids [rich in 16:0 and 18:1 ($n-9$) fatty acids]	Ranjith and Arumughan (2009)
Carotenoids: α-, β-, γ-, δ-carotene, β-cryptoxanthin, lutein, lycopene, and zeaxanthin	Ranjith and Arumughan (2009), Sayegh et al. (2014)
Fatty acids: Linoleic (18:2), alpha-linolenic (18:3), behenic (22:0), stearic (18:0), myristic (14:0), heneicosanoic (21:0), pentadecanoic (15:0), and palmitoleic (16:1) acid	Kalia et al. (2011), Pavlović et al. (2016), Ranjith and Arumughan (2009), Šnē et al. (2013), Stobdan et al. (2013), Zheng et al. (2017)
Sterols: 24-Methylenecykloartanol, β-sitosterol, campesterol, citrostadienol, ergosterol, isofucosterol, lanosterol, sitosterol, stigmasterol, stigmastanol squalene, and uvaol	Pavlović et al. (2016), Ranjith and Arumughan (2009), Stobdan et al. (2013), Teleszko et al. (2015)
Triterpenoids saponins: 3-*O*-[β-ᴅ-glucopyranosyl(1→2)-β-ᴅ-glucopyranosyl-(1→3)]-[α-l-rhamnopyranosyl-(1→2)]-α-l-arabinopyranosyl-13-ene-19-one-28-oic acid 28-*O*-β-ᴅ-glucopyranosyl ester; 3-*O*-[β-ᴅ-glucopyranosyl(1→2)-β-ᴅ-glucopyranosyl-(1→3)]-[α-l-rhamnopyranosyl-(1→2)]-α-l-arabinopyranosyl-13-ene-19-one-30-hydroxyolean-28-oic acid 28-*O*-β-ᴅ-glucopyranosyl ester; 3-*O*-[β-ᴅ-glucopyranosyl (1→2)-β-ᴅ-glucopyranosyl-(1→3)]-[α-l-rhamnopyranosyl-(1→2)]-β-ᴅ-glucopyranosyl-13-ene-19-one-28-oic acid 28-*O*-β-ᴅ-glucopyranosyl ester; and 3-*O*-[β-ᴅ-glucopyranosyl(1→2)-β-ᴅ-glucopyranosyl-(1→3)]-[α-l-rhamnopyranosyl-(1→)] -β-ᴅ-glucopyranosyl-13-ene-19-one-30-hydroxyolean-28-oic acid 28-*O*-β-ᴅ-glucopyranosyl ester	Chen et al. (2014)
Triterpenic acids: oleanolic and ursolic acids	Hu et al. (2015)
Tocotrienols	Stobdan et al. (2013)

H. rhamnoides berry contains an average of 35% of the rare and valuable palmitoleic acid (16:1n-7) (Kalia et al., 2011; Ranjith and Arumughan, 2009). This rare omega-7 series fatty acid is a component of skin fat and is known to support cell, tissue, and wound healing (Kalia et al., 2011; Ranjith and Arumughan, 2009). It is practically absent in seed oils (Ranjith and Arumughan, 2009).

3.5.3 Organic Acids and Other Volatile Compounds

Among the organic acids, *H. rhamnoides* berries mainly contain malic acid and quinic acid, together constituting around 90% of all the fruit acids in different origins (Beveridge et al., 2002; Stobdan et al., 2011a, b). Others include citric, oxalic, tartaric, and succinic acid (Stobdan et al., 2011a, b). Volatile aroma compounds composition analysis in 13 cultivars of *H. rhamnoides* berries using gas chromatography (GC) with the solid phase micro-extraction method during two consecutive years (2012 − 2013) identified 69 volatile compounds comprising 26 alcohols, 12 aldehydes, 11 ketones, 9 acids, and 11 esters (Vítová et al., 2015). Structural elucidation using spectroscopic and chemical methods carried for fruits of *H. rhamnoides* led to isolation of 10 volatile compounds. These compounds were hippophae cerebroside (1), oleanolic acid (2), ursolic acid (3), 19-α-hydroxyursolic acid (4), dulcioic acid (5), 5-hydroxymethyl-2-furancarbox-aldehyde (6), cirsiumaldehyde (7), octacosanoic acid (8), palmitic acid (9), and 1-*O*-hexadecanolenin (10). Among them, hippophae cerebroside was claimed as a novel compound (Zheng et al., 2009).

3.5.4 Polyphenolic Compounds

Several analytical and preparative chromatographic techniques, including countercurrent chromatography, HPLC-DAD, HPLC-ECD, HPLC with chemi-luminescence detection, HPLC-ESI-MS/MS, Sephadex LH-20 gel chromatography, NMR, and FTIR, have been applied for isolation, purification, and structural elucidation of flavonoid and phenolic constituents from crude extracts of *H. rhamnoides* berries; its juice and pulp as well (Ranjith and Arumughan, 2009; Rösch et al., 2004). Flavonols are found to be the predominating polyphenols, with phenolic acids and catechins found in minor amounts (Ranjith and Arumughan, 2009). Polyphenolic compounds, namely isorhamnetin and its glycosides, typically are found in the largest quantities, together with quercetin and small amounts of kaempferol (Sayegh et al., 2014; Zhao et al., 2013). The flavonoid content usually is higher in leaves than in berries (Šnē et al., 2013).

On comparison with other wild berries, *H. rhamnoides* berries have the second-highest total flavonol content and therefore have excellent antioxidant potential (Sayegh et al., 2014). Phenolics, including flavonols, flavones, phenolic acids, proanthocyanidins, and hydrolysable tannins are reported as the major

Table 3.2 Principal Polyphenolic Compounds in Different Parts of *Hippophae rhamnoides*

Polyphenolic Compounds	References
Phenolic acids: Caffeic, chlorogenic, cinnamic, ellagic, ferulic, gallic, gentisic, *p*-coumaric, *p*-hydroxybenzoic, protocatechuic, sinapic, and syringic acids	Ranjith and Arumughan (2009), Sayegh et al. (2014), Šnē et al. (2013), Fatima et al. (2015), Li and Wardle (2003)
Flavonoid compounds: Isorhamnetin, kaempferol, myricetin, quercetin, rutin, and their glycosides (isorhamnetin-3-*O*-glycoside, kaempferol-3-*O*-glycoside, quercetin 3-*O*-glycoside, and 7-*O*-rhamnosides of isorhamnetin, kaempferol, and quercetin, 1-feruloyl-β-d-glucopyranoside, kaempferol-3-*O*-β-D-(6′′-*O*-coumaryl) glycoside, isorhamnetin-3-*O*-rutinoside, quercetin 3-*O*-β-D-glucopyranoside and quercetin 3-*O*-β-D-glucopyranosyl-7-*O*-α-l-rhamnopyranoside)	Ranjith and Arumughan (2009), Zhao et al. (2013), Kim et al. (2010), Upadhyay et al. (2010)
Tannins: Hippophaenins A and B; *Catechins* such as (−)-epicatechin, (+)-gallocatechin, (−)-epi-gallocatechin, (+)-gallocatecholgallate, (−)-epicatecholgallate, and (−)-gallocatechin and; proanthocyanidins like *Procyanidin*, prodelphinidin, coumarin, catechin-(4α-8)-catechin and catechin-(4α-8)-epicatechin	Kallio et al. (2014), Ranjith and Arumughan (2009), Šnē et al. (2013), Upadhyay et al. (2010)

contributors to the antioxidant, cytoprotective, and antibacterial activities of *H. rhamnoides* berries (Ranjith and Arumughan, 2009; Šnē et al., 2013; Upadhyay et al., 2010). The flavonoid content in the fruit of *H. rhamnoides* has been reported to range from 120 to 1000 mg/100 g fresh fruit; the total phenolic content of *H. rhamnoides* berries ranges from 114 to 244 mg/100 g fruit (Ranjith and Arumughan, 2009).

The total flavones of *H. rhamnoides* (TFH) have been extracted from its leaves and fruits using HPLC (Zhao et al., 2013). It has been reported that THF includes 12 compounds, with the major ones being quercetin, kaempferol, isorhamnetin, quercetin-3-*O*-glucoside, isorhamnetin 3-*O*-rutinoside. Animal and human studies suggest that THF exhibit antioxidant, anti-ulcerogenic, hypoglycemic, and hepato-protective actions and could lower blood viscosity, lower blood pressure, enhance cardiac function, and suppress platelet aggregation (Xie et al., 2010; Zhao et al., 2013). We have listed some of the many polyphenolic compounds detected in *H. rhamnoides* berries, leaves, pomace, pulp, and seeds in Table 3.2.

3.5.5 Proteins and Amino Acids

Protein is one of the essential chemical components in *H. rhamnoides* leaves that has value in animal feed and can be used as a source of unconventional protein for human food. Essential amino acids such as lysine, threonine, valine, methionine, and phenylalanine are present in *H. rhamnoides* leaves (Li and

Wardle, 2003; Šnē et al., 2013). They also contain two semiessential amino acids and eight non-essential amino acids (Qin et al., 2013).

Serotonin, a derivative of tryptophan and a neurotransmitter in humans is well-known for its antidepressive, energizing, and immune-induction effects. It has been quantified in the leaves, fruits, offshoots, and offshoot bark of *H. rhamnoides* (Brad et al., 2007).

3.5.6   Vitamins and Minerals

Vitamin C is one of the major vitamins found in *H. rhamnoides* (Zakynthinos and Varzakas, 2015). The leaves and fruits of *H. rhamnoides* contain vitamins C and E and folic acid, however, the concentration of vitamin C is higher in the berries (Ranjith and Arumughan, 2009). In smaller amounts, vitamin C also occurs in *H. rhamnoides* flowers, while the highest concentrations are in roots, and the concentration in plants grown at high altitudes can be twice that found in freshly squeezed juice (Šnē et al., 2013). The total free tocopherol content in oil from whole berries has been reported to be 1.01–1.28 mg/g of oil, with α-tocopherol (62.5%–67.9%) being predominant. While in seed oil, it is 1.1–2.3 mg/g (Kalia et al., 2011). Vitamin E is known to promote blood coagulation, and certain studies have indicated its potential in the treatment of AIDS (Kalia et al., 2011). The berries also contain δ-tocopherols (32.1%–37.5%) and traces of γ-tocopherol. Green berries, however, usually have marked amount of γ-tocopherol, but its content decreases to traces as the berries ripen (Ranjith and Arumughan, 2009; Sayegh et al., 2014).

The fruit of *H. rhamnoides* also exhibits a good mineral element profile as revealed through the phytochemical analysis by Sabir et al. (2005) and Gutzeit et al. (2008). *H. rhamnoides* berries, leaves, and juice are reported to contain calcium, iron, magnesium, phosphorus, potassium, sodium, and zinc (Gutzeit et al., 2008; Sabir et al., 2005). In a study higher levels of calcium, iron, manganese, sodium, and zinc were found in *H. rhamnoides* leaves as compared to the berries (Šnē et al., 2013). Berries and juices of *H. rhamnoides* also contain arsenic, boron, chromium, copper, manganese, molybdenum, nickel, and selenium as determined by atomic absorption spectroscopy (AAS) and inductively coupled plasma-mass spectrometry (ICP-MS) (Gutzeit et al., 2008). In *H. rhamnoides*, the microelements are combined as complex chelatic derivatives, which makes them bioavailable (Brad et al., 2007).

3.6   BIOACTIVITIES OF *H. RHAMNOIDES*

H. rhamnoides possess numerous bioactivities because of numerous metabolites with health-promoting benefits. This section includes some of these activities, especially in context to its bioefficacy against HA pathophysiology.

3.6.1 Antihypoxic Effects

Several studies advocate the protective efficacy of *H. rhamnoides* in ameliorating the ill-effects of hypoxia. For example, Saggu and Kumar studied the antioxidant potential of *H. rhamnoides* leaf aqueous extract to reduce oxidative stress in liver and muscle of rats on attaining a rectal temperature (T_{rec}) of 23°C during cold (5°C), hypoxia (428 mmHg), and restraint (C-H-R) exposure and after recovery (T_{rec} 37°C) from C-H-R-induced hypothermia. The C-H-R is a multiple stress animal model that considers attainment of T_{rec} of 23°C as a measure of endurance (Sharma et al., 2015). The aqueous extract was administered orally in rats at a dose of 100 mg/kg in both single and five doses, 30 min before exposure, and its effects were examined in terms of catalase (CAT), glutathione-S-transferase (GST), MDA, reduced glutathione (GSH) and superoxide dismutase (SOD) levels in liver and muscle. Single- and five-dose extract treatment restricted increment in liver and muscle MDA levels, and five doses of extract treatment further improved the levels of liver antioxidants, viz. GSH, on recovery of T_{rec} 37°C, increased SOD during exposure and recovery; normalized CAT activity in liver during C-H-R exposure and an increase on recovery of T_{rec} 37°C. Whereas, GST levels in both single-dose and five-dose extract-treated rats were similar to that of untreated rats. Therefore, the results indicated efficacy of that *H. rhamnoides* aqueous extract in reducing oxidative stress in liver and muscle of rats during C-H-R exposure and poststress recovery (Saggu and Kumar, 2007a).

Similar observations were made during evaluation of *H. rhamnoides* fruit aqueous extract's adaptogenic activity in rats under simulated conditions of C-H-R, along with its toxicity studies. The dose of 75 mg/kg body weight was considered as the maximal effective adaptogenic dose. The toxicity studies revealed that there were no significant changes in body weight, organ/body weight ratios, hematological and biochemical variables in any subacute or subchronically extract-treated rats compared to respective controls, when administered either with 1 g/kg and 2 g/kg for 14 days or 75 mg/kg and 500 mg/kg for 30 days. Taking into account the findings of this study, it can be concluded that the extract possessed potent adaptogenic activity and is nontoxic in rats at its maximal effective dose administered for 30 days (Tulsawani et al., 2010).

The effect of freeze-dried *H. rhamnoides* leaf extract powder incorporated into biscuits also has been studied recently in C-H-R animal models. There was 37% increase in time to attain T_{rec} 23°C and 42% faster recovery of T_{rec} 37°C, in comparison to control rats fed normal feed and in comparison to animals fed on diet containing placebo biscuits. These biscuits are claimed to be highly nutritive and rich in polyphenols and flavonoids, further supporting their usage as health supplement in diet (Khanum et al., 2017).

In 2005, Ti-juan et al. studied the acute toxicity of compound *H. rhamnoides* granule (CHRG), its effects on the thrombosis formation and the influence on the whole brain lacking oxygen and energy metabolism in mice. The whole brain lacking oxygen model was made by injecting $NaNO_2$ in mice and the contents of lactic acid (LD) and ATPase in brain were determined. Compared with the normal control group, CHRG at a dose 160 g/kg significantly delayed the persistent time of gasping, increased ATPase activity, and decreased LD contents. It also supported survival of mice on thrombosis formation and lengthened existent time in mice whole brain lacking oxygen otherwise. Overall, the study demonstrated the inhibitory action for thrombosis formation and protective effect for the whole brain lacking oxygen by CHRG (Ti-juan et al., 2005). The potential role of *H. rhamnoides* in preventing high altitude-induced polycythemia in rats has been studied by Zhou et al. The changes in hemodynamic, hematologic parameters, and erythropoietin content were considered and the results showed that administration of total flavonoids (35, 70, 140 mg/kg, ig) from *H. rhamnoides* aided in averting HA-induced polycythemia in rats (Zhou et al., 2012).

Studies pertaining to the neuroprotective effect of *H. rhamnoides* against hypobaric hypoxia also have been conducted. Narayanan et al. (2005) demonstrated that treatment of C-6 glioma cells with the alcoholic leaf extract of *H. rhamnoides* (200 μg/mL) significantly repressed cytotoxicity, reactive oxygen species (ROS) production, and maintained antioxidant levels similar to that of control cells. Further, the leaf extract restored the mitochondrial integrity and averted DNA damage induced by hypoxia (Narayanan et al., 2005). Another exemplary study suggesting the protective efficacy of *H. rhamnoides* against hypobaric hypoxia (equivalent at an altitude of 7600 m) and reoxygenation-induced neuronal injury in brain includes supplementation of its leaf extract (SBTLE; 100 mg/kg body weight). It was found that the extract reduced free radical generation and, subsequently, lipid peroxidation by upregulating the expression of the key biosynthetic enzyme: γ-glutamylcysteine synthetase, thereby increasing GSH level in hippocampal region of the brain. This group claimed that affluence of the extract in antioxidant compounds could be a contributing factor for this bioactivity (Baitharu et al., 2016).

Several researchers have carried out studies comprising elucidation of the mechanisms underlying the protective effect of *H. rhamnoides* against hypobaric hypoxia. Saggu and Kumar in 2007 proposed that *H. rhamnoides* extract treatment resulted in a shift from anaerobic to aerobic metabolism during C-H-R exposure and post-stress recovery, which otherwise are suppressed along with hexose monophosphate (HMP) pathway. Such an effect was attributable to the ability of the extract in limiting the decrease in tissue glycogen and enzyme activities: citrate synthase (CS), glucose-6-phosphate dehydrogenase (G-6-PD), hexokinase (HK), and phosphofructokinase (PFK) in blood, liver, and muscle during C-H-R exposure and recovery (Saggu and Kumar, 2007b). The same group in 2008 suggested another mechanism for adaptogenic activity of *H. rhamnoides* leaf aqueous extract. According to this study, the extract facilitated efficient use of free fatty acids (FFA) for energy production and conserved cell membrane permeability by decreasing blood MDA levels and increasing GSH and CAT levels while lowering FFA and lactate dehydrogenase (LDH) levels in serum on attaining T_{rec} 23°C during C-H-R exposure (Saggu and Kumar, 2008).

Molecular approaches also have been employed to comprehend the mechanisms responsible for the antihypoxic effects of *H. rhamnoides*. In 2014, we designed a study to elucidate protective action of aqueous extract of *H. rhamnoides* in hippocampal neurons against hypoxia (0.5% O_2) in vitro, with the hypothesis that the effect was mediated via Janus-activated kinase/signal transducers and activators of transcription (JAK/STATs) pathway. Exposure of neuronal cells to hypoxia (0.5% O_2) displayed higher ROS with compromised antioxidant status compared to unexposed control cells. Also observed were hypoxic exposure, elevated levels of pro-inflammatory cytokines, tumor necrosis factor α (TNF-α), interleukin 6 (IL-6), and nuclear factor kappa B (NF-κB). Besides, there was an increased expression of JAK1 with phosphorylation of STAT3 and STAT5. Cells treated with JAK1, STAT3, and STAT5 specific inhibitors resulted in more cell death compared to hypoxic cells. However, treatment with *H. rhamnoides* aqueous extract inhibited hypoxia induced oxidative stress by altering cellular JAK1, STAT3, and STAT5 levels thus augmenting cellular survival response to hypoxia and proposed plausible use of aqueous extract of *H. rhamnoides* in assisting tolerance to hypoxia (Manickam and Tulsawani, 2014).

In vivo, Keshri et al. (2015) investigated the potential effect of supplementation with aqueous *H. rhamnoides* extract in adult rats categorized as susceptible (<10 min), normal (10–25 min) and tolerant (>25 min) on the basis of time taken for onset of gasping (indicator for hypoxic tolerance) when exposed to a simulated altitude of 9754 m (~205 mmHg). Supplementation of the extract distinctly improved the hypoxic gasping time in animals. The researchers suggested that this might be because the *H. rhamnoides* mediated increased in the citrate synthase (CS), G-6-PD activity and AMP-activated protein kinase-α (AMPKα), and glucose transporter4 (GLUT4) expression in the treated group compared to hypoxia group. Thereby, aqueous *H. rhamnoides* extract supplementation imparted tolerance to hypoxia in susceptible animals by facilitating intracellular energy content and augmenting antioxidants under acute hypoxia (Keshri et al., 2015). The above investigation was supported by a study conducted by Jain et al. (2016), in which the efficacy of *H. rhamnoides* was evaluated using the onset of gasping time (GT) at an altitude of 9754 m that was 100% increased. It also curtailed hypoxia-induced cardiac damage and free radical production. It upregulated hypoxia inducible factor (HIF)-1α and increased heme oxygenase-1 along with 100% increase in nitric oxide (NO) levels. *H. rhamnoides* reduced protein carbonylation; potentiated antiinflammatory effects by downregulating nuclear factor kappa B (NF-κB) and tumor necrosis factor-alpha (TNF-α), enhanced heat shock protein (HSP) 70 levels, and significantly lowered the expression of endoplasmic reticulum (ER) stress markers: glucose regulated protein (GRP78), protein kinase R (PKR)-like endoplasmic reticulum kinase (PERK), and C/EBP homologous protein (CHOP) (Jain et al., 2016).

Earlier studies also indicate potential of *H. rhamnoides* leaf extract (SBTLE) in curtailing hypoxia-induced transvascular permeability in the lungs. SBTLE provided significant protection against hypoxia-induced transvascular permeability by stabilizing the levels of reduced glutathione and antioxidant enzymes. The extract also attenuated hypoxia-induced increase in vascular endothelial growth factor (VEGF) and catecholamine levels (Purushothaman et al., 2011). With a similar approach, Purushothaman et al. in 2008 demonstrated protective role of *H. rhamnoides* seed oil against hypoxia-induced transvascular fluid leakage in brain of hypoxia-exposed rats (Purushothaman et al., 2008).

3.6.2 Antioxidant Activity

The clinical complications of HA could develop oxidative stress on exposure. Detoxification with adaptogens to curb the imbalance between the free radical generation and body's defense mechanism could be of immense benefit in this context. Considering this, antioxidant activity is an additional factor to the adaptogenic property of any herbal formulation. Sharma et al. (2015) provided an illustrative study, wherein they demonstrated that in vitro antioxidant activity of various subspecies of *H. rhamnoides* and *H. salicifolia* influenced their adaptogenic potential during exposure to simulated cold and hypoxia environment. The results indicated that the aqueous extracts of both *Hippophae* species had higher efficacy than their alcoholic counterparts as adaptogens under multiple stresses of C-H-R exposure.

Recently, shade-dried leaves of 10 *H. rhamnoides* populations of Kargil were evaluated for their phenolic content and antioxidant activity. During the study, the samples were found to exhibit substantial antioxidant potential. The values for total phenolic content and antioxidant activity measured by DPPH assay varied from 42.8–72.4 mg GAE (gallic acid content) and 37.4%–89%, respectively (Gupta and Kaul, 2017). Besides, a study by Bai et al. (Bai et al., 2015) showed that the dried leaves, even the pomace of *H. rhamnoides* also have shown strong reducing power and potent scavenging effect against superoxide anion radical and hydroxyl radical.

In another study, the antioxidant activities of *H. rhamnoides* leaf tea extracts (S) were compared with that of green tea (G) extracts. The extracts were prepared using water (SW, GW) and ethanol at room temperature (SE, GE), respectively, and at 80°C (SWH, GWH, SEH, and GEH, respectively). GEH, GWH, SE, and SEH contained more antioxidant compounds (phenolic, flavonoid, and ascorbic acid content) and had higher activities, whereas SWH, SEH, GWH, and GEH had elevated antioxidant enzyme activity levels in H_2O_2-treated RAW264.7 cells. The cells treated with SWH and SEH demonstrated increased expression of nuclear factor (erythroid-derived 2)-like 2 and maintained the cellular levels of glutathione (GSH)/glutathione disulfide (GSSG) ratio similar to H_2O_2-untreated controls (Cho et al., 2014).

Various studies have found that the antioxidant activity of *H. rhamnoides* is attributable to its antioxidant metabolites composition. For example, a study conducted by Fan et al. (2013) found a positive correlation between proanthocyanidin content and antioxidant activities of all the *H. rhamnoides* seed extracts. *H. rhamnoides* seed extracts (250 µg/mL) containing substantial amount of proanthocyanidins inhibited formation of conjugated diene hydroperoxides and thiobarbituric acid reactive substance (TBARS) by 0%–90% and 0%–88.6%, respectively, when evaluated via copper-catalyzed lecithin liposome oxidation assay for their antioxidant potential (Fan et al., 2013). Another significant study suggested that the antioxidant compounds β-sitosterol (Saeidnia et al., 2014) and ursolic acid (Checker et al., 2012) were the major contributors for the antioxidant activity of male and female *H. rhamnoides* leaf extracts (Singh et al., 2014).

3.6.3 Cardioprotective Action

The flavonoids and unsaturated fatty acids in *H. rhamnoides* berry oil have shown to improve cardiovascular functions, with promising effect on cardiac rhythm and the strength of heart muscle contraction (Krejcarová et al., 2015; Suryakumar and Gupta, 2011). Basu et al. (2007) have provided evidence for the antiatherogenic and cardioprotective activity of supercritical CO_2 extracted *H. rhamnoides* seed oil. This group found that feeding *H. rhamnoides* seed oil to normal rabbits for 18 days caused a significant decline in plasma cholesterol, LDL-cholesterol (LDL-C), atherogenic index (AI) and LDL/HDL ratio along with significant increase in HDL-cholesterol (HDL-C) levels, HDL-C/TC ratio (HTR) and vasorelaxant activity of the aorta. Similar results were observed in cholesterol-fed hypercholestromic animals when administered *H. rhamnoides* seed oil. Additionally, a decline in the blood total cholesterol (TC), triglyceride (TG), LDL-C, and AI was observed. Moreover, the increase in HDL-C was more pronounced in seed oil-treated hypercholesterolemic animals than in normal oil-fed animals (Basu et al., 2007).

Another investigation accentuating the defensive role of *H. rhamnoides* oil against isoproterenol (ISO)-induced cardiotoxicity in rats was conducted by Malik et al. (2011). They found *H. rhamnoides* oil administered at a dosage of 20 mL/kg per day alleviated myocardial damage induced by ISO by maintaining biochemical, hemodynamic, histopathological, and ultrastructural perturbations because of its free radical scavenging and antioxidant activities (Malik et al., 2011).

The fruit and leaves of *H. rhamnoides* also have been found to improve the functioning of the cardiovascular system. Their flavonoids have been acclaimed as contributing metabolites for their cardioprotective activity. The total flavonoids of *Hippophae* (THF) extracted from its leaves and fruits protect against myocardial ischemia in addition to being effective against reperfusion, tumors, oxidative injury, and aging (Suryakumar and Gupta, 2011). *H. rhamnoides* fruits also exhibit antihypertensive potential as described by Pang et al. (2008). This group designed a study to investigate the antihypertensive effect of total flavones extracted from seed residues of *H. rhamnoides* (TFH-SR) and its underlying mechanism in chronic sucrose-fed rats. The results indicated that TFH-SR (at the dose of 150 mg/kg/day) exerted its antihypertensive effects partly by improving insulin sensitivity and blocking angiotensin signal pathway. Overall, the study advocated potential use of TFH-SR in the management of hyperinsulinemia in a nondiabetic state with cardiovascular diseases (Pang et al., 2008). The work conducted by Koyama et al. (2009) also advocate the antihypertensive activity of *H. rhamnoides*. In their study, the powder of *H. rhamnoides* fruits reduced hypertensive stress on the ventricular microvessels in hypertensive stroke-prone rats when they were fed a concentration of 0.7 g/kg of feed for 60 days. These rates showed significant decrease in the mean arterial blood pressure, heart rate, total plasma cholesterol, triglycerides, and glycated hemoglobin when their diet was supplemented with the powder. A decreased alkaline phosphatase expression by the arteriolar capillary portions of microvessels and increase in total capillary density also was noted (Krejcarová et al., 2015).

Another study highlighting the cardioprotective effect of polyphenol extracts from *H. rhamnoides* (PESB) against myocardial ischemia reperfusion injury (MIRI) also was conducted. Overall, this study demonstrated that PESB protected rat hearts against I/R injury by inhibiting autophagy. In order to explain how PESB mediated I/R-induced autophagy, the expression levels of Beclin-1 and the ratio of microtubule-associated protein 1 light chain 3 (LC3) I and II, which have been regarded as autophagy markers, were measured. Autophagy is impaired following myocardial I/R because of upregulation of Beclin-1 while, an increased ratio of LC3-II/I indicates autophagosome formation. The study revealed that treatment with PESB caused significant ($P < 0.01$) reduction in Beclin-1 expression levels and LC3-II/I ratios compared with I/R control group, demonstrating inhibition of autophagy by PESB treatment. PESB treatment also decreased the coronary effluent of lactate dehydrogenase (LDH) and creatine kinase isoenzyme (CK-MB) in isolated rat hearts, indicating decrease in the extent of myocardial injury. The authors proposed substantial antioxidant activity of PESB in vitro for this activity (Tang et al., 2016).

3.6.4 Hepatoprotective Action

The liver is a vital organ involved in various physiological processes such as metabolism, secretion, storage and detoxification. Hepatic disease has become one of the serious concerns in human health today. Treatment of hepatic diseases with herbal-based products have gained attention of pharmaceutical industries because they are less toxic than modern medicines and have no or minimal side effects (Madrigal-Santillán et al., 2014). *H. rhamnoides* also exhibit hepatoprotective action as evident from several studies.

The study conducted by Geetha and her coworkers evaluated the hepatoprotective activity of *H. rhamnoides* on carbon tetrachloride (CCl_4)-induced liver injury in male albino rats. Pretreatment of leaf extract at concentrations of 100 and 200 mg/kg body weight significantly ($P < 0.05$) protected the animals from CCl_4-induced liver injury. The extract significantly restricted the CCl_4-induced increase of glutamate oxaloacetate transferase (GOT), glutamate pyruvate transferase (GPT), alkaline phosphatase (ALP), and bilirubin, and better maintained protein levels in the serum. It also enhanced GSH and decreased hepatic MDA levels (Geetha et al., 2008). Following this work, Maheshwari et al. in 2011 assessed the hepatoprotective activity of phenolic rich fraction (PRF) of *H. rhamnoides*. The fraction was found to be rich in phenolic compounds: gallic acid, isorhamentin, kaempferol, myricetin, and quercetin. Its administration (25–75 mg/kg body weight) significantly decreased CCl_4-induced increase in aspartate aminotransferase (AST), alanine aminotransferase (ALT), c-glutamyl transpeptidase (GGT), and bilirubin in serum, elevation in hepatic lipid peroxidation, hydroperoxides, and protein carbonyls, and depletion of hepatic GSH. It also enhanced the activities of hepatic antioxidant enzymes: CAT, GPX, glutathione reductase (GR), GST, and SOD. Histopathological changes such as hepatocytic necrosis, fatty changes, and vacuolation also were limited by PRF (Maheshwari et al., 2011).

Another study examined the hepatoprotective activity of oil from *H. rhamnoides* berries against toxicity induced by aflatoxin B1 (AFB1) in broiler chickens. Simultaneous administration of AFB1 and *H. rhamnoides* berries oil resulted in significant reduction of hepatic necrosis and fatty liver formation compared with chickens treated with only AFB1. A significant decrease in liver expression levels of cyclooxygenase (COX)-2, B-cell lymphoma (Bcl)-2, and p53 also were observed with *H. rhamnoides* oil supplementation. These findings showed that *H. rhamnoides* oil reduces the concentration of aflatoxins in liver and lessens their adverse effects, and therefore has potential hepatoprotective activity (Solcan et al., 2013). A study in 2014 determined the protective efficacy of *H. rhamnoides* in the treatment of human patients with nonalcoholic fatty liver disease (NAFLD). This group measured serum lipids, transaminase, and serum liver fibrosis indices along with liver computed tomography (CT) and Fibroscan examination at baseline and after *H. rhamnoides* treatment (orally, 1.5 g thrice per day for 3 months). Results demonstrated that *H. rhamnoides* treatment significantly decreased serum levels of ALT, LDL-C, hyaluronic acid, and collagen type IV. The CT liver/spleen ratio of the treated patients also was increased considerably with a significant reduction in the liver stiffness measurement (LSM) of the treated patients as compared with control patients or baseline data (Gao et al., 2014). Zhang et al. (2017) demonstrated the potential role of *H. rhamnoides* berry polysaccharide (SP) against carbon tetrachloride (CCl4)-induced hepatotoxicity. Oral pretreatment with SP (once daily for 14 consecutive days) prior to CCl_4 challenge reduced ALT, AST, total bilirubin levels, and MDA content. An increase in the levels of pre-albumin, glutathione peroxidase (GSH-Px), SOD activities, and GSH levels, along with diminished liver injuries, also were observed. SP pretreatment distinctly reduced the CCl_4-induced expression of TNF-α, interleukin-1 beta (IL-1β), inducible nitric oxide synthase (iNOS and NO). Other observations made with SP pretreatment were decrease in hepatic Toll-like receptor 4 (TLR4) expression and inhibition of p38 MAPK phosphorylation, extracellular signal-regulated kinase (p-ERK), c-Jun N-terminal kinase (p-JNK), and NF-κB in CCl_4-challenged mice. Therefore, this study illustrated that the antioxidative and antiinflammatory activities of SP contributed toward its hepatoprotective activity against CCl_4-induced liver damage (Zhang et al., 2017).

3.6.5 Anticarcinogenic Activity

H. rhamnoides is enriched with a plethora of bioactive constituents with potential antioxidant potential. Antioxidants prevent oxidative damages of the mitochondrial system and neutralize free-radicals, which can induce cancer of the respective organ (Zeb, 2006). *H. rhamnoides* contains quercetin, an anticarcinogenic compound, which induces apoptosis in cancer cells and has been reported to be effective in the treatment of patients with colon cancer, leukemia, and prostatic carcinoma (Patel et al., 2012). The oil from *H. rhamnoides* is described to alleviate hematological damage caused by leukemia (Yang and Kallio, 2002). Yasukawa et al., in 2009, isolated three phenolic compounds, (+)-catechin, (+)-gallocatechin, and (−)-epigallocatechin, and a triterpenoid, ursolic acid from the active fraction of the 70% ethanol extract of *H. rhamnoides*, demonstrating a significant antitumor activity (Yasukawa et al., 2009). The inhibition of factors causing stomach cancer in humans by *H. rhamnoides* also have been cited (Li and Beveridge, 2003). Induction of the apoptotic activity and apoptotic morphological changes of the nucleus, including chromatin condensation, also have been observed in HL-60 cells treated with flavonols isolated from *H. rhamnoides*, such as quercetin, kaempherol, and myricetin (Hibasami et al., 2005). However, Grey et al. (2010) observed that *H. rhamnoides'* ethyl acetate fraction that is rich in ursolic acid had more pronounced antiproliferative effects on the Caco-2 cell line than its ethanol:water extract, which had strong inhibitory effects on Hep G2 cells proliferation. The authors claimed that ursolic acid was more significant than the polyphenols in inhibiting the cancer cell proliferation. *H. rhamnoides* leaf extract was studied for its stimulatory apoptotic and inhibitory effect on proliferation of rat glioma C6 cells. The results revealed that *H. rhamnoides* leaf extract demonstrated such effects because of its ability to decrease ROS production and promote nuclear localization of proapoptotic protein Bcl-2-associated X (Bax) and upregulation of Bax (Kim et al., 2017).

3.6.6 Antimicrobial Effect

H. rhamnoides contains the phenolic compounds as major group of phytochemicals that possess proven antibacterial and antiviral effects. These compounds suppress Gram-negative bacteria (Khan et al., 2010) and reduce Gram-positive bacterial growth (Kumar and Sagar, 2007). Hipporamin, a purified fraction of *H. rhamnoides* polyphenols fraction containing monomeric hydrolysable gallo-ellagi-tannins (strictinin, isostrictinin, casuarinin, casuarictin pedunculagin, stachyurin) exhibits a broad-spectrum antibacterial as well as antiviral activity (Krejcarová et al., 2015). Antimicrobial activities have been reported not only for *H. rhamnoides* berries Michel et al., 2012), but also for its seeds (Chauhan et al., 2012; Michel et al., 2012) and leaves (Upadhyay et al., 2010; Michel et al., 2012). Michel et al. (2012) reported that *H. rhamnoides* mainly inhibits *Bacillus cereus, Enterococcus faecalis, Pseudomonas aeruginosa, Staphylococcus aureus,* and *Yersinia enterocolitica.* Oil obtained by pressing is a very effective inhibitor of bacterial growth, especially of *Escherichia coli* (Michel et al., 2012).

H. rhamnoides also exhibits biological properties against viruses, such as those that cause influenza and herpes. It inhibits viral neuraminidase of the influenza virus and has reported in vitro anti-HIV as well (Shipulina et al., 2005, Michel et al., 2012). *H. rhamnoides* bud dry extract, at a concentration of 50 μg/mL, was found to be able to reduce the growth of the Influenza-A H1N1 virus in vitro (Torelli et al., 2015). Jain et al., in 2008, suggested that *H. rhamnoides* leaf extract has a significant antidengue activity (Jain et al., 2016). Taken together, these investigations describe substantial potential of *H. rhamnoides* against microbes.

3.6.7 Antiinflammatory Effect

Inflammation holds a vital role in several diseases, such as asthma, atherosclerosis, and rheumatoid arthritis. Although inflammation is primarily a protective response, inflammation that is chronic and uncontrolled becomes detrimental to tissues. An inflammatory response involves proinflammatory cytokines, adhesion molecules, growth factors, and NF-κB, and is antagonized by antiinflammatory cytokines. NF-κB regulates the expression of various genes encoding proinflammatory cytokines, chemokines, growth factors, and inducible enzymes such as cyclooxygenase-2 (COX-2) and inducible iNOS, both of which stimulate the production of proinflammatory mediators (Mueller et al., 2010). Treatment with *H. rhamnoides* leaf extract has caused a significant decrease in NO production induced by lipopolysaccharide (LPS) via iNOS activation in murine macrophage cell line RAW 264.7 (Padwad et al., 2006). This study is supported by the 2017 investigation by Tanwar et al. In their study, the antiinflammatory activity of fractions (parallel methanol, PM; successive chloroform, SC; and successive methanol, SM) obtained from leaves of *H. rhamnoides* in mouse peritoneal macrophages were evaluated. The results revealed that the cells treated with PM fraction significantly suppressed LPS-induced NO production and proinflammatory cytokines such as TNF-α, IL-6, and IFN-γ as compared to SC and SM treatment. The iNOS and COX-2 expressions also were reduced significantly after treatment with PM fraction. The PM fraction was characterized for the bioactive compounds such as carbohydrate, proteins, and tannins groups. In conclusion, the study emphasized the anti-inflammatory effect of *H. rhamnoides* via its immunomodulatory action (Tanwar et al., 2017).

H. rhamnoides leaf extract also has demonstrated promising antiinflammatory action in adjuvant induced arthritis (AIA) rat model via its immunomodulatory effect (Ganju et al., 2005).

3.6.8 Dermatological Activity

H. rhamnoides oil traditionally has been used to improve skin injuries and support the process of dermal healing. In modern medicine, it upholds the same application and is widely used to cure various skin conditions, including burns, eczema, scalds, ulcerations, infections, wounds, and skin-damaging effects of solar and therapeutic radiations (Kumar et al., 2011). *H. rhamnoides* oil has UV-blocking activity as well as emollient properties and it is an aid in promoting regeneration of tissues (Zeb, 2004). It absorbs strongly in the UV-B range (290–320 nm) and is an active ingredient of natural sunscreen absorber creams (Suryakumar and Gupta, 2011). Topical administration of *H. rhamnoides* flavone in wounded rats enhanced the healing process as indicated by improved rate of wound contraction and decreased time taken for epithelialization (16.3 days vs 24.8 days in controls). Moreover, there was a noteworthy increase in hydroxyproline (26.0%), hexosamine (30.0%) content along with increment in vitamin C (70.0%) content, catalase (20.0%), reduced glutathione (55.0%) activities, and decrease in lipid peroxide levels (39.0%) in wound granulation tissue. The study suggested that the *H. rhamnoides* flavone promotes wound healing activity (Gupta et al., 2006).

In addition to the flavones, *H. rhamnoides* procyanidins (SBPC) also have shown healing action on acetic acid-induced lesions in the rat stomach. SBPC was found to reduce the size of the ulcers at day 7 and day 14. The SBPC-treated group, in comparison to a control group, had a lowered ulcer index (UI) and a higher epidermal growth factor (EGF) in plasma. The authors suggested increased expression of epidermal growth factor receptor (EGFR) and proliferating cell nuclear antigen (PCNA) around the ulcer in high-dose SBPC stomach as the plausible mechanism behind the healing role of SBPC of acetic acid-induced gastric lesions (Xu et al., 2007).

Topical application of *H. rhamnoides* leaf extract at effective concentration of 5.0% (w/w), in comparison to control and silver sulfadiazine (SSD) ointment-treated groups, accelerated neovascularization, collagen synthesis, and stabilization at the wound site, because of up-regulated expression of VEGF, collagen type-III, matrix metalloproteinases, and increased levels of hydroxyproline and hexosamine (Upadhyay et al., 2011). A similar molecular mechanism has been reported for the protective role in healing of burn wounds in rats by supercritical CO_2-extracted *H. rhamnoides* seed oil. *H. rhamnoides* seed oil was administered by two routes at a dose of 2.5 mL/kg body weight (p.o.) and 200 μL (topical) for seven days. This investigation signified the safety and efficacy of supercritical CO_2-extracted *H. rhamnoides* seed oil on a burn wound model (Upadhyay et al., 2009).

Palmitoleic acid, an important metabolite of *H. rhamnoides* oil, is a component of skin and is regarded a valuable agent in treating burns and healing wounds. This fatty acid also nourishes the skin when taken orally and is considered effective in treating systemic diseases, such as atopic dermatitis (Zeb, 2004). A study by Hou et al. (2017), involved evaluation of topical effects of *H. rhamnoides* oil on atopic dermatitis (AD)-like lesions in a mouse model generated by repeated topical administration of 2,4-dinitrochlorobenzene (DNCB) in BALB/c mice. Topically applied *H. rhamnoides* oil in DNCB-treated mice improved the severity score of dermatitis, reduced epidermal thickness, decreased spleen and lymph node weights, and prevented mast cell infiltration. In IFN-γ/TNF-α-activated HaCaT cells, *H. rhamnoides* oil suppressed Th2 chemokines TARC and MDC via dose-dependent inhibition of NF-κB, JAK2/STAT1, and p38-MAPK signaling pathways. Overall, the results advocated potential role of *H. rhamnoides* oil in the treatment of atopic dermatitis (Hou et al., 2017).

3.6.9 Platelet Aggregation Activity

H. rhamnoides flavonoids suppress platelet aggregation induced by collagen, probably by inhibition of the tyrosine kinase activity (Patel et al., 2012). Another important component of *H. rhamnoides* that facilitates platelet aggregation is sitosterol (Johansson et al., 2000). The leaves, fruit, and seed (oil) of *H. rhamnoides* all exhibit ability to inhibit platelet aggregation (Vij et al., 2010).

3.6.10   Radioprotective Effect

The polyphenols in the *H. rhamnoides* berries are suggested to impart radioprotective potential, probably because of their antioxidant activity. A study conducted in 2006 by Shukla et al. revealed the radioprotective potential of a polyphenolic rich fraction, REC-1001 from *H. rhamnoides* berries. REC-1001 (250 µg/mL) protected mitochondrial and genomic DNA from radiation-induced damage in rat thymocytes (Shukla et al., 2006). These results are corroborated by a study conducted by Chawla et al. (2007). In another study, a preparation of *H. rhamnoides* leaf extract named SBL-1 also has shown its radioprotective efficacy. A single dose of SBL-1 rendered a survival rate of >90% when administered 30 min prior to irradiation and 80%–90% survival when administered 1–4 h before irradiation. It activated hemopoietic stem cell proliferation, increased total thiol (T-SH), decreased plasma and liver free radicals and lipid peroxidation, and normalized the liver ALP activity (Bala et al., 2009). The radioprotective SBL-1 also has demonstrated its ameliorative effects in improving microbiota dysbiosis (Bala et al., 2014) and renal damage caused by ^{60}Co-gamma-radiation (Saini et al., 2014). Research by Bala et al. in 2015 suggested that the presence of polyphenols, such as quercetin, ellagic acid, gallic acid; tannins; and thiols identified in SBL-1 might have contributed to radiation protection by stimulating cryptal stem cells proliferation, regulating apoptosis, and countering radiation-induced chromosomal damage through their antioxidant activity in jejunum and bone marrow of ^{60}Co-gamma-irradiated mice (Bala et al., 2015).

Oxidative photodamage caused by UV radiation results in serious skin damage characterized by laxity, pigmentation, roughness, and wrinkling. *H. rhamnoides* has shown its therapeutic effect against oxidative photodamage by UV radiation. The efficacy of an *H. rhamnoides* fruit blend (SFB) comprising *H. rhamnoides* fruit extract, blueberry extract, and collagen on UV-induced skin aging were examined by treating hairless mice for six weeks with UV irradiation and SFB administered orally. The advantages of SFB intake included a decrease in wrinkle formation, lowered thickness of the epidermis, and an increase in skin moisture content in the damaged skin of UV-irradiated mice. SFB application decreased MMP-1 and MMP-9 expression levels compared with those in the vehicle-treated group, describing the mechanism underlying its potential against UV radiation-induced skin damage (Hwang et al., 2012).

3.7 HEALTH AND NUTRITIONAL PRODUCTS FROM *H. RHAMNOIDES*

The literature provides substantial evidence from traditional uses as well as modern scientific investigations regarding numerous medicinal and nutritional benefits of *H. rhamnoides*. Several *H. rhamnoides*-based products are being marketed in China, Finland, Germany, India, Mongolia, and the Scandinavian countries, among others (Bhartee et al., 2014). The fruit, seed oil, pulp, and leaves of *H. rhamnoides* all have been exploited for their nutritional and therapeutic potentials (Bhartee et al., 2014). A wide range of products, such as food, beverages, health supplements, cosmetic products (bath soap, body lotions, hair oil, shampoo), baby food, candy, vitamin C tablets, and ice cream are available throughout the globe. About 10 different drugs have been isolated from *H. rhamnoides* in Asia and Europe and are available in various forms, such as aerosols, films, liniments, liquids, powders, plasters, pastes, pills, and suppositories (Bhartee et al., 2014). Some of the commercially available *H. rhamnoides*-based products are shown in Fig. 3.2.

Remarkable efforts have been made by various laboratories of India's Defence Research Development Organization (DRDO) in developing products from *H. rhamnoides* with established health benefits, especially with reference to survival of Indian armed forces and civilians under the extremes of high altitude environments (Bhartee et al., 2014; Stobdan et al., 2011a, b). These DRDO laboratories include Defence Institute of High Altitude Research (DIHAR), Leh; Defence Food Research Laboratory (DFRL), Mysore; and Defence Institute of Physiology and Allied Sciences DIPAS (DRDO), Delhi. DIHAR has major role, contributing since the early 1990s in the development of various *H. rhamnoides*-based products (Stobdan et al., 2011a, b). Significant contributions made by these institutions are detailed in the following subsections.

3.7.1 Contributions by DIHAR, Leh

DIHAR, Leh, has developed a patented process for preparation of an *H. rhamnoides* fruit-based beverage from the highly acidic berries. This beverage is commercially available in the Indian market under the brand names Leh Berry, Ladakh Berry, Power Berry, and Sindhu Berry. The process of preparation maintains all the preventive and curative properties of *H. rhamnoides* berries (Stobdan et al., 2011a, b). Seapricot, another beverage, has been developed by blending *H. rhamnoides* and apricot fruit pulp. Seapricot is suggested to exhibit antioxidant, nutraceutical, and health-refreshing properties because of its richness in vitamins A, B1, B2, B3, B6,

FIG. 3.2

Some of the commercially available products based on *Hippophae rhamnoides* berries, fruit pulp, leaves, and seed oil. Below each image, a company website link is given for each product.

B9, B12, C, and E, unsaturated fatty acids, carbohydrate, proteins, and minerals. (Stobdan et al., 2011a, b). A patented *H. rhamnoides* jam contains vitamins (thiamine, riboflavin, niacin, pantothenic acid, B6, B12, C, A, and E) and minerals (calcium, copper, iron, magnesium, phosphorus, potassium, selenium, sodium, and zinc) as the major constituents (Stobdan et al., 2011a, b).

Herbal tea also has been developed using *H. rhamnoides* leaves as a major constituent. It is claimed to possess high quantities of polyphenols, which detoxify free radicals. The other constituents are locally available cost-effective species and herbs. It has refreshing, stimulating, soothing, and stress-relieving effects, as well as being a thirst quencher with fine aroma. Studies have revealed that the tea mix has the potency to reduce the oxidative stress induced by cholesterol-induced hypercholestrolemia in experimental animals. It also has demonstrated its protective effect against oxidative stress-induced DNA damage (Khanum et al., 2017). The tea is rich in flavonoids and vitamins. The technology has been transferred to commercial firms and is available in the Indian market under trade name Si-beri (Stobdan et al., 2011a, b). DIHAR also has developed an antioxidant herbal food supplement from *H. rhamnoides* and other fruit pulp. This product is rich in vitamin C (124 mg/100 g), vitamin A (121 IU/100 mL), vitamin-B complex, vitamin E, minerals, and unsaturated fatty acids (Stobdan et al., 2011a, b). A soft gel capsule containing *H. rhamnoides* oil is being marketed as being rich in omega fatty acids, vitamin E, and carotene. This capsule, at a dose of 500 mg per day, is claimed to better oxidative stress, improve cardiovascular circulation, and enhance cognitive functions (Stobdan et al., 2011a, b).

An oil from *H. rhamnoides* to protect the skin against harmful UV radiations has been developed with a patented extraction process. This oil includes standardized proportions of oil extracted from *H. rhamnoides*, *Prunus armeniaca* and *Seasemae indicum* along with other high altitude medicinal herbs containing UV protective compounds. It has been found to be rich in unsaturated fatty acids and carotenoids, and has a topical antioxidant, as well as UV reflective, skin nourishing, and moisturizing properties (Stobdan et al., 2011a, b). An herbal adaptogenic appetizer is composed of *H. rhamnoides*, apricot, and rhodiola by DIHAR. It is an appetizer that is rich in vitamins, minerals, unsaturated fatty acids, and isorhmnoids and has adaptogenic properties (Stobdan et al., 2011a, b).

3.7.2   Contributions by DFRL, Mysore

DFRL has developed numerous bakery products emphasizing the nutritional benefits of *H. rhamnoides*. It has prepared biscuits from lyophilized *H. rhamnoides* pulp and leaf extracts. These biscuits are are marketed as being rich in dietary fiber, flavonoids, and other antioxidant phytoconstituents with substantial adaptogenic activity as evaluated by C-H-R studies. Other bakery products such as buns, bread, cakes, and rusks also have been developed with antioxidant-rich *H. rhamnoides* leaves (Khanum et al., 2017; Stobdan et al., 2011a, b). DFRL also has formulated yogurt with *H. rhamnoides* and *Lactobacillus bifido*, which is easily digestible, antioxidant enriched, and has the ability to protect and maintain normal microbial flora (Khanum et al., 2017; Stobdan et al., 2011a, b). *H. rhamnoides* fruit pulp jelly blended with pulp of other fruits (papaya, watermelon, and grapes) with good organoleptic characteristics and antioxidant content is another contribution by DFRL as well as health drinks such as spiced squash and squash blended with other fruits (Khanum et al., 2017; Stobdan et al., 2011a, b).

3.7.3 Contributions by DIPAS, Delhi

DIPAS has developed *H. rhamnoides*-based products of pharmaceutical importance for combating the ill-effects of high altitude-induced hypoxia.

DIPAS has produced DIP-LIP, a supercritical CO_2 extracted *H. rhamnoides*-based product rich in omega-3, omega-6, and omega-9 fatty acids, tocopherols and β-sitosterol. It is marketed as a nontoxic vasorelaxant with considerable antiatherogenic and cardioprotective potential (Rathor et al., 2014; Stobdan et al., 2011a, b). DIP-HIP, another significant contribution by DIPAS, is an herbal adjuvant developed from supercritical CO_2 extracted *H. rhamnoides*-based product. It is marketed to enhance immunogenicity of vaccines by accelerating both the cell mediated and humoral immunogenic responses. It is safe, efficacious, and comparable to commercially available adjuvants such as Freund's adjuvant and alum. It has been found to be compatible with different antigens and requires fewer boosters for its effectiveness. It has shelf life of more than three years in its crude form, while in formulation with antigens it is stable for four months at 4°C (Rathor et al., 2014: Stobdan et al., 2011a, b). DIPAS also has produced DIPANTOX, an herbal antioxidant with high vitamin C, vitamin E, flavonoids, and phytosterols contents of *H. rhamnoides* origin. It has potent adaptogenic, antioxidant activity and promotes hypoxic survival. The preparation is marketed as an antioxidant in stress-induced disorders (Stobdan et al., 2011a, b). DIPAS has also contributed Herbo-healer, a potent wound healer. Its treatment causes increase in endogenous antioxidants and increase in lipid peroxidation levels in the wound granulation tissue. It enhances blood flow and reduces inflammation during cold injury. The preparation has wide application for incision and excision type injuries, diabetic wounds, burn wounds, and surgical dressings. It also has antibacterial activity and is safe to use (Rathor et al., 2014; Stobdan et al., 2011a, b).

3.8 CONCLUSIONS

High altitude environments induce numerous health ailments by distorting the normal oxidative status and its related bioenergetics and biochemical physiological processes. In order to restore and/or to prevent such physiological imbalances, a potent herbal remedy with substantial antioxidant and adaptogenic activity such as *Hippophae* could be of immense benefit. *H. rhamnoides* is enriched with several nutrients, including antioxidants, and has been found to be more efficacious than other herbal plants. This chapter has described most of the significant research regarding the potentials of *H. rhamnoides* formulations and health supplements, especially in context to hypobaric hypoxia. Commercialization of these products, however, needs serious consideration to make them available globally for their potential nutritional and medicinal applications. Target-based studies with concentration on their mechanisms of action, lethal dose/effective dose, and bioavailability also are required to establish the traditional usage of *H. rhamnoides* based-products in modern medicine.

Acknowledgments

Authors are thankful to Dr. Kshipra Misra, additional director, head of the Department, DIPAS, DRDO, for her constant support and constructive suggestions. Dr. Manimaran M. acknowledges the research support from PathGene Healthcare Private Limited, Tirupathi-517 501, A.P., India.

References

Acharya, S., Stobdan, T., Singh, S.B., 2010. Seabuckthorn (Hippophae sp. L.): new crop opportunity for biodiversity conservation in cold arid trans-Himalayas. J. Soil Water Conserv. 9 (3), 201–204.

Bai, S., Tang, C., Tian, J., Yan, H., Xu, X., Fan, H., 2015. Extraction and antioxidant activity of total flavonoids from Sea Buckthorn Pomace. Food Sci. 10, 014.

Baitharu, I., Deep, S.N., Ilavazhagan, G., 2016. Neuroprotective effects of sea buckthorn leaf extract against hypobaric hypoxia and post-hypoxic reoxygenation induced hippocampal damage in rats. Int. J. Phytomed. 8 (1), 47–57.

Bala, M., Prasad, J., Singh, S., Tiwari, S., Sawhney, R.C., 2009. Whole-body radioprotective effects of SBL-1: a preparation from leaves of Hippophae rhamnoides. J. Herbs Spices Med. Plants 15 (2), 203–215.

Bala, M., Beniwal, C.S., Tripathi, R.P., Prasad, J., 2014. Effects of SBL-1 on jejunal microbiota in total body 60 cobalt gamma-irradiated mice-a metagenomic study with implications toward radioprotective drug development. Balance 1, 3.

Bala, M., Gupta, M., Saini, M., Abdin, M.Z., Prasad, J., 2015. Sea buckthorn leaf extract protects jejunum and bone marrow of 60 cobalt-gamma-irradiated mice by regulating apoptosis and tissue regeneration. Evid. Based Complement. Alternat. Med. 2015, 765705.

Banjade, M.R., 1999. Sea Buckthorn: Gift for the Fragile Mountains. A Project Paper Submitted to the Partial Fulfillment of the Requirement of B. Sc. Forestry Degree. Institute of Forestry, Pokhara, Nepal.

Basistha, B.C., 2001. Vivipary in sea buckthorn (*H. salicifolia* D. Don). J. Hill Res. 14 (1), 67.

Basu, M., Prasad, R., Jayamurthy, P., Pal, K., Arumughan, C., Sawhney, R.C., 2007. Anti-atherogenic effects of seabuckthorn (*Hippophaea rhamnoides*) seed oil. Phytomedicine 14 (11), 770–777.

Beveridge, T., Harrison, J.E., Drover, J., 2002. Processing effects on the composition of sea buckthorn juice from *Hippophae rhamnoides* L. cv. Indian summer. J. Agric. Food Chem. 50 (1), 113–116.

Bhartee, M., Basistha, B.C., Pradhan, S., 2014. Sea buckthorn: a secret wonder species. SMU Med. J 1 (2), 102–115.

Brad, I., Vlasceanu, G.A., Brad, I.L., Manea, S.T., 2007. Characterization of Sea buckthorn fruits and copses in terms of serotonin and microelements. Innov. Roman. Food Biotechnol. 1, 24.

Buzoianu, A.D., Ioan, V., Socaciu, C., 2014. Untargeted metabolomics for sea buckthorn (*Hippophae rhamnoides* ssp. *carpatica*) berries and leaves: Fourier Transform Infrared spectroscopy as a rapid approach for evaluation and discrimination. Notulae Bot. Horti Agrobot. Cluj-Napoca 42 (2), 545–550.

Chakraborty, M., Karmakar, I., Haldar, S., Das, A., Bala, A., Haldar, P.K., 2016. Amelioration of oxidative DNA damage in mouse peritoneal macrophages by *Hippophae salicifolia* due to its proton (H+) donation capability: ex vivo and in vivo studies. J. Pharm. Bioallied Sci. 8 (3), 210.

Chauhan, S., Varshneya, C., 2012. The profile of bioactive compounds in Seabuckthorn: berries and seed oil. Int. J. Theoret. Appl. Sci. 4, 216–220.

Chauhan et al., 2007, Michel, T., Destandau, E., Le Floch, G., Lucchesi, M.E., Elfakir, C., 2012. Antimicrobial, antioxidant, and phytochemical investigations of sea buckthorn (*Hippophaë rhamnoides L.*) leaf, stem, root, and seed. Food Chem. 131 (3), 754–760.

Chawla, R., Arora, R., Singh, S., Sagar, R.K., Sharma, R.K., Kumar, R., Sharma, A., Gupta, M.L., Singh, S., Prasad, J., Khan, H.A., 2007. Radioprotective and antioxidant activity of fractionated extracts of berries of *Hippophae rhamnoides*. J. Med. Food 10 (1), 101–109.

Checker, R., Sandur, S.K., Sharma, D., Patwardhan, R.S., Jayakumar, S., Kohli, V., Sethi, G., Aggarwal, B.B., Sainis, K.B., 2012. Potent anti-inflammatory activity of ursolic acid, a triterpenoid antioxidant, is mediated through suppression of NF-κB, AP-1, and NF-AT. PLoS One 7 (2), e31318.

Chen, C., Gao, W., Cheng, L., Shao, Y., Kong, D.Y., 2014. Four new triterpenoid glycosides from the seed residue of *Hippophae rhamnoides* subsp. *sinensis*. J. Asian Nat. Prod. Res. 16 (3), 231–239.

Chiej, R., 1984. The Macdonald Encyclopedia of Medicinal Plants. Macdonald & Co (Publishers) Ltd.

Chirila, E., Oancea, E., Oancea, I.A., 2014. Physico-chemical characterisation of sea buckthorn extracts for cosmetic use. Anal. Unive. Ovidius Constant. Ser. Chimi. 25 (2), 75–80.

Cho, H., Cho, E., Jung, H., Yi, H.C., Lee, B., Hwang, K.T., 2014. Antioxidant activities of sea buckthorn leaf tea extracts compared with green tea extracts. Food Sci. Biotechnol. 23 (4), 1295–1303.

Dhar, P., Bajpai, P.K., Tayade, A.B., Chaurasia, O.P., Srivastava, R.B., Singh, S.B., 2013. Chemical composition and antioxidant capacities of phytococktail extracts from trans-Himalayan cold desert. BMC Complem. Alternat. Med. 13 (1), 259.

Fan, J., Zhang, J., Song, H., Zhu, W. and Liu, Y., Antioxidant activity and phenolic components of sea buckthorn (*Hippophae rhamnoides*) seed extracts. (In Proceedings of the 2013 International Conference on Advanced Mechatronic Systems). 2013.

Fatima, T., Kesari, V., Watt, I., Wishart, D., Todd, J.F., Schroeder, W.R., Paliyath, G., Krishna, P., 2015. Metabolite profiling and expression analysis of flavonoid, vitamin C, and tocopherol biosynthesis genes in the antioxidant-rich sea buckthorn (*Hippophae rhamnoides L.*). Phytochemistry 118, 181–191.

Ganju, L., Padwad, Y., Singh, R., Karan, D., Chanda, S., Chopra, M.K., Bhatnagar, P., Kashyap, R., Sawhney, R.C., 2005. Anti-inflammatory activity of sea buckthorn (*Hippophae rhamnoides*) leaves. Int. Immunopharmacol. 5 (12), 1675–1684.

Gao, Z., Zhang, C., Jin, L., Yao, W., 2014. Efficacy of sea buckthorn therapy in patients with non-alcoholic fatty liver disease. Chin. Med. 5 (04), 223.

Geetha, S., Jayamurthy, P., Pal, K., Pandey, S., Kumar, R., Sawhney, R.C., 2008. Hepatoprotective effects of sea buckthorn (*Hippophae rhamnoides L.*) against carbon tetrachloride induced liver injury in rats. J. Sci. Food Agric. 88 (9), 1592–1597.

Grey, C., Widén, C., Adlercreutz, P., Rumpunen, K., Duan, R.D., 2010. Antiproliferative effects of sea buckthorn (*Hippophae rhamnoides L.*) extracts on human colon and liver cancer cell lines. Food Chem. 120 (4), 1004–1010.

Gunenc, A., Khoury, C., Legault, C., Mirrashed, H., Rijke, J., Hosseinian, F., 2016. Sea buckthorn as a novel prebiotic source improves probiotic viability in yogurt. LWT Food Sci. Technol. 66, 490–495.

Gupta, R., Deswal, R., 2012. Low temperature stress modulated secretome analysis and purification of antifreeze protein from *Hippophae rhamnoides*, a Himalayan wonder plant. J. Proteome Res. 11 (5), 2684–2696.

Gupta, D., Kaul, V., 2017. Antioxidant activity vis-a-vis phenolic content in leaves of sea buckthorn from Kargil District (J&K, India): a preliminary study. Natl. Acad. Sci. Lett. 40 (1), 53–56.

Gupta, A., Kumar, R., Pal, K., Singh, V., Banerjee, P.K., Sawhney, R.C., 2006. Influence of sea buckthorn (*Hippophae rhamnoides L.*) flavone on dermal wound healing in rats. Mol. Cell. Biochem. 290 (1), 193–198.

Gupta, S.M., Pandey, P., Grover, A., Ahmed, Z., 2011. Breaking seed dormancy in *Hippophae salicifolia*, a high value medicinal plant. Physiol. Mol. Biol. Plants 17 (4), 403.

Gutzeit, D., Winterhalter, P., Jerz, G., 2008. Nutritional assessment of processing effects on major and trace element content in sea buckthorn juice (*Hippophaë rhamnoides L.* ssp. rhamnoides). J. Food Sci. 73 (6), H97–102.

Hibasami, H., Mitani, A., Katsuzaki, H., Imai, K., Yoshioka, K., Komiya, T., 2005. Isolation of five types of flavonol from sea buckthorn (*Hippophae rhamnoides*) and induction of apoptosis by some of the flavonols in human promyelotic leukemia HL-60 cells. Int. J. Mol. Med. 15 (5), 805–809.

Hou, D.D., Di, Z.H., Qi, R.Q., Wang, H.X., Zheng, S., Hong, Y.X., Guo, H., Chen, H.D., Gao, X.H., 2017. Sea buckthorn (Hippophaë rhamnoides L.) oil improves atopic dermatitis-like skin lesions via inhibition of NF-κB and STAT1 activation. Skin Pharmacol. Physiol. 30 (5), 268–276.

Hu, N., Suo, Y., Zhang, Q., You, J., Ji, Z., Wang, A., Han, L., Lv, H., Ye, Y., 2015. Rapid, selective, and sensitive analysis of triterpenic acids in *Hippophae rhamnoides L.* using HPLC with precolumn fluorescent derivatization and identification with postcolumn APCI-MS. J. Liq. Chromatogr. Relat. Technol. 38 (4), 451–458.

Hwang, I.S., Kim, J.E., Choi, S.I., Lee, H.R., Lee, Y.J., Jang, M.J., Son, H.J., Lee, H.S., Oh, C.H., Kim, B.H., Lee, S.H., 2012. UV radiation-induced skin aging in hairless mice is effectively prevented by oral intake of sea buckthorn (*Hippophae rhamnoides L.*) fruit blend for 6 weeks through MMP suppression and increase of SOD activity. Int. J. Mol. Med. 30 (2), 392–400.

Jain, K., Suryakumar, G., Prasad, R., Ganju, L., Singh, S.B., 2016. Enhanced hypoxic tolerance by Seabuckthorn is due to upregulation of HIF-1α and attenuation of ER stress. J. Appl. Biomed. 14 (1), 71–83.

Jia, D.R., Wang, Y.J., Liu, T.L., Wu, G.L., Kou, Y.X., Cheng, K., Liu, J.Q., 2016. Diploid hybrid origin of *Hippophaë gyantsensis* (*Elaeagnaceae*) in the western Qinghai–Tibet plateau. Biol. J. Linn. Soc. 117 (4), 658–671.

Johansson, A.K., Korte, H., Yang, B., Stanley, J.C., Kallio, H.P., 2000. Sea buckthorn berry oil inhibits platelet aggregation. J. Nutr. Biochem. 11 (10), 491–495.

Junzeng, Z., Yanze, L. and Zhimin, W., 2015. Dietary Chinese Herbs: Chemistry, Pharmacology, and Clinical Evidence, Springer-Verlag Wien.

Kalia, R.K., Singh, R., Rai, M.K., Mishra, G.P., Singh, S.R., Dhawan, A.K., 2011. Biotechnological interventions in sea buckthorn (*Hippophae L.*): current status and future prospects. Trees 25, 559–575.

Kallio, H., Yang, W., Liu, P., Yang, B., 2014. Proanthocyanidins in wild sea buckthorn (*Hippophae rhamnoides*) berries analyzed by reversed-phase, normal-phase, and hydrophilic interaction liquid chromatography with UV and MS detection. J. Agric. Food Chem. 62 (31), 7721–7729.

Kapoor, D.N., 2017. A review on pharmacognostic, phytochemical and pharmacological data of various species of *Hippophae* (Sea buckthorn). Int. J. Green Pharm 11 (01), S62–S75.

Kaushal, M., Sharma, P.C., Sharma, R., 2013. Formulation and acceptability of foam mat dried sea buckthorn (*Hippophae salicifolia*) leather. J. Food Sci. Technol. 50 (1), 78–85.

Keshri, G.K., Gupta, A., Jain, K., Suryakumar, G., Sharma, P., Gola, S., Yadav, A., Verma, S., Ganju, L., 2015. Sea buckthorn (*H. rhamnoides L.*) mediated acute hypoxic tolerance in the skeletal muscle of rats by differential activation of energy metabolism and enhanced antioxidants. Int. J. Pharm. Sci. Res. 6 (12), 5259.

Khan, B.A., Akhtar, N., Mahmood, T., 2010. A comprehensive review of a magic plant, *Hippophae rhamnoides*. Pharmacogn. J. 2 (16), 65–68.

Khanum, F., Anand, T., Ilaiyaraja, N., Patil, M.M., Singsit, D., Shiromani, S., Sharma, R.K., 2017. Health food for soldiers. Defence Life Sci. J. 2 (2), 111–119.

Kim, J.S., Kwon, Y.S., Sa, Y.J., Kim, M.J., 2010. Isolation and identification of sea buckthorn (*Hippophae rhamnoides*) phenolics with antioxidant activity and α-glucosidase inhibitory effect. J. Agric. Food Chem. 59 (1), 138–144.

Kim, S.J., Hwang, E., Yi, S.S., Song, K.D., Lee, H.K., Heo, T.H., Park, S.K., Jung, Y.J., Jun, H.S., 2017. Sea buckthorn leaf extract inhibits glioma cell growth by reducing reactive oxygen species and promoting apoptosis. Appl. Biochem. Biotechnol. 182 (4), 1663–1674.

Korekar, G., Stobdan, T., Singh, H., Chaurasia, O., Singh, S., 2011. Phenolic content and antioxidant capacity of various solvent extracts from sea buckthorn (*Hippophae rhamnoides L.*) fruit pulp, seeds, leaves, and stem bark. Acta Aliment. 40 (4), 449–458.

Kou, Y.X., Wu, Y.X., Jia, D.R., Li, Z.H., Wang, Y.J., 2014. Range expansion, genetic differentiation, and phenotypic adaption of *Hippophaë neurocarpa* (*Elaeagnaceae*) on the Qinghai–Tibet plateau. J. Syst. Evol. 52 (3), 303–312.

Koyama, T., Taka, A., Togashi, H., 2009. Effects of a herbal medicine, *Hippophae rhamnoides*, on cardiovascular functions and coronary microvessels in the spontaneously hypertensive stroke-prone rat. Clin. Hemorheol. Microcirc. 41 (1), 17–26.

Krejcarová, J., Straková, E., Suchý, P., Herzig, I., Karásková, K., 2015. Sea buckthorn (Hippophae rhamnoides L.) as a potential source of nutraceutics and its therapeutic possibilities: a review. Acta Vet. Brno 84 (3), 257–268.

Kumar, S., Sagar, A., 2007. Microbial associates of Hippophae rhamnoides (Seabuckthorn). Plant Pathol. J. 6 (4), 299–305.

Kumar, R., Kumar, G.P., Chaurasia, O.P., Singh, S.B., 2011. Phytochemical and pharmacological profile of sea buckthorn oil: a review. Res. J. Med. Plant 5, 491–499.

Li, T.S., Beveridge, T.H., 2003. Sea Buckthorn (Hippophae rhamnoides L.) Production and Utilization. NRC Research Press.

Li, T.S., Schroeder, W.R., 1996. Sea buckthorn (Hippophae rhamnoides L.): a multipurpose plant. HortTechnology 6 (4), 370–380.

Li, T.S.C., Wardle, D., 2003. Effect of harvest period on the protein content in sea buckthorn leaves. Can. J. Plant Sci. 83 (2), 409–410.

Li, F.L., Wang, R., Du, Q.Y., Sun, W., Han, B.L., 2011. Antioxidant effect of water soluble polysaccharides from commercially available *Hippophae rhamnoides* leaf tea in aging model mice [J]. Food Sci. 5, 063.

Liu, Y., Li, J., Fan, G., Sun, S., Zhang, Y., Zhang, Y., Tu, Y., 2016. Identification of the traditional Tibetan medicine "Shaji" and their different extracts through tristep infrared spectroscopy. J. Mol. Struct. 1124, 180–187.

Liu, Y., Fan, G., Zhang, J., Zhang, Y., Li, J., Xiong, C., Zhang, Q., Li, X., Lai, X., 2017. Metabolic discrimination of sea buckthorn from different *Hippophaë* species by 1H NMR based metabolomics. Sci. Rep. 7, 1585.

Madrigal-Santillán, E., Madrigal-Bujaidar, E., Álvarez-González, I., Sumaya-Martínez, M.T., Gutiérrez-Salinas, J., Bautista, M., Morales-González, Á.y., González-Rubio, M.G.L., Aguilar-Faisal, J.L., Morales-González, J.A., 2014. Review of natural products with hepatoprotective effects. World J. Gastroenterol. 20 (40), 14787.

Maheshwari, D.T., Kumar, M.Y., Verma, S.K., Singh, V.K., Singh, S.N., 2011. Antioxidant and hepatoprotective activities of phenolic rich fraction of Seabuckthorn (Hippophae rhamnoides L.) leaves. Food Chem. Toxicol. 49 (9), 2422–2428.

Malik, S., Goyal, S., Ojha, S.K., Bharti, S., Nepali, S., Kumari, S., Singh, V., Arya, D.S., 2011. Sea buckthorn attenuates cardiac dysfunction and oxidative stress in isoproterenol-induced cardiotoxicity in rats. Int. J. Toxicol. 30 (6), 671–680.

Manickam, M., Tulsawani, R., 2014. Survival response of hippocampal neurons under low oxygen conditions induced by Hippophae rhamnoides is associated with JAK/STAT signaling. PLOS One 9(2).

Matthews, V., 1994. The new plantsman. R. Hortic. Soc. 1, 1352–4186.

Michel, T., Destandau, E., Le Floch, G., Lucchesi, M.E., Elfakir, C., 2012. Antimicrobial, antioxidant, and phytochemical investigations of sea buckthorn (*Hippophaë rhamnoides L.*) leaf, stem, root and seed. Food Chem. 131 (3), 754–760.

Mueller, M., Hobiger, S., Jungbauer, A., 2010. Anti-inflammatory activity of extracts from fruits, herbs, and spices. Food Chem. 122 (4), 987–996.

Narayanan, S., Ruma, D., Gitika, B., Sharma, S.K., Pauline, T., Ram, M.S., Ilavazhagan, G., Sawhney, R.C., Kumar, D., Banerjee, P.K., 2005. Antioxidant activities of sea buckthorn (*Hippophae rhamnoides*) during hypoxia induced oxidative stress in glial cells. Mol. Cell. Biochem. 278 (1), 9–14.

Ni, W., Gao, T., Wang, H., Du, Y., Li, J., Li, C., Wei, L., Bi, H., 2013. Antifatigue activity of polysaccharides from the fruits of four Tibetan plateau indigenous medicinal plants. J. Ethnopharmacol. 150 (2), 529–535.

Olas, B., 2016. Sea buckthorn as a source of important bioactive compounds in cardiovascular diseases. Food Chem. Toxicol. 97, 199–204.

Padwad, Y., Ganju, L., Jain, M., Chanda, S., Karan, D., Banerjee, P.K., Sawhney, R.C., 2006. Effect of leaf extract of Seabuckthorn on lipopolysaccharide induced inflammatory response in murine macrophages. Int. Immunopharmacol. 6 (1), 46–52.

Pang, X., Zhao, J., Zhang, W., Zhuang, X., Wang, J., Xu, R., Xu, Z., Qu, W., 2008. Antihypertensive effect of total flavones extracted from seed residues of *Hippophae rhamnoides* L. in sucrose-fed rats. J. Ethnopharmacol. 117 (2), 325–331.

Patel, C.A., Divakar, K., Santani, D., Solanki, H.K., Thakkar, J.H., 2012. Remedial prospective of Hippophae rhamnoides Linn. (Sea Buckthorn). ISRN Pharmacol. 2012 (2012), 436857.

Pavlović, N., Valek Lendić, K., Miškulin, M., Moslavac, T., Jokić, S., 2016. Supercritical CO_2 extraction of sea buckthorn. Food Health Disease: Sci. Professional J. Nutr. Dietetics 5 (2), 55–61.

Purushothaman, J., Suryakumar, G., Shukla, D., Malhotra, A.S., Kasiganesan, H., Kumar, R., Sawhney, R.C., Chami, A., 2008. Modulatory effects of sea buckthorn (Hippophae rhamnoides L.) in hypobaric hypoxia induced cerebral vascular injury. Brain Res. Bull. 77 (5), 246–252.

Purushothaman, J., Suryakumar, G., Shukla, D., Jayamurthy, H., Kasiganesan, H., Kumar, R., Sawhney, R.C., 2011. Modulation of hypoxia-induced pulmonary vascular leakage in rats by sea buckthorn (*Hippophae rhamnoides L.*). Evid. Based Complement. Alternat. Med., 574524.

Qin, L., Cheng, W.J., Pan, X.L., 2013. Measurement and analysis of amino acids in different parts of Hippophae rhamnoides' branches and leaves. Acta Ecol. Anim. Domast. 6.

Raina, S.N., Jain, S., Sehgal, D., Kumar, A., Dar, T.H., Bhat, V., Pandey, V., Vaishnavi, S., Bhargav, A., Singh, V., Rani, V., 2012. Diversity and relationships of multipurpose sea buckthorn (Hippophae L.) germplasm from the Indian Himalayas as assessed by AFLP and SAMPL markers. Genet. Resour. Crop Evol. 59 (6), 1033–1053.

Ramu, S., Krishnaraj, K., Devika, A., Murali, A., 2014. Protective effects of Hippophae salicifolia D. Don fruit pulp extract in aluminum toxicity. Spatula DD Peer Rev. J. Complem. Med. Drug Discov. 4 (4), 207–212.

Ranjith, A., Arumughan, C., 2009. Phytochemical Investigations on Sea Buckthorn (Hippophae rhamnoides) Berries. Doctoral dissertation, National Institute for Interdisciplinary Science and Technology.

Ranjith, A., Kumar, K.S., Venugopalan, V.V., Arumughan, C., Sawhney, R.C., Singh, V., 2006. Fatty acids, tocols, and carotenoids in pulp oil of three sea buckthorn species (Hippophae rhamnoides, H. salicifolia, and H. tibetana) grown in the Indian Himalayas. J. Am. Oil Chem. Soc. 83 (4), 359–364.

Rathor, R., Sharma, P., Suryakumar, G., Gupta, A., Himashree, G., Bhaumik, G., 2014. Herbs for mitigating high altitude maladies. In: Translational Research in Environmental and Occupational Stress. Springer, India, pp. 255–266.

Rathor, R., Sharma, P., Suryakumar, G., Ganju, L., 2015. A pharmacological investigation of Hippophae salicifolia (HS) and Hippophae rhamnoides *turkestanica* (HRT) against multiple stress (CHR): an experimental study using rat model. Cell Stress Chaperon. 20 (5), 821–831.

Rongsen, L., Ahani, H., Shaban, M., Esfahani, M.N., Alizade, G., Rostampour, M., Moazeni, N., Javadi, S., Mahdavian, S.E., 2013. The genetic resources of Hippophae genus and its utilization. Int. J. Scholar. Res. Gate 1, 15–21.

Rösch, D., Krumbein, A., Mügge, C., Kroh, L.W., 2004. Structural investigations of flavonol glycosides from sea buckthorn (Hippophaë rhamnoides) pomace by NMR spectroscopy and HPLC-ESI-MS n. J. Agric. Food Chem. 52 (13), 4039–4046.

Sabir, S.M., Maqsood, H., Hayat, I., Khan, M.Q., Khaliq, A., 2005. Elemental and nutritional analysis of sea buckthorn (Hippophae rhamnoides ssp. *turkestanica*) berries of Pakistani origin. J. Med. Food 8 (4), 518–522.

Saeidnia, S., Manayi, A., Gohari, A.R., Abdollahi, M., 2014. The story of beta-sitosterol—a review. Eur. J. Med. Plants 4 (5), 590–609.

Saggu, S., Kumar, R., 2007a. Modulatory effect of sea buckthorn leaf extract on oxidative stress parameters in rats during exposure to cold, hypoxia, and restraint (C-H-R) stress and post stress recovery. J. Pharm. Pharmacol. 59 (12), 1739–1745.

Saggu, S., Kumar, R., 2007b. Possible mechanism of adaptogenic activity of sea buckthorn (Hippophae rhamnoides) during exposure to cold, hypoxia and restraint (C–H–R) stress induced hypothermia and post stress recovery in rats. Food Chem. Toxicol. 45 (12), 2426–2433.

Saggu, S., Kumar, R., 2008. Effect of sea buckthorn leaf extracts on circulating energy fuels, lipid peroxidation and antioxidant parameters in rats during exposure to cold, hypoxia and restraint (C–H–R) stress and post stress recovery. Phytomedicine 15 (6), 437–446.

Saini, M., Madhu, B., Prasad, J., Farooqi, H., Abdin, M.Z., 2014. A standardized sea buckthorn leaf extract (SBL-1) counters radiation induced renal histopathology, oxidative stress as well as changes in mRNA levels. Proceedings of the International Conference on Radiation Biology:

Frontiers in Radiobiology–Immunomodulation, Countermeasures and Therapeutics: Abstract book, souvenir, and scientific programme.

Sayegh, M., Miglio, C., Ray, S., 2014. Potential cardiovascular implications of sea buckthorn berry consumption in humans. Int. J. Food Sci. Nutr. 65 (5), 521–528.

Selvamurthy, W., Basu, C.K., 1998. High altitude maladies: recent trends in medical management. Int. J. Biometeorol. 42 (2), 61–64.

Sharma, P., Suryakumar, G., Singh, V., Misra, K., Singh, S.B., 2015. In vitro antioxidant profiling of sea buckthorn varieties and their adaptogenic response to high altitude-induced stress. Int. J. Biometeorol. 59 (8), 1115–1126.

Sharma, A., Sharma, S., Khattri, S., Garg, H., 2016. Role of sea buckthorn oil in management of chronic periodontitis: follow-up study. Int. J. Dental Res. 4 (2), 33–37.

Sherpa, L.Y., Stigum, H., Chongsuvivatwong, V., Nafstad, P., Bjertness, E., 2013. Prevalence of metabolic syndrome and common metabolic components in high altitude farmers and herdsmen at 3700 m in Tibet. High Alt. Med. Biol. 14 (1), 37–44.

Shipulina, L.D., Tolkachev, O.N., Krepkova, L.V., Bortnikova, V.V., Shkarenkov, A.A., 2005. Antiviral anti-microbial and toxicological studies on Seabuckthorn (Hippophae rhamnoides). In: Singh, V. (Ed.), Seabuckthorn (Hippophae L.): A Multipurpose Wonder Plant. In: 2, Daya Publishing House, New Delhi, pp. 471–483.

Shukla, S.K., Chaudhary, P., Kumar, I.P., Samanta, N., Afrin, F., Gupta, M.L., Sharma, U.K., Sinha, A.K., Sharma, Y.K., Sharma, R.K., 2006. Protection from radiation-induced mitochondrial and genomic DNA damage by an extract of *Hippophae rhamnoides*. Environ. Mol. Mutagen. 47 (9), 647–656.

Singh, K.N., 2012. Traditional knowledge on ethnobotanical uses of plant biodiversity: a detailed study from the Indian western Himalaya. Biodivers. Res. Conserv. 28, 63–77.

Singh, R., Mishra, S.N., Dwivedi, S.K., Ahmed, Z., 2006. Genetic variation in sea buckthorn (Hippophae rhamnoides *L.*) populations of cold arid Ladakh (India) using RAPD markers. Curr. Sci. 91 (10), 1321–1322.

Singh, A.K., Deep, P., Dubey, S., Attrey, D.P., Naved, T., 2014. Comparative studies on antioxidant activity, total phenol content and high performance thin layer chromatography analysis of sea buckthorn (Hippophae rhamnoides *L.*) leaves. Pharmacogn. J 6 (5), 5–7.

Šnē, E., Galoburda, R. and Segliņa, D., 2013, August. Sea buckthorn vegetative parts: a good source of bioactive compounds. In Proceedings of the Latvian Academy of Sciences. Section B. Natural, Exact, and Applied Sciences (Vol. 67, No. 2, pp. 101–108).

Solcan, C., Gogu, M., Floristean, V., Oprisan, B., Solcan, G., 2013. The hepatoprotective effect of sea buckthorn (Hippophae rhamnoides) berries on induced aflatoxin B1 poisoning in chickens. Poult. Sci. 92 (4), 966–974.

Stobdan, T., Korekar, G., Chaurasia, O.P., Balaji, B., Yadav, A., Dwivedi, S.K., Targais, K., Mundra, S., Srivastava, R.B., 2011a. Sea Buckthorn Production for Greening and Sustainable Income Generation in Cold Desert of India, pp. 71–86.

Stobdan, T., Yadav, A., Mishra, G.P., Chaurasia, O.P., Srivastava, B.R., 2011b. Sea Buckthorn: The Super Plant. Defence Institute of High Altitude Research, Defence Research and Development Organisation, Leh, Ladakh, Jammu, and Kashmir, India.

Stobdan, T., Korekar, G.B., Srivastava, R., 2013. Nutritional attributes and health application of sea buckthorn (Hippophae rhamnoides L.): a review. Curr. Nutr. Food Sci. 9 (2), 151–165.

Sun, K., Chen, X.L., Ma, R.J., Li, C., Wang, Q., Ge, S., 2002. Molecular phylogenetics of Hippophae L. (Elaeagnaceae) based on the internal transcribed spacer (ITS) sequences of nrDNA. Plant Syst. Evol. 235 (1), 121–134.

Suryakumar, G., Gupta, A., 2011. Medicinal and therapeutic potential of Sea buckthorn (Hippophae rhamnoides L.). J. Ethnopharmacol. 138 (2), 268–278.

Tang, L., Lv, H., Li, S., Bi, H., Gao, X., Zhou, J., 2016. Protective effects of polyphenol extracts from sea buckthorn (Hippophaë rhamnoides L.) on rat hearts. Open J. Mol. Integr. Physiol. 6 (01), 10.

Tanwar, H., Singh, D., Singh, S.B., Ganju, L., 2017. Anti-inflammatory activity of the functional groups present in Hippophae rhamnoides (Seabuckthorn) leaf extract. Inflammopharmacology, 26 (1), 291–301.

Tayade, A.B., Dhar, P., Sharma, M., Chauhan, R.S., Chaurasia, O.P. and Srivastava, R.B., 2013. Antioxidant capacities, phenolic contents, and GC/MS analysis of Rhodiola imbricata Edgew. Root extracts from Trans-Himalaya. J. Food Sci. 78(3), C402–10.

Teleszko, M., Wojdyło, A., Rudzińska, M., Oszmiański, J., Golis, T., 2015. Analysis of lipophilic and hydrophilic bioactive compounds content in sea buckthorn (Hippophae rhamnoides L.) berries. J. Agric. Food Chem. 63 (16), 4120–4129.

Thakur, S., Chilikuri, P., Pulugurtha, B., Yaidikar, L., 2015. Hippophae salicifolia D. Don berries attenuate cerebral ischemia reperfusion injury in a rat model of middle cerebral artery occlusion. J. Acute Dis. 4 (2), 120–128.

Ti-juan, C., Shu-chang, W., Jun-ying, Y., Jungang, L., Honglin, Y., 2005. The influence of compound Hippophae rhamnoides L. on the whole brain lacking oxygen and energy metabolism. Pharmacol. Clin. Chin. Mater. Med. 1, 018.

Torelli, A., Gianchecchi, E., Piccirella, S., Manenti, A., Piccini, G., PASTOR, E.L., Canovi, B., Montomoli, E., 2015. Sea buckthorn bud extract displays activity against cell-cultured influenza virus. J. Prevent. Med. Hyg. 56 (2), E51.

Tulsawani, R., Sharma, P., Divekar, H.M., Meena, R.N., Singh, M., Kumar, R., 2010. Supplementation of fruit extract of Hippophae rhamnoides speeds adaptation to simulated high altitude stressors in rats. J. Complem. Integr. Med. 7 (1). https://doi.org/10.2202/1553-3840.1323. Article 30.

Upadhyay, N.K., Kumar, R., Mandotra, S.K., Meena, R.N., Siddiqui, M.S., Sawhney, R.C., Gupta, A., 2009. Safety and healing efficacy of Sea buckthorn (Hippophae rhamnoides *L*.) seed oil on burn wounds in rats. Food Chem. Toxicol. 47 (6), 1146–1153.

Upadhyay, N.K., Kumar, M.Y., Gupta, A., 2010. Antioxidant, cytoprotective, and antibacterial effects of Sea buckthorn (Hippophae rhamnoides L.) leaves. Food Chem. Toxicol. 48 (12), 3443–3448.

Upadhyay, N.K., Kumar, R., Siddiqui, M.S., Gupta, A., 2011. Mechanism of wound-healing activity of Hippophae rhamnoides L. leaf extract in experimental burns. Evid. Based Complem. Alternat. Med 2011, 659705.

Uprety, Y., Asselin, H., Boon, E.K., Yadav, S., Shrestha, K.K., 2010. Indigenous use and bioefficacy of medicinal plants in the Rasuwa District, Central Nepal. J. Ethnobiol. Ethnomed. 6 (1), 3.

Usha, T., Middha, S.K., Goyal, A.K., Karthik, M., Manoj, D.A., Faizan, S., Goyal, P., Prashanth, H.P., Pande, V., 2014. Molecular docking studies of anticancerous candidates in Hippophae rhamnoides and Hippophae salicifolia. J. Biomed. Res. 28 (5), 406.

Vij, A.G., Krishna, K., Joginder, D., Karan, P., Minakshi, B., Sawhney, R.C., 2010. Inhibitory effect of sea buckthorn (Hippophea rhamnoides) on platelet aggregation and oxidative stress. J. Complem. Integr. Med. 7 (1), 1553–3840. ISSN (Online).

Vítová, E., Sůkalová, K., Mahdalová, M., Butorová, L., Melikantová, M., 2015. Comparison of selected aroma compounds in cultivars of sea buckthorn (Hippophae rhamnoides L.). Chem. Pap. 69 (6), 881–888.

Wang, A., Schluetz, F., Liu, J., 2008. Molecular evidence for double maternal origins of the diploid hybrid Hippophae goniocarpa (Elaeagnaceae). Bot. J. Linn. Soc. 156 (1), 111–118.

Wang, H., Gao, T., Du, Y., Yang, H., Wei, L., Bi, H., Ni, W., 2015. Anticancer and immunostimulating activities of a novel homogalacturonan from Hippophae rhamnoides L. berry. Carbohyd. Polym. 131, 288–296.

Wang, H., Bi, H., Gao, T., Zhao, B., Ni, W., Liu, J., 2018. A homogalacturonan from Hippophae rhamnoides L. berries enhance immunomodulatory activity through TLR4/MyD88 pathway mediated activation of macrophages. Int. J. Biol. Macromol. 107 (Pt A), 1039–1045

Wani, T.A., Wani, S.M., Ahmad, M., Ahmad, M., Gani, A., Masoodi, F.A., 2016. Bioactive profile, health benefits, and safety evaluation of sea buckthorn (Hippophae rhamnoides L.): A review. Cogent Food Agric. 2 (1), 1128519.

Xie, Y., Zeng, X., Li, G., Cai, Z., Ding, N., Ji, G., 2010. Assessment of intestinal absorption of total flavones of Hippophae rhamnoides L. in rat using in situ absorption models. Drug Dev. Ind. Pharm. 36 (7), 787–794.

Xu, X.Y., Xie, B.J., Pan, S.Y., Liu, L., Wang, Y.D., Chen, C.D., 2007. Effects of sea buckthorn procyanidins on healing of acetic acid-induced lesions in the rat stomach. Asia Pac. J. Clin. Nutr. 16 (S1), 234–238.

Xu, L., Wang, H., La, Q., Lu, F., Sun, K., Fang, Y., Yang, M., Zhong, Y., Wu, Q., Chen, J., Birks, H.J.B., 2014a. Microrefugia and shifts of Hippophae tibetana (Elaeagnaceae) on the north side of Mt. Qomolangma (Mt. Everest) during the last 25,000 years. PlOS One 9(5).

Xu, M., Wang, N., Zhang, W., Wang, J., 2014b. Protective effects of sea buckthorn seed oil against acute alcoholic-induced liver injury in rats. J. Food Nutr. Res. 2 (12), 1037–1041.

Yadav, V.K., Sah, V.K., Singh, A.K., Sharma, S.K., 2006. Variations in morphological and biochemical characters of sea buckthorn (Hippophae salicifolia D. Don) populations growing in Harsil area of Garhwal Himalaya in India. Trop. Agric. Res. Ext. 9, 1–7.

Yang, B., Kallio, H., 2002. Composition and physiological effects of sea buckthorn (Hippophae) lipids. Trends Food Sci. Technol. 13 (5), 160–167.

Yang, B., Linko, A.M., Adlercreutz, H., Kallio, H., 2006. Secoisolariciresinol and matairesinol of sea buckthorn (Hippophaë rhamnoides L.) berries of different subspecies and harvesting times. J. Agric. Food Chem. 54 (21), 8065–8070.

Yasukawa, K., Kitanaka, S., Kawata, K., Goto, K., 2009. Antitumor promoters phenolics and triterpenoid from Hippophae rhamnoides. Fitoterapia 80 (3), 164–167.

Yongshan, L., Xuelin, C., Hong, L., 2003. Taxonomy of Seabuckthorn (Hippophae L.). In: Singh, V. et al., (Ed.), Sea Buckthorn (Hippophae L.): A Multipurpose Wonder Plant. Indus Publishing Group, New Delhi, pp. 34–46.

Zakynthinos, G., Varzakas, T., 2015. Hippophae rhamnoides: safety and nutrition. Curr. Res. Nutr. Food Sci. J. 3 (2), 89–97.

Zeb, A., 2004. Important therapeutic uses of sea buckthorn (Hippophae): a review. J. Biol. Sci. 4 (5), 687–693.

Zeb, A., 2006. Anticarcinogenic potential of lipids from Hippophae; Evidence from the recent literature. Asian Pac. J. Cancer Prevent. 7 (1), 32.

Zeb, A., Khan, I., 2008. Pharmacological Applications of Sea Buckthorn *(Hippophae)*. Advances in Phytotheraphy ResearchResearch Signpost, New Delhi.

Zeb, A., Ullah, S., 2015. Sea buckthorn seed oil protects against the oxidative stress produced by thermally oxidized lipids. Food Chem. 186, 6–12.

Zhang, W., Zhang, X., Zou, K., Xie, J., Zhao, S., Liu, J., Liu, H., Wang, J., Wang, Y., 2017. Sea buckthorn berry polysaccharide protects against carbon tetrachloride-induced hepatotoxicity in mice via anti-oxidative and anti-inflammatory activities. Food Funct. 8 (9), 3130–3138.

Zhao, G., Duan, J., Xie, Y., Lin, G., Luo, H., Li, G., Yuan, X., 2013. Effects of solid dispersion and self-emulsifying formulations on the solubility, dissolution, permeability and pharmacokinetics of isorhamnetin, quercetin, and kaempferol in total flavones of Hippophae rhamnoides L. Drug Dev. Ind. Pharm. 39 (7), 1037–1045.

Zheng, R.X., Xu, X.D., Tian, Z., Yang, J.S., 2009. Chemical constituents from the fruits of Hippophae rhamnoides. Nat. Prod. Res. 23 (15), 1451–1456.

Zheng, L., Shi, L.K., Zhao, C.W., Jin, Q.Z., Wang, X.G., 2017. Fatty acid, phytochemical, oxidative stability, and in vitro antioxidant property of sea buckthorn (Hippophaë rhamnoides L.) oils extracted by supercritical and subcritical technologies. LWT Food Sci. Technol. 86, 507–513.

Zhou, X., Ma, J., Wang, W., Gong, N., Liu, J., 2010. Genome size of the diploid hybrid species Hippophae goniocarpa and its parental species, H. rhamnoides ssp. sinensis and H. neurocarpa ssp. neurocarpa (Elaeagnaceae). Acta Biol. Cracov. Ser. Bot. 52 (2), 12–16.

Zhou, J.Y., Zhou, S.W., Du, X.H., Zeng, S.Y., 2012. Protective effect of total flavonoids of sea buckthorn (Hippophae rhamnoides) in simulated high-altitude polycythemia in rats. Molecules 17 (10), 11585–11597.

Further Reading

Farinazzi-Machado, F.M.V., 2017. Can the consumption of seeds, leaves and fruit peels avoid the risk factors for cardiovascular disorders? Int. J. Nutrol. 10 (2), 37–45.

Shi, J., Wang, L., Lu, Y., Ji, Y., Wang, Y., Dong, K., Kong, X., Sun, W., 2017. Protective effects of sea buckthorn pulp and seed oils against radiation-induced acute intestinal injury. J. Radiat. Res. 58 (1), 24–32.

CHAPTER 4

Valeriana sp.: The Role in Ameliorating High-Altitude Ailments

Jigni Mishra*, Kshipra Misra†

*Chemistry Division, Department of Biochemical Sciences (DBCS), Defence Institute of Physiology and Allied Sciences (DIPAS), Delhi, India, †Department of Biochemical Sciences (DBCS), Defence Institute of Physiology and Allied Sciences (DIPAS), Delhi, India

Abbreviations

AMS	acute mountain sickness
ATP	adenosine triphosphate
CAT	catalase
CNS	central nervous system
DNA	deoxy ribonucleic acid
DPPH	2,2-diphenyl-1-picrylhydrazyl
GABA	gamma aminobutyric acid
GC	gas chromatography
HACE	high-altitude cerebral edema
HAPE	high-altitude pulmonary edema
HPLC	high-performance liquid chromatography
LDH	lactate dehydrogenase
MDA	malondialdehyde
MES	modulating electroshock seizures
MRC-5	medical research council strain 5
MS	mass spectrometry
SOD	superoxide dismutase
sp.	species
STZ	streptozotocin
VW	*Valeriana wallichii* DC

4.1 INTRODUCTION

Recent research about healthcare improvement has seen the use of plant-derived drugs as vital therapeutics because of their minimal side effects as compared to undesired bioconversion reactions arising from conventional allopathic medicines. This has culminated in the paradigm of phytomedicines. The bioactive constituents in such phytomedicines range from conjugated sugars, flavonoids, lipids, proteins, phenolics, and terpenoids

69

Management of High Altitude Pathophysiology. https://doi.org/10.1016/B978-0-12-813999-8.00004-5

to other similar classes of biomolecules. These active constituents isolated from various parts of plant—fruits, leaves, rhizomes, roots, or stems—have been used to treat various diseases, including cancer, epilepsy, infectious diseases, and inflammatory diseases. Plant-derived bioactive drugs also impart pharmaceutical effects such as antioxidant, antiinflammatory, antiproliferative, and immunostimulatory.

In this context, the genus *Valeriana*, encompassing about 200 species worldwide, is noteworthy. These species abound in various types of pharmaceutical properties (Devi and Rao, 2014). Among all these species, *Valeriana wallichii* is unique for its numerous medicinal properties and ameliorative effects that have been described in ancient medical literature. *V. wallichii* also has been extensively described in Unani and Ayurvedic texts because of its sedative, neurostimulant, and antidepressant effects. This chapter briefly describes studies performed on different formulations of *V. wallichii* that have been used to treat health disorders and ailments, with special focus on alleviating effects arising from exposure to hypobaric hypoxia.

4.2 DESCRIPTION OF THE PLANT

V. wallichii DC (henceforth referred to as VW), also known as *Valeriana jatamansi Jones* is a small, perennial, pubescent herb belonging to the family *Valerianaceae*, order *Dipsacales*, that is 14–45 cm tall (Fig. 4.1). Commonly referred to as Indian valerian or tagar-ganthoda, VW has horizontal, thick rootstock with thick descending fibrous roots (Singh et al., 2013). The roots are usually yellowish brown in color, measuring 1.5–7 cm long and 1–2 mm high, and constitute one of the medicinally most important parts of the VW plant (Fig. 4.2). The roots have piliferous layer of papillosed cells, some of which develop into root hairs. Exodermis is a single layer of quadrangular to polygonal cells with suberized walls, containing globules of volatile oil. The root cortex is parenchymatous with several starch granules. Cells in the outermost layer also have abundant globules of volatile oil (Mhaske et al., 2011). The rhizome is yellowish to brownish, subcylindrically shaped; having a length of 4–7 cm and thickness of about 1 cm. Leaves are radical, persistent, stalked, cordate-ovate, acute, and toothed. Flowers of VW are pinkish-white or whitish in color, arranged in a terminal corymb having an approximate diameter of 3–8 cm. The flowers are unisexual, with male and female flowers on different plants. Fruits of VW plant are small and smooth, sometimes hairy. VW odor is typically valerianaceous and taste is largely bitter (Bos et al., 1997; Devi and Rao, 2014).

FIG. 4.1
Valeriana wallichii herb thriving in Himalayan valleys.

FIG. 4.2
Roots of *Valeriana wallichii* herb.

4.3   GEOGRAPHICAL DISTRIBUTION AND HABITAT

VW is predominantly distributed in cold and temperate regions of the Himalayas at altitudes of 1500–3000 m and is spread across Himalayan ranges in Afghanistan, Bhutan, Burma, India, Nepal, and Pakistan. VW is the counterpart of European *Valeriana officinalis*, which has been recognized as a medicinal drug in British and European pharmacopoeias (Sah et al., 2010). Habitations of VW typically include moist and damp regions, as well as ditches and streams. Flowering and fruiting of VW takes place during March-June. Different subspecies of VW, with morphological and genetic diversities, are found in different geographical locations (Jugran et al., 2013). Because of increasing awareness regarding its herbal properties, cultivation of VW also has begun in Belgium, England, France, Germany, India (Himachal Pradesh and Uttarakhand), the Netherlands, the Russian Federation, and the United States. The geographical distribution of the major *Valeriana* species is depicted in Fig. 4.3.

4.4   ACTIVE CONSTITUENTS IN *V. WALLICHII*

Identification of active phytoconstituents in VW has been accomplished using chromatographic techniques such as high-performance liquid chromatography (HPLC), gas chromatography (GC), and mass spectrometric (MS) studies. The most prominent chemical constituents in VW imparting various bioactivities are valepotriates (iridoids) (Becker and Chavadej, 1985), dihydrovaltrate (Bounthanh et al., 1981), and isovalerinate and Linan-isovalerianate (Thies, 1968). These compounds are characteristic to *Valeriana* and thus often act as biomarkers for analytical and quality assurance purposes. Rhizomes and roots

Valeriana atypifolia ▬▬
Valeriana occidentalis ▬▬
Valeriana officinalis ▬▬
Valeriana wallichii ▬▬

FIG. 4.3
Geographical distribution of major *Valeriana* species.

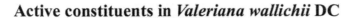

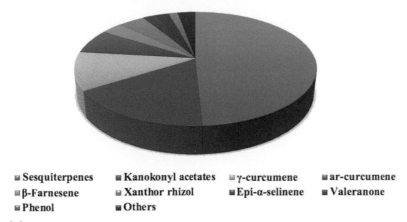

FIG. 4.4

Major bioactive constituents in *Valeriana wallichii*.

contain essential volatile oil (valerianic oil), which is composed of alkaloids, boryl isovalerianate, 1-camphene, chatinine, formate, glucoside, isovalerenic acid, 1-pinene, resins, terpineol, valeranone, valerenic acid, and valerianine (Gilani et al., 2005). Rhizomes have been established as source of organic acids, including citric acid, malic acid, succinic acid, and tartaric acid. Volatile components isolated from VW rhizomes include sesquiterpenes, kanokonyl acetate, γ-curcumene, ar-curcumene, (Z)-β-farnesene, xanthor rhizol, 7-epi-α-selinene, valeranone, and curcuphenol (Devi and Rao, 2014). VW also contains other bioactive flavonoids such as 6-methylapigenin, hesperidin, and sesquiterpenoids (Katoch et al., 2012). Fig. 4.4 provides a quick insight into the proportions of major bioactive constituents in VW.

4.5 BIOLOGICAL ACTIVITIES OF *V. WALLICHII*

The basis for nomenclature of the term *Valerianaceae* is derived from the Latin word *valere*, which means "being in a state of good health" (Syed et al., 2014). The herb VW has been used extensively for centuries in traditional Unani and Ayurvedic systems of medicines. Although all the plant parts possess richness in terms of phytoconstituents, the roots and rhizomes of VW have the most elaborate medicinal properties (Table 4.1).

VW roots have been used as sedatives to cure sleep-related disorders and insomnia. Several neurological anomalies, such as epilepsy, imbalanced reflexes, nervous unrest, hysteria, anxiety, neurosis, spasmolytic attacks, memory dysfunctions, neurodegenerative disorders, cognitive dysfunctions, and central

Table 4.1 List of Biological Activities of *Valeriana wallichii* Contributing to Improvement of Health Under Hypoxia

Biological activity	Mechanism of action	Reference (s)
Adaptogenic	a. Improvement in malondialdehyde level b. Modulation of lactate dehydrogenase level c. Lowering of superoxide dismutase and catalase values	Sharma et al. (2012)
Analgesic	Inhibition of prostaglandin synthesis	Sah et al. (2010)
Anticonvulsant	Minimization of electroshock seizures by manipulating ion channels	Joseph et al. (2015)
Antidepressant	a. Influence of 6-methylapigenin and 2(S)-hesperidin on central nervous system b. Activation of gamma amino butyric acid receptor complexes	a. Marder et al. (2003) b. Subhan et al. (2010)
Antimicrobial	Membrane damage	a. Patil et al. (2011) b. Sati et al. (2011) c. Thind and Suri (1979)
Antioxidant	a. Prevention of oxidative damage to nucleic acid b. Scavenging of free radicals c. Chelation of transition metal ions	a. Kalim et al. (2010) b. Sudhanshu et al. (2012) c. Thusoo et al. (2014)
Antispasmodic	Activation of ATP-dependent K^+ channel	Gilani et al., 2005
Perambulatory and locomotor activity improvement	Borneol as active constituent	Bhandarkar et al., 2014
Sleep improvement	a. Attenuation of monoamine levels b. Activation of iridoid esters	a. Sahu et al. (2012) b. Thies and Funke (1966)

nervous system-related problems, have been treated successfully using VW (Gilani et al., 2005; Shalam et al., 2007). VW also acts as an effective tranquilizer, antispasmodic, hypotensive, and anticonvulsant (Ortiz et al., 2006).

VW also aids in revitalizing vital organs, including the heart, liver, and kidney. The hepatoprotective effects of VW were studied by Syed et al. (2014) in which ethanolic extract of VW reduced CCl_4-induced hepatotoxicity by as much as 73%. An ayurvedic preparation called *Dhanya Panchaya Kashaya*, containing VW as one of its essential ingredients, has been proved to alleviate dyspepsia (Tripathi et al., 1982). VW roots also have been effective as a diuretic and a carminative agent, potential emmenagogue agent, and in curing jaundice (Nadkami and Nadkarni, 1976; Sah et al., 2010). Essential oils from VW are capable of reducing heart rate and blood pressure, chloroform and ethanolic extracts control arrhythmia and ischemia, and hexane extracts contribute to vasorelaxation. (Chen et al., 2015). The essential oils and iridoid compounds of VW were found to enhance microcirculatory blood perfusion in kidneys of experimental rats (Yang et al., 1998).

VW also has been reported to exert appreciable antimicrobial properties. Aqueous, hexane, and methanolic extracts of aerial parts of VW could inhibit the growth of bacterial pathogens such as *Staphylococcus aureus* and *Bacillus subtilis* (Sati et al., 2011). Rhizome extracts of VW have been reported to inhibit *S. aureus, Staphylococcus epidermidis, Escherichia coli*, and *Pseudomonas aeruginosa* (Patil et al., 2011). Antifungal activities of VW also are reported in literature, however, there is less evidence for this claim (Thind and Suri, 1979).

4.6 APPLICATIONS OF *V. WALLICHII* IN IMPROVING HEALTH AT HIGH ALTITUDES

Ascent to altitudes higher than 3000 m can be debilitating to individuals, causing several physiological and psychological health effects. The major repercussion arising from exposure to high altitudes is hypobaric hypoxia, a decreased oxygen supply to brain and tissues. Consequences of hypobaric hypoxia vary from moderate hypoxic symptoms, such as fatigue and nausea, to more severe health concerns, such as acute mountain sickness, characterized by decreased blood oxygenation, dizziness, loss of appetite, and insomnia (Shukitt-Hale et al., 1994); high-altitude cerebral edema. A vasogenic edema leading to accumulation of fluid in brain giving rise to ataxia, loss of consciousness, lassitude, and altered mental health (Bärtsch and Swenson, 2013); high-altitude pulmonary edema, a noncardiogenic edema of lungs, marked by fluid accumulation in alveoli (Yang et al., 2014). Chronic hypobaric hypoxia also triggers cognitive and neurological ailments, insomnia, convulsions, inflammatory problems, and gastrointestinal complications.

The subsequent sections describe the ameliorative role of VW in treating these health conditions, with an objective of curtailing the severity of hypobaric hypoxia-induced health effects.

4.6.1 Anxiolytic and Antidepressant

Extensive literature shows the importance of VW in treating hysteria and anxiety disorders. In previously reported studies, VW was found to exhibit anxiolytic activity through flavonoids such as anxiolytic benzodiazepine binding-site ligand 6-methylapigenin, and a sedative compound, 2S($-$)hesperidin (Marder et al., 2003; Wasowski et al., 2002). In another study (Subhan et al., 2010), the aqueous-ethanolic and methanolic extracts of VW showed antidepressant effects, possibly because of activation of $GABA_A$ receptor complexes. Ethanolic extract of VW rhizome ingested in the form of capsules reduce depression index as measured by Hamilton's brief psychiatric rating scale (Bhattacharyya et al., 2007).

4.6.2 Insomnia

Aqueous extracts of VW root have been reported to cure sleep latency, thus providing a lead to treatment of insomnia (Sahu et al., 2012). Improvement in sleep parameters were attributed to levels of monoamines in brain, where it was found that VW supplementation lessened the amounts of norepinephrine and serotonin in the brain cortex and stem, and decreased dopamine levels in the brain cortex.

Underground root organs of VW long have been used for their potential tranquillizing role. Presence of iridoid esters called valepotriates have been shown to impart this kind of tranquilizer-like effects (Thies and Funke, 1966). Valerian continues to be a safe sedative or hypnotic choice for patients with mild to moderate insomnia, mainly because of the influence of 2S(−)hesperidin on central nervous system (CNS) (Marder et al., 2003).

4.6.3 Antispasmodic Activity

Muscle spasms and cramps are common problems faced by both healthy and susceptible individuals during ascent to higher elevations. In this context, studies that report the antispasmodic activities of VW are noteworthy.

Aqueous and chloroform extracts of VW can lower spasms and blood pressure, causing complete relaxation of the muscular contractions. The underlying mechanism behind antispasmodic activity of VW was determined to be ATP-dependent K^+ channel activation, which facilitates vasodilation. VW exhibited efficiency in activating K^+ channels comparable to that of cromakalin, a prototypical K_{ATP} channel opener (Gilani et al., 2005).

4.6.4 Antipsychotic Activity

In one study, aqueous extract of VW markedly attenuated ischemia-reperfusion induced neuronal injury in mice, revealing its neuroprotective effects (Rehni et al., 2007).

In streptozotocin (STZ)-induced animal model of diabetes, the effects of VW on learning and memory impairment were measured by a Morris maze and elevated plus maze. Methanolic extracts of VW showed recovery effects by countering cholinergic dysfunction and reducing lipid peroxidation, which were attributed to the phenolic contents in the methanolic extracts of VW (Nabi et al., 2013). Petroleum ether extract of VW was investigated for perambulatory and spontaneous locomotor activity. Borneol was the active constituent that possibly improved ambulatory and locomotor activities (Bhandarkar et al., 2014).

4.6.5 Anticonvulsant Activity

Hydroethanolic extract of VW was found to decrease convulsions, while modulating electroshock seizures (MES) in Swiss albino mice. This substantiated the known sedative-like properties of VW, with added observations that VW had no side effects, which advocated its use in clinical treatment of seizures (Joseph et al., 2015).

4.6.6 Antioxidant Activity

The antioxidant activity of VW root extracts and essential oils was evaluated by DPPH method (Sudhanshu et al., 2012; Thusoo et al., 2014). For both types of samples under study, results showed appreciable free radical scavenging and metal chelating activities. Total phenolic content, total flavonoids content, total ascorbic acid, free radical scavenging activity, hydroxyl radical and peroxynitrite scavenging activity, and prevention of oxidative DNA damage also were studied in methanolic extracts of VW where substantially effective antioxidant values were observed (Kalim et al., 2010).

4.6.7 Analgesic Activity

Essential oil and alcohol extract of VW exerted good peripheral analgesic action. The inhibitory mechanism was attributed to inhibition of prostaglandin synthesis on acetic acid-induced writhing (Sah et al., 2010). Pure oil of valerian can be used topically for spinal rubs to lessen the spinal cord's sensibility to pain and stimulation (Kalim et al., 2010).

4.6.8 Antiinflammatory Activity

By using *in vitro* lipoxygenase inhibition assay, methanolic extract and ethyl acetate fractions of VW leaves were investigated for antiinflammatory activities. Carrageenan-induced hind paw edema tests on acute and chronic phase inflammation models were carried out in male Wistar rats. The best antiinflammatory activity was found with the ethyl acetate fraction (Khuda et al., 2013).

4.6.9 Radioprotective Activity

Aqueous extract of VW demonstrated ameliorative effect against radiation injury in plasmid DNA in MRC-5 cultured human embryonic lung fibroblast cells. This protective effect was attributed to hesperidin, one of the major constituents in aqueous extract of VW under study (Katoch et al., 2012).

4.6.10   Adaptogenic Activity

Sharma et al. in 2012 reported the role of VW as an effective adaptogen in preventing the deleterious effects of multiple stresses, such as cold, hypoxia, and restraint under an oxygen-limited hypobaric hypoxic environment. MDA and LDH levels were improved and SOD and CAT levels were decreased, thus providing effective protection against oxidative stress generated by hypobaric hypoxia. Soldiers, tourists, and other people exposed to such an environment suffer from severe maladies as a result of compromised blood flow and immunity (Sharma et al., 2012).

4.7   CONCLUSIONS

Even though *V. wallichii* has been for many centuries in Ayurveda to treat insomnia, as an antileishmanic agent, and as an antioxidant, the herb has not been explored for treatment of hypoxia-induced maladies. Because *V. wallichii* already possesses several pharmacological properties capable of restorative functions suitable for treatment of hypoxia, such as antidepressant, neuroprotective, antiinflammatory, and immunostimulant, further scope lies for exploring its benefits as prophylactic and as herbal medicine in solving these hypoxia-related ailments.

Acknowledgements

The authors would like to express their gratitude to the director of DIPAS for his constant support and encouragement. One of the authors, Miss Jigni Mishra, thanks Defense Research and Development Organization, India, for her doctoral fellowship.

References

Bärtsch, P., Swenson, E.R., 2013. Acute high-altitude illnesses. N. Engl. J. Med. 368 (24), 2294–2302.

Becker, H., Chavadej, S., 1985. Valepotriate production of normal and colchicine-treated cell suspension cultures of *Valeriana wallichii*. J. Nat. Prod. 48 (1), 17–21.

Bhandarkar, A.V., Shashidhara, S., Deepak, M., 2014. Pharmacognostic investigation of *Valeriana hardwickii* Wall: a threatened herb. Pharmacogn. J. 6 (5), 7.

Bhattacharyya, D., Jana, U., Debnath, P.K., Sur, T.K., 2007. Initial exploratory observational pharmacology of *Valeriana wallichii* on stress management: a clinical report. Nepal Med. Coll. J. 9 (1), 36–39.

Bos, R., Woerdenbag, H.J., Hendriks, H., Smit, H.F., Wikström, H.V., Scheffer, J.J., 1997. Composition of the essential oil from roots and rhizomes of *Valeriana wallichii* DC. Flavour Fragr. J. 12 (2), 123–131.

Bounthanh, C., Bergmann, C., Beck, J.P., Haag-Berrurier, M., Anton, R., 1981. Valepotriates, a new class of cytotoxic and antitumor agents. Planta Med. 41 (01), 21–28.

Chen, H.W., Wei, B.J., He, X.H., Liu, Y., Wang, J., 2015. Chemical components and cardiovascular activities of *Valeriana* species. Evid-Based Complement. Altern. Med., 1–11.

Devi, V.S., Rao, M.G., 2014. *Valeriana Wallichii*—a rich aroma root plant: a review. World J. Pharm. Pharmaceut. Sci. 3 (9), 1516–1525.

Gilani, A.H., Khan, A.U., Jabeen, Q., Subhan, F., Ghafar, R., 2005. Antispasmodic and blood pressure lowering effects of *Valeriana wallichii* are mediated through K+ channel activation. J. Ethnopharmacol. 100 (3), 347–352.

Joseph, L., Puthillath, R.E., Rao, S.N., 2015. Acute and chronic toxicity study of *Valeriana wallichii* rhizome hydroethanolic extract in Swiss albino mice. Asian J. Med. Sci. (E-ISSN 2091-0576; P-ISSN 2467-9100) 7 (2), 49–54.

Jugran, A.K., Bhatt, I.D., Rawal, R.S., Nandi, S.K., Pande, V., 2013. Patterns of morphological and genetic diversity of *Valeriana jatamansi* Jones in different habitats and altitudinal range of West Himalaya, India. Flora-Morphol. Distrib. Funct. Ecol. Plants 208 (1), 13–21.

Kalim, M.D., Bhattacharyya, D., Banerjee, A., Chattopadhyay, S., 2010. Oxidative DNA damage preventive activity and antioxidant potential of plants used in Unani system of medicine. BMC Complement. Alt. Med. 10 (1), 77.

Katoch, O., Kaushik, S., Kumar, M.S.Y., Agrawala, P.K., Misra, K., 2012. Radioprotective property of an aqueous extract from *Valeriana wallichii*. J. Pharm. Bioallied Sci. 4 (4), 327–332.

Khuda, F., Iqbal, Z., Khan, A., Nasir, F., Shah, Y., 2013. Anti-inflammatory activity of the topical preparation of *Valeriana wallichii* and *Achyranthes aspera* leaves. Pak. J. Pharm. Sci. 26 (3), 451–454.

Marder, M., Viola, H., Wasowski, C., Fernández, S., Medina, J.H., Paladini, A.C., 2003. 6-Methylapigenin and hesperidin: new *valeriana* flavonoids with activity on the CNS. Pharmacol. Biochem. Behav. 75 (3), 537–545.

Mhaske, D.K., Patil, D.D., Wadhawa, G.C., 2011. Antimicrobial activity of methanolic extract from rhizome and roots of *Valeriana wallichii*. Int. J. Pharmaceut. Biomed. Res. 2 (4), 107–111.

Nabi, N.U., Neeraj, K., Preeti, K., 2013. Effect of *Valeriana wallichii* DC on learning memory impairment in streptozotocin-induced animal model of diabetes. Int. J. Universal Pharm. Biosci. 2 (6), 476–493.

Nadkami, K.M., Nadkarni, A.K., 1976. Dr K.M. Nadkarmi's Indian Materia Medica: With Ayurvedic, Unani-Tibbi, Siddha, Allopathic, Homeopathic, Naturopathic and Home Remedies, Appendices and Indexes, third ed. Popular Prakashan, Bombay.

Ortiz, J.G., Rassi, N., Maldonado, P.M., González-Cabrera, S., Ramos, I., 2006. Commercial valerian interactions with [3H] Flunitrazepam and [3H] MK-801 binding to rat synaptic membranes. Phytother. Res. 20 (9), 794–798.

Patil, D.D., Mhaske, D.K., Wadhawa, G.C., 2011. Antibacterial and antioxidant study of *Ocimum basilicum Labiatae* (sweet basil). J. Adv. Pharm. Education Res. 2, 104–112.

Rehni, A.K., Pantlya, H.S., Shri, R., Singh, M., 2007. Effect of chlorophyll and aqueous extracts of *Bacopa monniera* and *Valeriana wallichii* on ischemia and reperfusion-induced cerebral injury in mice. Indian J. Exp. Biol. 45 (9), 764–769.

Sah, S.P., Mathela, C.S., Chopra, K., 2010. Elucidation of possible mechanism of analgesic action of *Valeriana wallichii* DC chemotype (patchouli alcohol) in experimental animal models. Indian J. Exp. Biol. 48 (3), 289–293.

Sahu, S., Ray, K., Kumar, M.Y., Gupta, S., Kauser, H., Kumar, S., Mishra, K., Panjwani, U., 2012. *Valeriana wallichii* root extract improves sleep quality and modulates brain monoamine level in rats. Phytomedicine 19 (10), 924–929.

Sati, S.C., Khulbe, K., Joshi, S., 2011. Antibacterial evaluation of the Himalayan medicinal plant *Valeriana wallichii* DC (Valerianaceae). Res. J. Microbiol. 6 (3), 289.

Shalam, M.D., Shantakumar, S.M., Narasu, M.L., 2007. Pharmacological and biochemical evidence for the antidepressant effect of the herbal preparation Trans-01. Indian J. Pharm. 39 (5), 231–234.

Sharma, P., Kirar, V., Meena, D.K., Suryakumar, G., Misra, K., 2012. Adaptogenic activity of Valeriana wallichii using cold, hypoxia and restraint multiple stress animal model. Biomed. Aging Pathol. 2 (4), 198–205.

Shukitt-Hale, B., Stillman, M.J., Welch, D.I., Levy, A., Devine, J.A., Lieberman, H.R., 1994. Hypobaric hypoxia impairs spatial memory in an elevation-dependent fashion. Behav. Neural Biol. 62 (3), 244–252.

Singh, S.K., Katoch, R., Kapila, R.K., 2013. Chemotypic variation for essential oils in *Valeriana jatamansi* Jones populations from Himachal Pradesh. J. Essent. Oil Res. 25 (2), 154–159.

Subhan, F., Karim, N., Gilani, A.H., Sewell, R.D., 2010. Terpenoid content of *Valeriana wallichii* extracts and antidepressant-like response profiles. Phytother. Res. 24 (5), 686–691.

Sudhanshu, N.R., Sandhya, M., Ekta, M., 2012. Evaluation of antioxidant properties of *Valeriana wallichii* to scavenge free radicals. Asian J. Pharm. Clin. Res. 5 (3), 238–240.

Syed, S.N., Rizvi, W., Kumar, A., Khan, A.A., Moin, S., Ahsan, A., 2014. Antioxidant and hepatoprotective activity of ethanol extract of *Valeriana wallichii* in CCl_4 treated rats. Br. J. Pharmaceut. Res. 4 (8), 1004–1013.

Thies, P.W., 1968. Linarin-isovalerianate, a currently unknown flavonoid from *Valeriana wallichii* DC 6. Report on the active substances of Valeriana. Plant. Med. 16 (4), 363.

Thies, P.W., Funke, S., 1966. Über die wirkstoffe des baldrians: 1. Mitteilung Nachweis und isolierung von sedativ wirksamen isovaleriansäureestern aus wurzeln und rhizomen von verschiedenen valeriana-und kentranthus-arten. Tetrahedron Lett. 7 (11), 1155–1162.

Thind, T.S., Suri, R.K., 1979. In vitro antifungal efficacy of four essential oils. Indian Perfume 23, 138–140.

Thusoo, S., Gupta, S., Sudan, R., Kour, J., Bhagat, S., Hussain, R., Bhagat, M., 2014. Antioxidant activity of essential oil and extracts of *Valeriana jatamansi* roots. Biomed. Res. Int. 2014.

Tripathi, S.V., Tiwari, S.K., Chaturvedi, G.N., 1982. Comparative clinical trial of certain indigenous drugs (Dhanya panchaka) in gastric secretion hyperchlorhydria. Naga 25, 170–174.

Wasowski, C., Marder, M., Viola, H., Medina, J.H., Paladini, A.C., 2002. Isolation and identification of 6-methylapigenin, a competitive ligand for the brain GABAA receptors, from *Valeriana wallichii* DC. Plant. Med. 68, 934–936.

Yang, J., Xue, C.K., Zhu, X.Z., 1998. Evaluate the effects of some TCMs extracts on improving microcirculation reperfusion volume of both cardiac and renal tissues by ^{86}Rb tracer. Chin. J. Microcircul. 8 (1), 15–17.

Yang, Y., Ma, L., Guan, W., Wang, Y., Du, Y., Ga, Q., Ge, R.L., 2014. Differential plasma proteome analysis in patients with high-altitude pulmonary edema at the acute and recovery phases. Exp. Ther. Med. 7 (5), 1160–1166.

CHAPTER 5

Rhodiola sp.: The Herbal Remedy for High-Altitude Problems

Priyanka Sharma*, Kshipra Misra†

*Cardiorespiratory Department, Defence Institute of Physiology and Allied Sciences (DIPAS), Delhi, India, †Department of Biochemical Sciences (DBCS), Defence Institute of Physiology and Allied Sciences (DIPAS), Delhi, India

Abbreviations

ABTS	2,2′-azino-bis(3-ethylbenzthiazoline-6-sulphonic acid)
AFI	antifatigue index
CAT	catalase
C-H-R	cold-hypoxia-restraint
DPPH	2,2′-diphenyl-1-picrylhydrazyl
GSH	glutathione
Gy	gray
IC50	half of maximum inhibitory concentration
IL-6	interleukin-6
LDH	lactate dehydrogenase
MTD	maximum tolerant doses
PBMC	peripheral blood mononuclear cell
sp.	species
SSRI	selective serotonin reuptake inhibitor
TNF	tumor necrosis factor

5.1 INTRODUCTION

Rhodiola plants commonly called golden root, rose root, or arctic root, belong to stone crop family, Crassulaceae. *Rhodiola* species have been popular as adaptogens in Asia and Northern Europe for a long time. Traditional Chinese medicine consider them to be a drug of "source of adaptation to environment" (Li et al., 2007). They are capable of improving human resistance to stress or fatigue and in promoting longevity (Mattioli et al., 2009, 2012; Tolonen et al., 2003). Some species of *Rhodiola* have been used to enhance the body's ability to survive in adverse environments, especially those faced by astronauts, divers, mountaineers, and pilots (Li et al., 2007).

81

Management of High Altitude Pathophysiology. https://doi.org/10.1016/B978-0-12-813999-8.00005-7

The extracts of *Rhodiola* plants are used widely throughout Asia, Europe, and the United States for antiallergic, antiinflammatory, and antistress activities, and have been reported to enhance mental alertness and physical vigor (Tolonen et al., 2003). Root extracts of *Rhodiola* have gained worldwide popularity as ingredients of food additives and other commercial pharmaceutical preparations. Extensive literature is available describing the medicinal properties of some *Rhodiola* species with regard to cancer therapy, immune response, memory and learning, and organ function. (Evstatieva et al., 2010).

Phytochemical investigation of *Rhodiola* sp. has been focused mainly on rosarin, rosavin, rosin, and salidroside (Tolonen et al., 2003). Salidroside has been reported in various Chinese studies as a sleep inducer (Li and Chen, 2001). Other constituents that are present in significant amounts in *Rhodiola* roots are essential oils, flavonoids, gallic acid and its esters, and tannins. (Evstatieva et al., 2010).

This chapter will present a comprehensive view of the various pharmacological applications of *Rhodiola* sp. because of the presence of myriad phytochemicals with special focus on the treatment of hypoxia.

5.2   GEOGRAPHICAL DISTRIBUTION OF *RHODIOLA* SP.

The genus *Rhodiola L.* (Crassulaceae) consists of about 96 species distributed in the alpine regions of Asia and Europe (Liu et al., 2013). The *Rhodiola* genus finds its origin in the mountain areas of Himalayas and Southeastern China (Darbinyan et al., 2000), growing in the alpine rocky abyss of tundra regions. The species of the *Rhodiola* genus are spread in a circumpolar manner. They cover Middle and Northern Asia from Altai Mountains over Mongolia, Kazakhstan, Uzbekistan, and in many regions of Siberia. In Europe, this plant encompasses Iceland and British Islands, Scandinavia, Pyrenees, Alps, Carpathians, and other peaks in the Balkan area. Table 5.1 displays the major geographical distribution of some *Rhodiola* species in various continents.

Plants belonging to *Rhodiola* sp. are popularly known as golden root or rose root. The genus sometimes is referred to as *Rhodiola rosea* (*R. rosea*). A few studies in traditional history attribute the name to the fact that the roots of this plant have a rose-like fragrance (Linnaeus, 1749). It also was presented as a bouquet to newlywed couples as a symbol of fertility and good health (Saratikov and Krasnov, 1987). Substantial literature is available that pertains to the use of *R. rosea* in pharmaceutical purposes and as a traditional medicine in France, Germany, Iceland, Norway, Russia, and Sweden (Brown et al., 2002). The Vikings used this species for a long time to increase physical strength and resistance, and *R. rosea* was included in the first edition of Pharmacopeia of Sweden in the year 1755.

Table 5.1 Species Distribution of Some of the *Rhodiola* Species

Continent	Country/States/Cities	Species
Asia	China (Hebei, Shanxi, Gansu etc.), Kazakhstan, Uzbekistan, Mongolia, Russia (Siberia, Altai etc.), India, Pakistan	*R. rosea, R. himalensis, R. crenulata, R. sachaliensis, R. kirilowii, R. serrata, R. fastigiata, R. crenulata, R. quadrifida, R. heterodonta, R. imbricata, R. sinuate, R. tibetica, R. wallichiana*
Europe	Austria, Bulgaria, Czechoslovakia, the United Kingdom, Italy, Norway, Poland	*R. rosea*
North America	Canada, the United States (Alaska, California, Colorado, New York, Nevada, Washington)	*R. rosea, R. kirilowii, R. yunnanensis, R. crenulata, R. sacra, R. sachalinensis*

R. rosea is a widely found species, distributed across 28 countries in Europe, from sea level to altitudes as high as 5000 m (Maftei, 2016). The plants preferably grow in dry sandy ground at high altitudes. Of all the *Rhodiola* sp., the most medicinal, pharmacological, and phytochemical research has been carried out on *R. rosea*.

In India, six species of *Rhodiola* are found: *R. heterodonta, R. imbricata, R. quadrifida, R. sinuate, R. tibetica,* and *R. wallichiana* (Chaurasia and Gurmet, 2003). *R. imbricata* is one of the most predominantly studied species and occurs primarily in the Himalayan region at a height of 2500–4000 m. The plant is perennial (Fig. 5.1A), has thick rhizomes (Fig. 5.1B), and is fragrant when freshly cut. It is commonly found in Indus and Leh valleys of Indian Trans-Himalayan regions (Chaurasia et al., 2007).

FIG. 5.1

Rhodiola plant (A) and its rhizomes (B).

5.3 BOTANICAL CLASSIFICATION OF *RHODIOLA* SP.

The taxonomical status of the genus *Rhodiola* is quite complex. Before World War II, taxonomists distinguished between different species of *Rhodiola* as independent genus (Hegi, 1963). Later, different researchers classified *Rhodiola* as subgenus, then again as a genus and so on. Because of morphological similarities, the classification of plants under this genus still remains a daunting task. Ghiorghita et al. (2015) detailed the systematic classification of *Rhodiola* sp.: Kingdom: *Plantae*; Phylum: *Magnoliophyta*; Class: *Magnoliopsida*; Order: *Rosales*; Family: *Crassulaceae*; Genus: *Rhodiola*; Species: *96* or more.

As mentioned in Section 5.2, majority of studies under genus *Rhodiola* are being conducted on *R. rosea*. This species approximately covers 51% of all animal studies and 94% of all human studies conducted. *R. rosea* has been found to be toxicologically safe for both animals and humans. This species is prevalent in three continents: Asia, Europe, and North America (Table 5.1). In general, *R. rosea* is a perennial, succulent herb with acylindrical or oblong, thick (diameter: 0.5–2.5 cm), fleshy and fragrant (characteristic rose) rhizome. Leaves are pale green in color, ovate or oblong (1–5 cm × 0.4–1.5 cm), fleshy and glabrous. Leaf margin is entire or dentate. The plants are dioecious. The species is diploid ($2n = 22$).

Similar to *R. rosea*, *R. imbricate* is a succulent herb with a thick rhizome (10–35 cm), golden outside, pink inside; leaves 1.3–3 cm long, oblong to narrow, margins nearly entire; flowers pale yellow in congested cluster, surrounded by an involucre of leaves (Chaurasia et al., 2007).

5.4 BIOACTIVE COMPONENTS IN *RHODIOLA* SP.

Phytochemical investigations have revealed that there are six important classes of bioactive constituents in *Rhodiola* rhizomes: flavonoids, monoterpernes, phenylethanolic derivatives, phenolic acids, phenylpropanoids, and triterpenes (Ali et al., 2008) (Table 5.2). Bioassay-guided fractionation of various extracts prepared from the *Rhodiola* plant has shown that the active components are mainly phenylethanolic derivatives and phenylpropanoids, including rosavin, salidroside, and tyrosol (Cifani et al., 2010; Kurkin and Zapesochnaya, 1986; Peschel et al., 2013; Wiedenfeld et al., 2007; Zapesochnaya et al., 1995). Phenylpropanoids, such as rosarin, rosavin, and rosin found in *Rhodiola* rhizomes are pharmacologically active as antioxidants and neurostimulants (Huang et al., 2009; Kucinskaite et al., 2004; Schriner et al., 2009). Compounds such as gallic acid and tyrosol have been proven to be good free radical scavengers (Lee et al., 2000; Sun et al., 2012). A recent study revealed that rhodionin and salidroside might have antitumor effects (Hu et al., 2010; Sun et al., 2012).

Table 5.2 Categories of Phytochemicals and Their Principle Bioactivity as Studied in Various *Rhodiola* Species

Phytochemicals	Examples	Bioactivities
Phenylpropanoides	Rosavin, rosin, rosarin	Antioxidants, neurostimulants, adaptogenic
Phenylethanol derivatives	Salidroside (rhodioloside), tyrosol	Sleep inducers, adaptogenic
Flavonoides	Rhodiolin, rhodionin, rhodiosin, acetylrodalgin, tricin	Learning and memory, antitumor, radioprotective
Phenolic acids	Chlorogenic acid, hydroxycinnamic acid, gallic acid	Antioxidants, free radical scavengers
Monoterpenes	Rosiridol, rosaridin	Antimicrobial, food additives
Triterpenes	Daucosterol, beta-sitosterol	antiinflammatory, immunomodulatory

Rhodionin also is recognized to be involved in learning and memory improvement (Choe et al., 2012; Kobayashi et al., 2008). Rhizomes are known to contain 1.5–4 times more salidroside (0.3%–0.4% dry wt.) and total rosavins (1.2%–3.0%), as compared to that in the roots (Wiedenfeld et al., 2007). The rosavins are specific solely to *R. rosea*; the salidroside also characterizes other species of the *Rhodiola* genus, other plant species, and some bacteria and yeasts (Ghiorghita et al., 2015).

5.5 PHARMACOLOGICAL AND MEDICINAL IMPORTANCE OF *RHODIOLA* SP.

The pharmacological and medicinal properties of *Rhodiola* are species-dependent. Some of the important medicinal properties are depicted in Fig. 5.2. *Rhodiola* sp. have been used as adaptogens for a very long time in Northern Europe and Russia (Mattioli et al., 2009; Mattioli et al., 2012; Tolonen et al., 2003). *Rhodiola* has been reported to increase the bioelectrical activity of the brain, stimulate norepinephrine, dopamine, serotonin, and nicotinic cholinergic effects in the central nervous system, and enhance the permeability of dopamine and serotonin to cross the blood-brain barrier.

Rhodiola sp. have been used in traditional Tibetan medicine for over 1000 years (Kylin, 2010). They have been found to cure hernias, headaches, hysteria, and leucorrhea (Alm, 2004; Brown et al., 2002; Galambosi, 2005; Kylin, 2010; Panossian et al., 2010). The plants have been known to improve longevity, work productivity, and resistance to altitude sickness (Maftei, 2016). The major active constituent of *Rhodiola* sp. is salidroside (Li and Chen, 2001), which could improve sleep and possess sedative and hypnotic action.

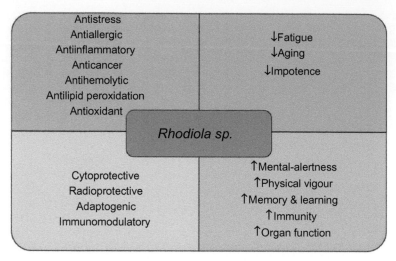

FIG. 5.2

Medicinal importance of *Rhodiola* species (↑, increases; ↓, decreases).

Rhodiola sp. have shown medicinal properties such as antiaging, anticancer, antiinflammatory, and antimutagenic activities (Saratikov and Krasnov, 1987). The species also have been reported to enhance physical endurance and to treat impotence, fatigue, and gastrointestinal, cardiac, and central nervous system disorders (Ohwi, 1984; Brown et al., 2002). The adaptogenic potential and therapeutic efficacy of *Rhodiola* sp. against high-altitude sickness, sexual dysfunction, and sleep disturbances (depression) has indicated its pharmacological importance in humans (Ganzera et al., 2001). Species such as *R. imbricata* have demonstrated significant bioefficacy by shifting anaerobic metabolism to aerobic metabolism after multiple-stress exposures (Gupta et al., 2010). The anticancerous potential of *R. imbricata* toward natural killer cell cytotoxicity (Mishra et al., 2008) and the wound healing potential toward an increase in antioxidant activity and a decrease in lipid peroxide levels in the granulation tissues have been extensively studied (Gupta et al., 2007).

Researchers also have reported the antiviral (Sun et al., 1993), antitumor (Wang et al., 1992), antiaging (Yi et al., 1992), antioxygen deficiency, antiradiation (Luo et al., 1996) effects, and other pharmacological properties of *Rhodiola* sp. (Xu et al., 2000).

The following section details some of the medicinal properties of *Rhodiola* sp. and the clinical studies that have been carried out on these plants.

5.5.1 Adaptation Under Exposure to Multiple Stresses [Cold, Hypoxia, and Restraint (C-H-R)] and Poststress Recovery

Gupta et al. (2010) examined the antioxidative potential of aqueous extract of *R. imbricate* root extract under multiple stressful conditions in rats, by orally administering a dose of 100 mg/Kg, as both single and multiple doses, 30 min prior to CHR exposure of cold (5°C), hypoxia (428 mmHg), and restraint. Multiple-dose treatment of the extract further increased glutathione (GSH) level in blood, liver and muscle; restricted increase in lactate dehydrogenase (LDH) level on attaining rectal temperature T_{rec} 23°C and recovery; and increased catalase (CAT) during recovery. The observations brought out the appreciable antioxidant potential of *R. imbricate* root extract during C-H-R exposure and poststress recovery, while effectively maintaining the cell membrane permeability (Gupta et al., 2007).

5.5.2 Anticancer and Immunomodulatory Effects

In a small pilot study involving 12 patients with superficial bladder carcinoma, treatment with minimum 0.8% salidroside and 3.0% rosavin improved parameters of leukocyte integrins and T cell immunity. The average frequency of relapse was reduced but did not reached statistical significance.

The *Rhodiola* sp. extract was found to stimulate production of interleukin-6 (IL-6) and tumor necrosis factor-alpha (TNF-alpha) in human PBMCs as well as RAW 264.7 cell line. The aqueous extract of *R. imbricate* rhizome was proved to be quite potent in upregulating immune response in immune-compromised patients (Mishra et al., 2008).

5.5.3 Antiproliferative Effect on HT-29 Human Colon Cancer Cells

The antiproliferative and in vitro free radical scavenging activities of acetone and methanol extracts of *Rhodiola* rhizome in HT-29 human colon cancer cells also have been reported (Senthilkumar et al., 2013). The extracts were further reported to exhibits 2,2'-diphenyl-1-picrylhydrazyl (DPPH), 2,2'-azino-bis-(3-ethylbenzthiazoline-6-sulphonic acid) (ABTS), superoxide anion, hydroxyl radical scavenging activities, and metal chelating ability because of the presence of natural antioxidants such as flavonoids, phenolics, and tannins. Moreover, the extracts also inhibited the proliferation of HT-29 human colon cancer upon treatment at higher concentration (200 μg/mL) (acetone and methanol, 84% each). On further investigations, acetone extract revealed antiproliferative potential in a dose-dependent manner; methanol extract showed both dose-dependent and time-dependent inhibitory potential.

5.5.4 Clinical Case Studies Using *Rhodiola rosea*

A visibly depressed 45-year-old writer regained her enthusiasm after consuming 100 mg of *R. rosea* extract (Rosavin, a preparation standardized to 1% salidroside and 3% rosavin) twice a day for six weeks (Brown et al., 2002).

A 45-year-old mental health professional, suffering from refractory depression and fibromyalgia for five years, recovered completely after an intake of 600 mg/day of a formulation containing *R. rosea* extract (Rosavin) with antidepressant sertraline (Zoloft, a selective serotonin reuptake inhibitor, SSRI). The formulation enabled her to return to normal enjoyment and full productivity in life after about two months (Brown et al., 2002).

Within 1 week of starting 300 mg *R. rosea* extract (Rosavin) twice daily, a patient with Parkinson's disease for 10 years began to recover with marked progressive improvements in his abilities to think, speak, read, and initiate independent activities (Brown et al., 2002).

5.5.5 Dermal Wound Healing Potential

Gupta et al. (2007) investigated the wound healing efficacy of extract of *R. imbricata* rhizome using a rat excision wound model. The extract was found to be rich in polyphenols, which possess significant wound healing activity. The results also were supported by histological examinations (Gupta et al., 2007).

5.5.6 Improvement in Mental Work Efficiency

Shevtsov et al. (2003) designed a randomized, double-blind, placebo-controlled, parallel-group clinical study with an extra nontreatment to measure the effect of a single dose of standardized SHR-5 *R. rosea* extract on capacity for mental work against fatigue and stress. Two doses were chosen. The first dose was a well-established psychostimulating dose; the second was 50% higher in concentration. For this purpose, uniformed cadets, 19–21 years old were chosen. The study showed a pronounced antifatigue effect reflected in a ratio called AFI (Anti-fatigue Index). The verum groups, when given two and three capsules, had AFI mean values of 1.0385 and 1.0195, respectively; the AFI value for the placebo group was 0.9046. This was statistically significant ($P < 0.001$) for both doses (verum groups). Although no significant difference between the two dosage groups was observed, there was a possible trend in favor of the lower dose in psychometric tests. No such trend was found in physiological tests (Shevtsov et al., 2003).

5.5.7   Radiomodulatory and Free Radical Scavenging Activity

Arora et al. (2008) studied the radio modifying effects, electron-donation potential, superoxide ion scavenging ($IC_{50} \leq 0.025$ mg/mL), nitric oxide (NO) scavenging potential ($IC_{50} = 0.5$ mg/mL), and antihemolytic activity of a fractionated extract. Reducing power, superoxide ion, and NO scavenging ability of the fractionated extract increased in a dose-dependent manner. This particular study showed that *R. imbricata* has immense potential for alleviation of biological damage in a radiation-prone environment (Arora et al., 2008).

5.5.8   *Rhodiola* sp. as a Radioprotector

Nuclear and radiological emergencies have been affecting human lives for a long time. Both synthetic and natural products have shown radioprotective potential; however, the uses of synthetic compounds have been found to be limited because of their higher toxicity than natural compounds, which are found to be safer and more reliable. In one of the studies by Goel et al. (2006), the radioprotective effects of both aqueous and aqueous-alcohol extracts of *R. imbricate* were reported. The maximum tolerant doses (MTD) for both the extracts were found to be more than 1000 mg/kg of body weight of mice. The maximum radioprotective effect was observed at doses one-fourth to one-third of the MTD. This kind of observation is contrary to the general observation that a medicinal plant offers radioprotection at doses that are quite close to their MTD. Thus, extracts of *Rhodiola* sp. are found to be nontoxic and exhibit quite safe margins. The chances of survival improved by about 83% in cases of whole body supralethal irradiation (10 Gy) (Goel et al., 2006).

5.6   CONCLUSIONS

Rhodiola species have demonstrated their myriad biological activities for almost 2000 years. Since its potential therapeutic potential was discovered, *Rhodiola* species have been thoroughly explored by athletes, divers, and even cosmonauts. Further research is needed to be carried out using controlled clinical trials to develop *Rhodiola*-based adaptogens and formulations against deadly diseases and ailments. At the same time, great care must be required to preserve these golden root species because they are very prone to environmental changes.

Acknowledgments

The authors are thankful to the director, Defense Institute of Physiology and Allied Sciences, Delhi, India, for the constant support and encouragement.

References

Ali, Z., Fronczek, F.R., Khan, I.A., 2008. Phenylalkanoids and monoterpene analogues from the roots of *Rhodiola rosea*. Plant. Med. 74 (02), 178–181.

Alm, T., 2004. Ethnobotany of *Rhodiola rosea* (Crassulaceae) in Norway. SIDA 21 (1), 321–344.

Arora, R., Singh, S., Sagar, R.K., Chawla, R., Kumar, R., Puri, S.C., Surender, S., Prasad, J., Gupta, M.L., Krishna, B., Siddiqui, M.S., 2008. Radiomodulatory and free radical scavenging activity of the fractionated aquoalcoholic extract of the adaptogenic nutraceutical (*Rhodiola imbricata*): a comparative in vitro assessment with ascorbate. J. Diet. Suppl. 5 (2), 147–163.

Brown, R.P., Gerbarg, P.L., Ramazanov, Z., 2002. *Rhodiola rosea*: a phytomedicinal overview. HerbalGram 56, 40–52.

Chaurasia, O.P., Ahmed, Z., Ballabh, B., 2007. Ethnobotany and Plants of Trans-Himalaya. Satish Serial Pub. House, Delhi.

Chaurasia, O.P., Gurmet, A., 2003. Checklist on Medicinal and Aromatic Plants of Trans-Himalayan Cold Deserts. FRL, Leh.

Choe, K.I., Kwon, J.H., Park, K.H., Oh, M.H., Kim, M.H., Kim, H.H., Cho, S.H., Chung, E.K., Ha, S.Y., Lee, M.W., 2012. The antioxidant and anti-inflammatory effects of phenolic compounds isolated from the root of *Rhodiola sachalinensis* A. BOR. Molecules 17 (10), 11484–11494.

Cifani, C., Di, B.M.V.M., Vitale, G., Ruggieri, V., Ciccocioppo, R., Massi, M., 2010. Effect of salidroside, active principle of *Rhodiola rosea* extract, on binge eating. Physiol. Behav. 101 (5), 555–562.

Darbinyan, V., Kteyan, A., Panossian, A., Gabrielian, E., Wikman, G., Wagner, H., 2000. *Rhodiola rosea* in stress-induced fatigue: a double blind cross-over study of a standardized extract SHR-5 with a repeated low-dose regimen on the mental performance of healthy physicians during night duty. Phytomedicine 7 (5), 365–371.

Evstatieva, L., Todorova, M., Antonova, D., Staneva, J., 2010. Chemical composition of the essential oils of *Rhodiola rosea L.* of three different origins. Pharmacogn. Mag. 6 (24), 256.

Galambosi, B., 2005. *Rhodiola rosea L.*, from wild collection to field production. Med. Plant Conserv. 11, 31–35.

Ganzera, M., Yayla, Y., Khan, I.A., 2001. Analysis of the marker compounds of *Rhodiola rosea L.* (golden root) by reversed phase high performance liquid chromatography. Chem. Pharm. Bull. 49 (4), 465–467.

Ghiorghita, G., Maftei, D.I., Maftei, D.E., 2015. *Rhodiola rosea L.*: a valuable plant for traditional and modern medicine. AN Stiint. Univ. AI I Cuza Iasi 61 (1/2), 5.

Goel, H.C., Bala, M., Prasad, J., Singh, S., Agrawala, P.K., Swahney, R.C., 2006. Radioprotection by *Rhodiola imbricata* in mice against whole-body lethal irradiation. J. Med. Food 9 (2), 154–160.

Gupta, A., Kumar, R., Upadhyay, N.K., Pal, K., Kumar, R., Sawhney, R.C., 2007. Effects of *Rhodiola imbricata* on dermal wound healing. Plant. Med. 73 (08), 774–777.

Gupta, V., Lahiri, S.S., Sultana, S., Tulsawani, R.K., Kumar, R., 2010. Antioxidative effect of *Rhodiola imbricata* root extract in rats during cold, hypoxia, and restraint (CHR) exposure and post-stress recovery. Food Chem. Toxicol. 48 (4), 1019–1025.

Hegi, G., 1963. *Rhodiola*, Rosenwurz. In: Hegi, G. (Ed.), Illustrierte Flora von Mitteleuropa. Band IV/2, Liefering 2/3. Zweite völlig neubearbeitete edn, Hamburg/Berlin, pp. 99–102.

Hu, X., Lin, S., Yu, D., Qiu, S., Zhang, X., Mei, R., 2010. A preliminary study: the antiproliferation effect of salidroside on different human cancer cell lines. Cell Biol. Toxicol. 26 (6), 499–507.

Huang, S.C., Lee, F.T., Kuo, T.Y., Yang, J.H., Chien, C.T., 2009. Attenuation of long-term *Rhodiola rosea* supplementation on exhaustive swimming-evoked oxidative stress in the rat. Chin. J. Physiol. 52 (5), 316–324.

Kobayashi, K., Yamada, K., Murata, T., Hasegawa, T., Takano, F., Koga, K., Fushiya, S., Batkhuu, J., Yoshizaki, F., 2008. Constituents of *Rhodiola rosea* showing inhibitory effect on lipase activity in mouse plasma and alimentary canal. Plant. Med. 74 (14), 1716–1719.

Kucinskaite, A., Briedis, V., Savickas, A., 2004. Experimental analysis of therapeutic properties of *Rhodiola rosea L.* and its possible application in medicine. Medicina (Kaunas, Lithuania) 40 (7), 614–619.

Kurkin, V.A., Zapesochnaya, G.G., 1986. The chemical composition and pharmacological properties of *Rhodiola* plants. Khim. Pharm. Z. 20 (10), 1231–1244.

Kylin, M., 2010. Genetic Diversity of Rose Root (*Rhodiola rosea L.*) from Sweden, Greenland, and Faroe Islands. Master's thesis. Swedish University of Agricultural Sciences, SLU, p. 58.

Lee, M.W., Lee, Y.A., Park, H.M., Toh, S.H., Lee, E.J., Jang, H.D., Kim, Y.H., 2000. Antioxidative phenolic compounds from the roots of *Rhodiola sachalinensis* a. Bor. Arch. Pharm. Res. 23 (5), 455.

Li, H.B., Chen, F., 2001. Preparative isolation and purification of salidroside from the Chinese medicinal plant *Rhodiola sachalinensis* by high-speed countercurrent chromatography. J. Chromatogr. A 932 (1), 91–95.

Li, T., Xu, G., Wu, L., Sun, C., 2007. Pharmacological studies on the sedative and hypnotic effect of salidroside from the Chinese medicinal plant *Rhodiola sachalinensis*. Phytomedicine 14 (9), 601–604.

Linnaeus, C., 1749. Materia medica: Liber I de plantis. IDC. p. 168.

Liu, Z., Liu, Y., Liu, C., Song, Z., Li, Q., Zha, Q., Lu, C., Wang, C., Ning, Z., Zhang, Y., Tian, C., 2013. The chemotaxonomic classification of *Rhodiola* plants and its correlation with morphological characteristics and genetic taxonomy. Chem. Central J. 7 (1), 118.

Luo, C.H., Shu, R., Gao, Y., 1996. Experiment study of antifatigue and ant-radiation effect of high-mountain plant *Rhodiola L.* Modern Apply Med. 13 (4), 5–7.

Maftei, D.E., 2016. *Rhodiola rosea L.*: preservation status in the Călimani national park. Biologie 25, 132–135.

Mattioli, L., Funari, C., Perfumi, M., 2009. Effects of *Rhodiola rosea L.* extract on behavioral and physiological alterations induced by chronic mild stress in female rats. J. Psychopharmacol. 23 (2), 130–142.

Mattioli, L., Titomanlio, F., Perfumi, M., 2012. Effects of *Rhodiola rosea L.* extract on the acquisition, expression, extinction, and reinstatement of morphine-induced conditioned place preference in mice. Psychopharmacology 221 (2), 183–193.

Mishra, K.P., Padwad, Y.S., Dutta, A., Ganju, L., Sairam, M., Banerjee, P.K., Sawhney, R.C., 2008. Aqueous extract of *Rhodiola imbricata* rhizome inhibits proliferation of an erythroleukemic cell line K-562 by inducing apoptosis and cell cycle arrest at G2/M phase. Immunobiology 213 (2), 125–131.

Ohwi, J., 1984. Flora of Japan. Smithsonian Institution, Washington, DC, p. 495.

Panossian, A., Wikman, G., Sarris, J., 2010. Rosen root (*Rhodiola rosea*): traditional use, chemical composition, pharmacology, and clinical efficacy. Phytomedicine 17 (7), 481–493.

Peschel, W., Prieto, J.M., Karkour, C., Williamson, E.M., 2013. Effect of provenance, plant part, and processing on extract profiles from cultivated European *Rhodiola rosea L.* for medicinal use. Phytochemistry 86, 92–102.

Saratikov, A.S., Krasnov, E.A., 1987. *Rhodiola rosea* is a Valuable Medicinal Plant (Golden Root). Tomsk State University, Tomsk, Russia.

Schriner, S.E., Abrahamyan, A., Avanessian, A., Bussel, I., Maler, S., Gazarian, M., Holmbeck, M.A., Jafari, M., 2009. Decreased mitochondrial superoxide levels and enhanced protection against paraquat in Drosophila melanogaster supplemented with *Rhodiola rosea*. Free Radic. Res. 43 (9), 836–843.

Senthilkumar, R., Parimelazhagan, T., Chaurasia, O.P., Srivastava, R.B., 2013. Free radical scavenging property and antiproliferative activity of *Rhodiola imbricata* Edgew extracts in HT-29 human colon cancer cells. Asian Pacific J. Trop. Med. 6 (1), 11–19.

Shevtsov, V.A., Zholus, B.I., Shervarly, V.I., Vol'skij, V.B., Korovin, Y.P., Khristich, M.P., Roslyakova, N.A., Wikman, G., 2003. A randomized trial of two different doses of a SHR-5 *Rhodiola rosea* extract versus placebo and control of capacity for mental work. Phytomedicine 10 (2), 95–105.

Sun, C., Wang, Z., Zheng, Q., Zhang, H., 2012. Salidroside inhibits migration and invasion of human fibrosarcoma HT1080 cells. Phytomedicine 19 (3), 355–363.

Sun, F., Wang, X.Q., Li, J.B., 1993. Experimental study the anti-AsQ B5 virus's effect of *Rhodiola*. Chin. Tradit. Herb Drugs 24 (10), 532–534.

Tolonen, A., Hohtola, A., Jalonen, J., 2003. Comparison of electrospray ionization and atmospheric pressure chemical ionization techniques in the analysis of the main constituents from *Rhodiola rosea* extracts by liquid chromatography/mass spectrometry. J. Mass Spectrom. 38 (8), 845–853.

Tolonen, A., Pakonen, M., Hohtola, A., Jalonen, J., 2003. Phenylpropanoid glycosides from *Rhodiola rosea*. Chem. Pharm. Bull. 51 (4), 467–470.

Wang, X.Q., Li, J.B., Zhang, H.Y., 1992. Experimental study the antitumor effect of *Rhodiola L*. JILIN Tradit.ional Chin. Med. 3, 40–42.

Wiedenfeld, H., Dumaa, M., Malinowski, M., Furmanowa, M., Narantuya, S., 2007. Phytochemical and analytical studies of extracts from *Rhodiola rosea* and *Rhodiola quadrifida*. Die Pharm. 62 (4), 308–311.

Xu, B.J., Zhen, Y.N., Li, X.G., 2000. New development of *Rhodiola L*. plant research. Chin. Med. 23 (9), 580–584.

Yi, G.S., Zhou, J.Q., Wang, G.X., 1992. Study of the prolonged life effect of extraction of *Rhodiola L*. root. Acta Nutr. Sin. 14 (1), 98–100.

Zapesochnaya, G.G., Kurkin, V.A., Boyko, V.P., Kolkhir, V.K., 1995. Phenylpropanoids promising biologically active compounds of medicinal plants. Khim. Farm. Zh. 29, 47–50.

CHAPTER 6

Cordyceps sp.: The Precious Mushroom for High-Altitude Maladies

Mamta Pal*, Kshipra Misra†

*Division of Forensic Science, School of Basic and Applied Sciences, Galgotias University, Greater Noida, India, †Department of Biochemical Sciences (DBCS), Defence Institute of Physiology and Allied Sciences (DIPAS), Delhi, India

Abbreviations

ABTS	2,2′-azino-bis (3-ethylbenzothiazoline-6-sulphonic acid)
AChE	acetylcholinesterase
AMS	acute mountain sickness
B. subtilis	*Bacillus subtilis*
C. sinensis	Cordyceps *sinensis*
CSE	*Cordyceps sinensis* extract
DPPH	1,1-diphenyl-2-picryl-hydrazyl
E. coli	*Escherichia coli*
ET-1	endothelin-1
FDA	Food and Drug Administration
H. armoricanus	*Hepialus armoricanus*
HACE	high-altitude cerebral edema
HAPE	high-altitude pulmonary edema
HATE	high-altitude thromboembolism
I/R	ischemia-reperfusion
LPS	lipopolysaccharide
O. sinensis	*Ophyiocordyceps sinensis*
PASMCs	pulmonary artery smooth muscle cells
PCNA	proliferating cell nuclear antigen
ROS	reactive oxygen species
S. aureus	*Staphylococcus aureus*
US	United States
VEGF	vascular endothelial growth factor

6.1 INTRODUCTION

Herbs fulfill important human requirements and are gaining importance in modern medicine because of their characteristic bioactivities that are responsible for mitigating physiological damages incurred from exposure to different

93

Management of High Altitude Pathophysiology. https://doi.org/10.1016/B978-0-12-813999-8.00006-9

stresses (Mamta et al., 2013b). Several traditional systems in India, such as Ayurveda and Unani, are known mainly for plant-based drugs. Materia Medica also describes the practice of indigenous herbs, and Vedic literature, such as Rig Veda and Atharva Veda, mentions the use of several plants as medicines. Charaka Samhita and Susruta Samhita also reveal the use of more than 700 herbs (Anandanayaki and Jegadeesan, 2010).

In addition to medicinal plants, medicinal mushrooms are also becoming popular in the pharmaceutical industry for the development of various drugs (Wasser, 2014). These mushrooms are known to be natural bioenhancers, similar to herbal plants, considering their substantial metabolite composition (Barros et al., 2009; Heleno et al., 2011). They also have been used as traditional medicines in many South Asian countries since ancient times. About 12,000 species of mushrooms exist in nature, out of which at least 2000 species are edible. Among the edible varieties about 35 species are being produced commercially to meet global demand (Chang, 1999). *Cordyceps sinensis* is one of these highly valuable medicinal mushrooms found at high altitudes (Holliday and Cleaver, 2008).

More than 680 species of *Cordyceps* have been documented (Holliday and Cleaver, 2004), but *C. sinensis* is most popular among scientists and has been used in traditional Chinese medicine for centuries (Singh et al., 2010). This medicinal mushroom, despite the harsh environment at high altitudes such as oxygen deprivation, low temperature, and intense exposure to UV radiation, thrives well probably because of its distinctive metabolite composition (Spehn et al., 2012). This chapter will review the chemical composition of *C. sinensis* and its health-promoting activities, particularly for high altitude-induced maladies.

6.2   VEGETATION AT HIGH ALTITUDE

C. sinensis is an entomogenous medicinal mushroom that is native to high-altitude regions. Several other varieties of *Cordyceps*, such as *C. militaris*, *C. sobolifera*, *C. subssesilus*, and *C. phioglossides* also exist in nature (Holliday and Cleaver, 2008), but *C. sinensis* is biologically more effective than the others (Singh et al., 2010). The ability of *C. sinensis* to sustain hostile high-altitude conditions likely has marked effects on its life cycle and its exclusive bioactive metabolites in comparison to the vegetation at plane regions (Spehn et al., 2012).

It is difficult for oxygen to enter into a human's vascular system at high altitudes because of decreased atmospheric pressure, which leads to insufficient oxygen in inspired air and an altered mitochondrial respiratory chain. Subsequently, people encounter various health-related issues and often end up with acute mountain sickness (AMS). AMS is manifested by headache, nausea, vomiting, breathlessness, dizziness, insomnia, muscle aches, fatigue, and loss of appetite and immunity (Tang et al., 2014). All the innate properties of this high-altitude fungus could help individuals tolerate high altitude-induced hypoxia, oxidative stress, and UV radiation.

6.3 ORIGIN AND GEOGRAPHICAL DISTRIBUTION OF *C. SINENSIS*

Cordyceps, the generic name of the mushroom, is derived from the Latin words *cord*, meaning "club," and *ceps*, meaning "head," which describe appearance of this fungus. Herdsmen first observed this fungus about 1500 years ago in the Tibetan mountains, where cattle stumbled upon it while grazing. After consuming this grass-like mushroom, cattle and livestock became more energetic and even older animals became dynamic (Singh et al., 2010). This miracle fungus was first recorded as Yarsagumba (caterpillar mushroom) in local Tibetan language in the 15th century (Childs and Choedup, 2014). In Chinese Tibetan medicine, it became famous as Yartsa gunbu, and in China, it became known as Dong Chong Xia Cao, meaning winter worm or summer grass, depending on the season (Garbyal et al., 2004). In India, it is referred as Keera ghaas (insect herb) throughout Kumaun Himalaya region (Negi et al., 2006). In some parts of Garhwal Himalaya, it is known as Keera jhar (Sharma, 2004).

C. sinensis belongs to *Ascomycetes* fungi, *Hypocreales* order, and is classified under *Clavicipataceae* family (Holliday and Cleaver, 2004). However, it has been placed in a new family, *Ophiocordycipitaceae*, based on the colony characteristics. Species of this family produce dark pigmented stroma, which can be either be soft or hard. Thus, *C. sinensis* has now been assigned to *Ophiocordyceps* genus and renamed as *Ophyiocordyceps sinensis* (*O. sinensis*) (Lo et al., 2013).

C. sinensis is found in the alpine and subalpine regions of Himalayas at an altitude from 3200 to 4500 m above the sea level, mainly on plateau areas such as Nepal, Tibet, Bhutan, and China. In India, it mainly grows at high altitudes of Arunachal Pradesh, in alpine meadows of Chiplakedar, Darma, Vyas, and Ralamdhura in Kumaun Himalaya (Negi et al., 2006). It also thrives in Chiplakot, Ultapara, Brahmkot, Najari, Nangnidhura, and Munshyari regions of Pithoragarh district. In some parts of Garhwal Himalaya, it is found in Niti and Mana valleys of the Chamoli district (Sharma, 2004).

6.4 LIFE CYCLE OF *C. SINENSIS*

The distribution of *Cordyceps* in nature is based on the distribution of its host. The fungus disperses through air, rain, and insects. It harbors only few species of insects that are listed in Table 6.1 (Li and Tsim, 2004).

Table 6.1 Various Insect Hosts of *Cordyceps*

Species			
Hepialus armoricanus (H. armoricanus)	*H. devidi*	*H. ganna*	*H. latitegumenus*
H. kangdingensis	*H. varians*	*H. nebulosua*	*H. damxungensis*
H. yushuensis	*H. zhangmoensis*	*H. yunlongensis*	*H. bibelteus*
H. oblifurcus	*H. zhayuensis*	*H. lijiangensis*	*H. bagingensis*
H. menyuanicus	*H. kangdingroides*	*H. macilentus*	*Hepialiscus nepalensis*
H. sichuanus	*H. yulongensis*	*H. meiliensis*	*Hepialiscus sylvinus*
H. baimaensis	*H. deqinensis*	*H. renzhiensis*	*Hepialiscus flavus*
H. pratensis	*H. yunnanensis*	*H. markamensis*	*Bipectilus yunnanensis*
H. ferrugineus	*H. jinshaensis*	*H. albipictus*	*Phassus giganodus*
H. jialangensis	*H. jianchuanensis*	*H. zaliensis*	*Napialus humanesis*
H. anomopterus	*H. zhongzhiensis*	*H. cingulatus*	*Forkalus xizangensis*
H. luquensis	*H. xunhuaensis*	*H. gonggaensis*	*Magnificus jiuzhiensis*

The life cycle of *Cordyceps* can be divided into three stages: infection, parasitism (before insect death), and saprophytism (after insect death). *C. sinensis* infects the host through its body wall, oral cavity, and other orifices during the larval stage. Another mode of infection is ingestion of food that is contaminated by the mycelia of *C. sinensis*. After infection, *Cordyceps* derives nutrients from the bowel of the host. This stage is known as the parasitic stage, wherein the host is alive. With time, the insect's surface fades from dark yellowish-brown to light yellow as the entire body of insect is covered by fungal mycelia. Soon after the host dies, the mycelia form hard tissue and, under suitable conditions, grow out through the oral cavity of the host, forming a fruiting body (Fig. 6.1). This is the saprophytic stage.

The entire medicinal mushroom, *C. sinensis*, consists of the fungal and larval parts (Fig. 6.2). The fungal portion arises as grassy stalk-like stroma (fruiting body) from the mummified insect's larvae (*H. armoricanus*) (Baral, 2017). The larval part, which is segmented throughout the body, is buried under the ground during the winter and eventually emerges on the ground level in the month of May (Li and Tsim, 2004). Because of this appearance of *C. sinensis*, as it grows on insect (keera) and looks like a grass (ghas/jhari), it is locally known as "Keera ghas" or "Keera jhari" in the Indian Himalaya regions. The lifecycle of *C. sinensis* results in a dried structure, which is composed of a unique complex of secondary metabolites. At this stage, the mushroom is collected to be sold in market.

6.5   MAJOR BIOACTIVE CONSTITUENTS ISOLATED FROM *C. SINENSIS* AND THEIR AMELIORATING EFFECTS

As evident from the literature, the bioactive compounds found in *C. sinensis* are amino acids, fatty acids, inorganic elements, nucleosides, peptides and

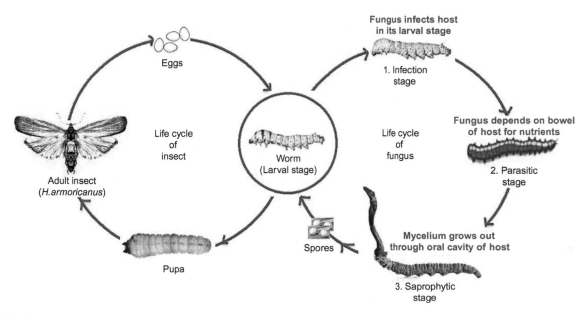

FIG. 6.1
Developmental stages of *Cordyceps sinensis* in the host caterpillar *Hepialus armoricanus* (*H. armoricanus*).

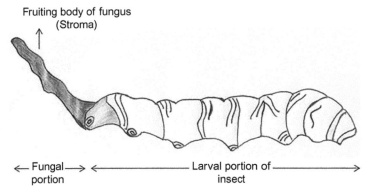

FIG. 6.2
After drying, mummified *Cordyceps sinensis* looks like "keera jhari."

proteins, phenolic compounds, polysaccharides, sterols, sugars, and vitamins. These compounds have been detected and isolated from the extracts and fractions of this high-altitude medicinal mushroom using various techniques. The numerous bioactive chemical constituents that have been confirmed in *C. sinensis* are given in Table 6.2.

Because *C. sinensis* is a repository of several bioactive principles, researchers also have studied its extracts, fractions, and isolated metabolites for their noticeable beneficial effects on biological systems through various experimental models (in vivo and in vitro), which are summarized in Table 6.3.

Table 6.2 Major Bioactive Compounds Confirmed in *Cordyceps sinensis*

Chemical Constituents	Examples	References
Amino acids	Phenylalanine, proline, histidine, valine, oxyvaline, arginine, glutamic acid	Sharma (2004)
	Glutamate, arginine and aspartic acid, tryptophan, and tyrosine	Zhou et al. (2009)
	Myriocin	Xiao et al. (2011)
Fatty acids and other organic acids	Lauric acid, myristic acid, penta decanoic acid, palmitoleic acid, linoleic acid, oleic acid, stearic acid, docosanoic acid, lignoceric acid, palmitic acid, succinic acid, linolenic acid	Li and Tsim (2004)
Fatty acid esters	Valeraldehyde dibutylacetal	Bhardwaj et al. (2015)
Nucleosides and related compounds	Uracil, guanosine, thymidine, tridine, dideoxyuridine, guanine, inosine	Zhu et al. (1998), Mamta et al. (2015), Zong et al. (2015)
	Adenine, hypoxanthine, adenosine, cordycepin	Huang et al. (2003), Khan et al. (2015), Mamta et al. (2015), Singh et al. (2013), Zong et al. (2015)
	Dideoxyadenosine, inosine, guanosine, uridine	Holliday and Cleaver (2008), Yu et al. (2006), Zong et al. (2015)
	Thymine, cytosine, inosine	Yang et al. (2011), Zong et al. (2015)
Peptide and proteins	*Cyclic dipeptides:* Cyclo-[Gly-Pro], cyclo-[Leu-Pro], cyclo-[Val-Pro], cyclo-[Ala-Leu], cyclo-[Ala-Val], and cyclo-[Thr-Leu]	Holliday and Cleaver (2008)
	Polyamines: 1,3-Diamino propane, Cadaverine, Spermidine, Spermine, Putrescine	Holliday and Cleaver (2008)
	Flazin, Perlolyrine	Yang et al. (2011)
	Cordymin, L-tryptophan	Yang et al. (2011), Qian et al. (2012), Zhang et al. (1991)
Polysaccharides and other sugar derivatives and related compounds	D-glucan, Cordysinocan, Mannoglucan, D-mannitol (cordycepic acid)	Yalin et al. (2006), Cheung et al. (2009), Wu et al. (2007), Li et al. (2001)
Sterols and their derivatives	Ergosterol, Cerevisterol, α-sitosterol, Ergosteryl-3-*O*-α-D-glucopyranoside, Ergosterol peroxide, 3-*O*-ferulylcycloartenol, Daucosterol, Stigmasterol, Stigmasterol 3-*O*-acetate, Fungisterol, Cholesterol, Campesterol, Dihydrobrassicasterol, △3 ergosterol, ergosteryl-3-*O*-β-D-glucopyranoside, β-sitosterol, cholesteryl palmitate	Chakraborty et al. (2014), Li and Tsim (2004)
Vitamins and inorganic elements	Vitamin-B1, B2, B12, and E and Minerals-K, Na, Ca, Mg, Fe, Cu, Mn, Zn, Pi, Se, Al, Si, Ni, Sr, Ti, Cr, Ga	Holliday and Cleaver (2008)
	C	Li and Tsim (2004)
	V and Zr	Chakraborty et al. (2014)

Table 6.3 Biological and Pharmacological Activities Shown by Various Isolated Bioactive Constituents From *Cordyceps sinensis*

Bioactive Constituents	Various Pharmacological and Biochemical Activities	References
Fatty acids and other organic acids	Cell signaling	Neitzel (2010)
Nucleosides	Antimetastatic activity Immunoregulatory activity	Nakamura et al. (2001) Zhou et al. (2008), Wang et al. (2016)
Peptide and proteins	Antibiotic activities Antitumor and immunopotentiation activity Antiinflammatory and antinocieptive effects	Xiao et al. (2011) Li and Tsim (2004) Qian et al. (2012)
Phenolic and flavonoid compounds	Antioxidant, antiinflammatory, antimicrobial, anticancerous activity	Cicerale et al. (2010), Ghasemzadeh and Ghasemzadeh (2011)
Phospholipids	Calcifying epiphyseal cartilage	Wuthier (1973)
Polysaccharides	Antitumor activity Antioxidant activity	Nie et al. (2013), Chen et al. (2006), Kiho et al. (1993)
Sterols	Antitumor activity	Bok et al. (1999)

6.6 CULTIVATION OF *C. SINENSIS*

Cordyceps is a storehouse of numerous metabolites exhibiting a wide spectrum of bioactivities. Consequently, an increased demand of *Cordyceps*-based nutraceutical and medicinal products have developed globally. *Cordyceps* is a very rare mushroom, however, and its natural variety is quite expensive. In order to meet market demands, various commercially viable culture methods have been established These vary considerably for each variety of *Cordyceps* (Holliday and Cleaver, 2004). The first successful attempt to cultivate *Cordyceps* was in 1982 by the Institute of Materia Medica, Chinese Academy of Medical Sciences. This organization cultivated mycelium of *Cordyceps* and isolated the strain CS-4, which was subjected to various studies and led to commercial production of *C. sinensis* (Gupta and Karkala, 2017). For both the mycelium and the fruiting body, a variety of media and substrates are being used currently. Mostly, insect larvae and various cereal grain-based substrates are employed (Holliday and Cleaver, 2004).

6.7 PRODUCTS BASED ON *C. SINENSIS*

Because *C. sinensis* accounts for remarkable health ameliorating effects, numerous products based on it are available in the market, such as Royale V^2 Plus, Cordizen, and Cordyzen (Paliwal et al., 2015; http://hyy.com.hk/eng/product/cordyzen/). Royale V^2 Plus formulation, composed of *Turnera diffusa* and *C. sinensis*, is available in the form of capsules and is promoted to increase both male and female vitality. Cordizen is a liquid oral supplement of the natural variety of *C. sinensis* (http://www.healthcare-exp.com/b2b/biotechnology_products/4/cordyceps_sinensis_oral_162.html). Cordyzen, which is based on cultivated variety of *C. sinensis*, is another product that is sold commercially as capsules. It is an ideal preparation for maintaining overall health, however, its emphasis is to provide a healthy respiratory system. It also is known to soothe lung and kidney functions (http://hyy.com.hk/product/cordyzen/). The United States Food and Drug Administration (FDA) has recognized *Cordyceps* as a "food" and classified it as "generally recognized as safe" (GRAS) and also regards it as a superfoods (Miller, 2009).

6.8   STUDIES SHOWING AMELIORATING EFFECTS

In traditional Chinese medicine, *Cordyceps* is valued for its bioactivity in restoring energy, promoting longevity, and improving quality of life (Holliday and Cleaver, 2004). It reinforces various biological systems, including respiratory, hepatobiliary, renal, and immune, and consequently affects numerous physiological functions. Several researchers have evaluated various extracts and fractions of *C. sinensis* for its protective efficacy under extreme environmental conditions, including high altitude.

6.8.1   Antihypoxic Activities

Several health problems emerge with the change in the environment, such as high altitude regions where hypoxia is considered to be the key factor affecting human physiology. Hypoxia with a low atmospheric pressure results in hypobaric hypoxia.

Hypobaric hypoxia often leads to AMS, which can result in severe complications such as high altitude cerebral edema (HACE), high altitude pulmonary edema (HAPE), and high altitude thromboembolism (HATE). In HACE, the brain swells and the sufferer can display incoherent behavior and a characteristic loss of coordination (Hackett and Roach, 2004). In HAPE, fluid accumulates in the lungs because of constriction of blood vessels (Yang et al., 2014).

In HATE, a clot forms in the blood vessel, breaks off, and is carried by the blood stream until it blocks a systemic vessel (Rathi et al., 2012). To overcome such maladies induced by high-altitude hypoxia, the protective efficacy of *C. sinensis* has been explored under simulated high-altitude environments using various models.

1. Gao et al., in 2010 conducted a study to assess the effect of water-based *C. sinensis* extract (CSE) on hypoxia-induced proliferation of pulmonary artery smooth muscle cells (PASMCs). CSE was found to inhibit

hypoxia-induced proliferation of PASMCs in male Sprague Dawley rats. Significant inhibition of the expression of PCNA, c-jun and c-fos in these cells after CSE supplementation was considered to be the plausible reason for this inhibitory effect. There also was a decrease in percentage of cells in the synthesis phase, second gap phase, and mitotic phase in cell cycle after CSE treatment. CSE, therefore, can be used as a therapeutic agent for the treatment of hypoxic pulmonary hypertension (Gao et al., 2010).

2. Our laboratory also is exploring the efficacy of *C. sinensis* against high-altitude ailments. We have investigated the protective action of aqueous and ethanolic *C. sinensis* extracts (CSEs) in murine hippocampal (HT22) cells under a simulated hypoxic environment (24 h duration). The hippocampus is thought to be one of the most vulnerable regions to neuronal cell damage from hypoxia-induced reactive oxygen species (ROS) generation. As a consequence, it can lead to serious neurodegenerative disorders, such as dementia, stroke, Huntington's disease, Parkinson's disease, and Alzheimer's disease (Brimson and Tencomnao, 2011).

   Our study suggested that treatment of CSEs in HT22 cells under hypoxia promoted survivability of cells and lowered hypoxia-induced ROS production. The levels of endogenous antioxidants (GSH, GPx, and SOD) in hypoxia-exposed cells were reduced drastically in comparison to normoxic control. These antioxidant markers significantly increased, however, in cells treated with CSEs in comparison to hypoxic control. Extent of lipid peroxidation and protein carbonylation were reduced in the hypoxia-exposed CSE-treated cells as compared to hypoxic control. Moreover, well-known inflammatory cytokines, IL-6 and TNF-α, and the transcription factor NF-kβ, were reduced significantly by CSE treatment under hypoxic condition as compared to hypoxic cells. Therefore, results are indicative of antihypoxic properties of CSEs and promising anti-inflammatory activities against hypoxia-induced inflammation (Pal et al., 2015).

3. Another study of our institute was carried out by Singh et al. (2013). In this study, aqueous extract of *C. sinensis* was evaluated for its defensive role against hypoxia in A549 (human lung epithelial) cells. Supplementation with *C. sinensis* extracts reduced hypoxia induced ROS generation and inhibited lipid and protein carbonylation. It also increased levels of cellular antioxidants (GSH, GPx, SOD, and GR) and antioxidant genes (HO1, MT, Nrf$_2$). *C. sinensis* treatment was associated with decreased levels of NF-kβ and TNF-α, which might be attributed to decreased levels of observed antioxidants genes. These findings clearly showed efficacy of CSEs in hypoxia-exposed cells (Singh et al., 2013).

4. In a similar study, liquid supplementation of *C. sinensis* was evaluated against oxygen and glucose deprivation (OGD) causing damage in human neuroblastoma cell line (SH-SY5Y). In this study, SH-SY5Y cells were exposed to different doses of *C. sinensis* and OGD condition was induced by transferring the cells from a high-glucose cell culture medium in a box gassed with air containing 5% CO_2 to a glucose-free cell culture medium in a box gassed with 94% N_2, 5% CO_2, and 1% O_2. On exposure to OGD environment, the cell viability on treatment with *C. sinensis* liquid (prepared from capsules) increased in a dose-dependent manner. The treatment of *C. sinensis* significantly inhibited both cell apoptosis and capase-3 activation in all groups in comparison to the model group (cultured with high-glucose cell culture medium and transferred to the OGD but received no dose of *C. sinensis* liquid). Treatment of *C. sinensis* significantly inhibited potential damage to the mitochondrial membrane. *C. sinensis*, therefore, could protect cells in both oxygen- and glucose-deprived environments by inhibiting the mitochondrial apoptosis pathway (Zou and Chu, 2016).

5. Another group of researchers evaluated the protective efficacy of a *C. sinensis* capsule (bailing capsule) in male Wistar rats. The levels of hypoxia induced proliferation of PASMCs were determined. The study was carried out by quantifying the hypoxia-induced ROS and endothelin-1 (ET-1) production. Additionally, expression of proliferating cell nuclear antigen (PCNA), c-fos, and c-jun in PASMCs also were resolved. Results showed that these capsules inhibited hypoxia-induced PASMC proliferation by lowering ROS and ET-1 levels, along with inhibiting the expression of PCNA, c-fos, and c-jun, suggesting antiproliferative properties of *C. sinensis* capsules under hypoxic environment (Li et al., 2016).

6.8.2 Performance Enhancing Activities

Because hypoxia is a pathophysiological condition wherein both mental and physical performances are hampered, various studies have been carried out to assess the potential of *C. sinensis* using different physical performance models.

1. To determine endurance enhancement properties, polysaccharides of *C. sinensis* were isolated and given to mice at different doses. The animals then were forced to undergo an exhaustive swimming exercise that resulted in oxidative stress. Results showed that exhaustive swimming time was significantly prolonged in all *C. sinensis*'s polysaccharides supplemented group in comparison to the control group. Other parameters associated with oxidative stress were also analyzed, including

SOD, GPx, and CAT, which were significantly higher, and MDA and 8-hydroxy-2′-deoxyguanosine (8-OHdG), which were significantly lower, in serum, liver, and muscles as compared to the control group. Therefore, this study supported defensive action of *C. sinensis* polysaccharides in combating exhaustive exercise-induced oxidative stress (Yan et al., 2014).

2. A similar study was conducted to elucidate the underlying mechanism of enhancement in exercise endurance activities by *C. sinensis* supplementation. Rats were given crude *C. sinensis* with or without swimming exercise. In both cases, *C. sinensis* resulted in significant increase in numerous metabolic regulators of skeletal muscles (AMPK, PGC-1α, and PPAR-δ) and performance-enhancing antioxidant genes (MCT1, MCT4, GLUT4, VEGF, NRF-2). Increased expression of oxidative stress responsive transcription factor NRF-2 and its associated targets SOD1 and thioredoxines in gastrocnemius muscle also were observed, which explained its molecular mechanism and implied that *C. sinensis* could be used as a potent performance enhancer (Kumar et al., 2011).

3. In another study, hot water fraction of *C. sinensis* mycelia was isolated. This fraction mainly constituted of carbohydrates. It was administered orally in animals in order to assess the antifatigue and antistress potential using the swimming model. The supplemented group exhibited a significant increase in swimming time from 75 min to 90 min with a decrease in fatigue. A index of stress, such as weight change of various organs (adrenal gland, spleen, thymus and thyroid), also was suppressed along with the inhibition of increased levels of total cholesterol and decreased alkaline phosphate in stressed animals during *Cordyceps* supplementation (Koh et al., 2003).

4. Apart from these physical performance studies, Gong et al. (2011) conducted a study that emphasized the learning memory of mice after supplementation with *C. sinensis* sporocarp extract. Locomotor activity also was assessed and no effect was found on *C. sinensis*-treated group. *C. sinensis* supplementation, however, shortened the escape latency and increased the time of come-crossing platform in Morris water maze animal model. Activity of acetylcholinesterase (AChE) also was found to be decreased with CSE supplementation. Thus, the study highlighted the potential of CSE in improving learning and memory impairment in mice attributable to its inhibitory effect on AChE activity.

6.8.3 Other Important Biological Activities

C. sinensis has several promising bioactivities that might have contributory effects in combating the pathophysiological complications arising from high-altitude maladies as discussed in the following paragraphs.

6.8.3.1 Antiaging Activities

Aging is a slow process, but it can be get accelerated under excessive oxidative stress as observed in nonacclimatized individuals such as soldiers, who often have to perform in hostile environments. As a result of such harsh conditions, their efficiency is affected and age-related activities such as learning and memory ability are impaired.

The antiaging activity of *C. sinensis* was evaluated in terms of brain functions, along with learning and memory ability in aged mice. The sexual functions also were tested in aged rodents. Aging was achieved by D-galactose treatment in mice and by castration in rats. On supplementation of CSE, brain functions and age-related enzyme's activities (SOD, GPx, and CAT) were enhanced along with declination of oxidation of lipid and monoamine oxidase in aged mice. Time taken for mount also was found to decrease in castrated rats after *C. sinensis* treatment. Thus, *C. sinensis* supplementation showed its antiaging activities by improving brain function, learning and memory ability, along with sexual functions (Ji et al., 2009).

6.8.3.2 Antimicrobial Activities

At high altitudes, human physiology is usually exposed to the potential hazards of weather extremes including cold, hypoxia, lower humidity, and UV radiation. The intricate interactions of these inevitable stressors obstruct immunity and consequently increase vulnerability to infections (Basnyat and Starling, 2016; Ericsson et al., 2001). Microbes can thrive in any environment, including hypoxic atmosphere and can colonize various biotic and abiotic substrates (Wu et al., 2005). The degree of severity of an infection depends on the occurrence of specific pathogens, the immune status of the host, and the duration of exposure (Basnyat and Starling, 2016). Studies have been conducted to evaluate the inhibitory action of *C. sinensis* against bacteria and viruses, however, no data have been reported concerning its antifungal activity.

Antibacterial Activities

1. Alves et al. (2012) isolated an antibacterial protein (CSAP) from *C. sinensis* and tested it against *Staphylococcus aureus* and *Bacillus subtilis*. CSAP exhibited effective antimicrobial activity against *S. aureus* but was less effective against *B. subtilis*. It also demonstrated antimicrobial activities against Gram-negative bacteria such as *Escherichia coli, Proteus vulgaris,* and *Salmonella typhi*.
2. In another study, the antimicrobial potential of CSEs (aqueous and ethanolic extracts with varying ratio of water and ethanol) were examined against *E. coli, P. aeruginosa,* and *B. subtilis*. All the ethanolic extracts showed relatively higher inhibitory effect against these bacterial pathogens than the aqueous extract (Mamta et al., 2015).

3. Miller (2009) reviewed that cordycepin, a nucleoside isolated from *C. sinensis* is effective against bacterial infections. This bioactive nucleoside also is suggested to exhibit antiviral effect that is believed to be because of its ability to inhibit viral DNA synthesis.

Antiviral Activities

Several research groups have reported that the marker compound of *Cordyceps*, cordycepin (an adenosine derivative), and its analogs have antiviral activity against the influenza virus, plant viruses, human immunodeficiency virus (HIV), murine leukemia virus, and Epstein-Barr virus (EBV) (Ryu et al., 2014). Some of the significant studies in this context are explained as follows.

1. Interferon-treated cells induce production of cellular enzymes, which affect viral replication thereby inhibiting cellular transformation by oncogenic viruses. The paucity and nonspecific targeting of these endogenous interferons, however, have necessitated development of antiviral agents such as nucleoside analogs with specificity against viral-infected mammalian cells with minimum cytotoxicity. Doetsch et al. (1981) demonstrated that the cordycepin analog of (2'-5')oligo(A) and the naturally occurring core (2'-5')oligo(A) were relatively more effective in preventing the morphological transformation of human lymphocytes after infection with Epstein-Barr virus (EBV) than human fibroblast interferons. It also was found that the synthetic core cordycepin analog was noncytotoxic to uninfected lymphocytes and proliferating lymphoblasts in comparison to the naturally occurring core (2'-5')oligo(A). Therefore, the study supported the use of core cordycepin analog as an effective antiviral agent against EBV infection (Doetsch et al., 1981).

2. Another study evaluated the antiviral activity of cordycepin against EBV infection of gastric epithelial cells. It was observed that treatment significantly downregulated most EBV genes and reduced both extracellular and intracellular EBV genome copy numbers up to 55% and 30%, respectively, at a concentration of 125 μM (Ryu et al., 2014).

3. Application of cordycepin as a potential adjuvant of the hepatitis B virus (HBV) vaccine has been validated by the results of in vitro and in vivo experiments, which showed that cordycepin significantly enhanced humoral and cellular immunity in BALB/c mice without any side effects. This immunity is a defensive mechanism that is necessary against HBV infection (Wang et al., 2017).

4. Infection of cells with human immunodeficiency virus type 1 (HIV-1) often leads to transient increment in the enzyme 2',5'-oligoadenylate synthetase and hence, increased cellular level of 2',5'-oligoadenylate. This oligoadenylate causes increased endoribonclease (RNAse L), which

might account for the inhibitory effects on viral RNA and viral protein synthesis. It has been observed that cordycepin (3'-deoxyadenosine) core trimer (Co3) and its 5'-monophosphate derivative (pCo3), the analogues of 2',5'-oligoadenylates (2-5A) synthesized from cordycepin (marker metabolite of *Cordyceps* species), have marked anti-HIV-1 activity in vitro. These compounds at a concentration of $10\,\mu M$ reduced the activity of HIV-1 reverse transcriptase (RT) by 90% by interfering with the binding site of $tRNA^{Lys.3}$ to RT (Mueller et al., 1991).

5. A patent of a pharmaceutical composition that has as an intervention for treating AIDS has been claimed. This preparation includes *C. sinensis*, *Radix rehmanniae*, *Mylabris*, and *Rhizoma anemarrhenae* in its composition. Clinical trials have showed considerable therapeutic effect of the pharmaceutical composition in the treatment of AIDS (Wang, 2017).

6.8.3.3 *Antioxidant Activities*

Numerous free radicals, generated under high altitude-mediated oxidative stress, are responsible for various inflammations and diseases and ultimately affect the whole biological system. Antioxidants are molecules that prevent cellular damage caused by these free radicals. They prevent the production of free radicals either by terminating the chain reactions that produce these free radicals by removing the intermediates or by inhibiting the chain reactions by oxidizing themselves (Mamta et al., 2013a). *C. sinensis* is well known for possessing a substantial amount of polyphenoloic compounds. Various extracts of *C. sinensis* have been evaluated for their antioxidant potential in both biological systems (in vitro studies) or in chemical systems (in vitro chemical assays).

1. In 2008, Dong and Yao investigated the antioxidant potential of *C. sinensis* mycelia by evaluating hot-water extracts from both the natural and cultured varieties. The researchers found that these extracts possessed potent inhibitory activity against linoleic acid peroxidation. These also were effective in scavenging superoxide anion, hydroxyl and DPPH (1,1-diphenyl-2-picryl-hydrazyl) radicals, and exhibited substantial reducing power and ferrous ion chelating activity (Dong and Yao, 2008).

2. Our laboratory also has contributed in elucidating the antioxidant activities of five different extracts of *C. sinensis* (CSEs). These consisted of aqueous and hydro-ethanolic extracts. We subjected them to phytochemical evaluation in terms of total phenolics and total flavonoids contents while their antioxidant activity was determined by free radical scavenging assays. All the extracts were found to possess substantial phenolic and flavonoid contents. They also demonstrated marked ABTS [2,2'-azino-bis (3-ethylbenzothiazoline-6-sulphonic acid)] and DPPH free radicals scavenging activities along with ferric ion

reducing antioxidant power and total reducing power. This study suggested substantial antioxidant potential of all the CSEs (Mamta et al., 2015). Additionally, the extracts also were evaluated for their hydroxyl, nitric oxide, and superoxide radical scavenging activities. They were found to exhibit considerable radical scavenging potential against these free radical species, further supporting their antioxidant behavior (Pal et al., 2015).

3. Polysaccharides (mannose and galactose) rich water-soluble fraction of *C. sinensis* mycelia also was examined for its antioxidative effects in mouse monocyte–macrophage (RAW264.7) cells to which oxidative stress was induced by H_2O_2 treatment. The polysaccharide fraction reduced H_2O_2-induced cell injury as indicated by diminished depolarization of mitochondrial membrane after the treatment. The probable mechanism of its antioxidative behavior was because of the inhibition of sphingomyelinase activity and reduction of ceramide levels (Wang et al., 2011).

6.8.3.4 Immune Enhancing, Antiinflammatory, and Anticancerous Activities

Hypoxia induces inflammation resulting from overproduction of ROS, which clearly shows reduction of immune power and might result in several chronic inflammatory diseases, including neurodegenerative diseases, rheumatoid arthritis, and cancer (Pham-Huy et al., 2008). Some of the studies pertaining to beneficial effects of *C. sinensis* against such diseases are detailed below.

1. C57BL/6 mice were implanted with EL-4 (mouse lymphoma cell lines) cells and supplemented with CSE to study its immune enhancing activities. Phagocytic activity of peritoneal macrophages and chemotaxis were suppressed in EL-4 transplanted mice as well as in cyclophosphamide (chemotheraputic agent used to treat cancer) treated group. With administration of CSE, these activities were improved, predicting its immune-enhancing effect (Yamaguchi et al., 1990).

2. In another study, aqueous extract of *C. sinensis* was examined in human monocytic cell lines (THP1). THP1 cells were treated with and without lipopolysaccharide (LPS) and aqueous extract of *C. sinensis* (with different doses). The aqueous extract significantly suppressed LPS-induced release of TNF-α and IL-1β in THP1 cells. It also suppressed the NO release in macrophage cells. This study demonstrated the potent antiinflammatory activity of *C. sinensis*, leading to the conclusion that it could be used to protect against inflammatory diseases (Rathor et al., 2014).

3. In an in vitro study, the anticancerous effects of *C. sinensis* were examined against B16-F10 melanoma cells, which were supplemented by a water

soluble polysaccharide fraction of *C. sinensis* mycelium. The polysaccharide fraction inhibited the expressions of MMP-1 in these cells either by downregulating the nuclear factor kappa-B (NF-kB) or ERK/p38 MAPK and therefore inhibited the migration of B16-F10 cells. This study suggested that the *C. sinensis* polysaccharide fraction could be used as therapeutic agent for the treatment of cancer metastasis (Jayakumar et al., 2014).

6.8.3.5 *Replenishment of Vital Organs*

Considering that ROS mediated oxidative stress induces inflammation and leads to malfunctioning of various vital organs, studies have been conducted that highlight protective efficacy of *C. sinensis* in replenishment of vital organs, such as lungs, liver, and kidneys.

1. The supplementation effect of *C. sinensis* was assessed against renal ischemia–reperfusion (I/R) injury in animal model. Rats supplemented with *C. sinensis* significantly improved the I/R-induced renal damage by increasing stromal cell-derived factor-1 in different parts of the kidney and by decreasing the CXC chemokine receptor 4 (CXCR4) and Ki-67 levels. Its treatment also lowered the elevated level of creatinine after renal (I/R) injury. This study revealed the protective action of *C. sinensis* against I/R induced renal damage (Wang et al., 2013).

2. In another study, the effect of CSE on glomerular sclerosis rat model was studied along with the underlying mechanism. CSE treatment improved the renal damage by reducing proteinuria and fibronectin, collagen type IV, connective tissue growth factor, and plasminogen activator inhibitor. Furthermore, there was increased expression of matrix metalloprotease-2 after *C. sinensis* treatment in model rats. Thus, *C. sinensis* might be an effective agent for improving functions of damaged kidneys (Song et al., 2010).

3. The protective effect of *C. sinensis* also has been examined on rat kidney with diabetic nephropathy along with the possible mechanism for its action. In this study, a group of rats with diabetic nephropathy was treated with *C. sinensis* and showed decreased expression of vascular endothelial growth factor (VEGF) in the kidney in comparison to the control group and an untreated diabetic nephropathy group. This implies that *C. sinensis* might protect against chronic injury in diabetic nephropathy (Yuan et al., 2013).

4. The curative effect of *C. sinensis* on dimethylnitrosamine-induced liver fibrosis also was studied in a rat model. The contents of hydroxyproline, tissue inhibitor of metalloproteinase-2, type IV collagen and type I collagen, were found to be higher in an untreated group than those in the normal control group. Levels of these markers, however, were

lowered in *C. sinensis* treated group relative to untreated group. The level of matrix metalloproteinases-2 was remarkably lower in untreated rats in comparison to the normal control rats, but in *C. sinensis* treated rats its level was considerably elevated as compared to the untreated group. These findings revealed that *C. sinensis* can convalescent liver fibrosis (Li et al., 2006).

5. *C. sinensis* also has been appraised for its beneficial effect against lung fibrosis. In this context, Chen et al. (2012) explored effect of *C. sinensis* in a rat model with bleomycin-induced lung fibrosis. Treatment with *C. sinensis* attenuated infiltration of inflammatory cells, deposition of fibroblastic loci and collagen, generation of ROS, and production of cytokines along with revival of MMP-9/TIMP-1 in fibrotic group of rats. This group further examined the underlying mechanism in a fibrotic cell model (human lung epithelial A549) with induction of transforming growth factor beta-1. The researchers concluded that cordycepin, one of the active components of *C. sinensis*, might have reversed the epithelial-mesenchymal transition by elevating vimentin expression in parallel to decrease in E-cadherin. These findings provide the mechanism whereby *C. sinensis* has significant preventive and therapeutic potential in the treatment of lung fibrosis (Chen et al., 2012).

6.9   CONCLUSION

In this chapter, we reviewed high altitude and endurance promoting activities of *C. sinensis*, which are evidently owed to its secondary metabolite composition. Considering the fact that the United States FDA also has recognized *Cordyceps* as a "superfood," there is a push for the development and promotion of *C. sinensis* based products to be used as beneficial therapeutics in ameliorating high altitude-induced pathology. Its rareness in nature makes it expensive, so the cultivated variety should be promoted as substitution to fulfill the global demand.

Acknowledgments

The authors thank the pro-vice chancellor of Galgotias University, Greater Noida, India, and director of the Defense Institute of Physiology and Allied Sciences, Delhi, India, for their constant support and encouragement.

References

Alves, M.J., Ferreira, I.C., Dias, J., Teixeira, V., Martins, A., Pintado, M., 2012. A review on antimicrobial activity of mushroom (Basidiomycetes) extracts and isolated compounds. Planta Med. 78 (16), 1707–1718.

Anandanayaki, S., Jegadeesan, M., 2010. Comparative pharmacognostical studies on selected plants (Pedalium murex roen ex. L. and Martyniaannua L.). Ph. D. Thesis. pp. 1–301. http://hdl.handle.net/10603/1026 or http://shodhganga.inflibnet.ac.in/bitstream/10603/1026/7/07_chapter%202.pdf.

Baral, B., 2017. Entomopathogenicity and biological attributes of Himalayan treasured fungus *Ophiocordyceps sinensis* (Yarsagumba). J. Fungi 3 (1), 4.

Barros, L., Dueñas, M., Ferreira, I.C., Baptista, P., Santos-Buelga, C., 2009. Phenolic acids determination by HPLC–DAD–ESI/MS in sixteen different Portuguese wild mushrooms species. Food Chem. Toxicol. 47 (6), 1076–1079.

Basnyat, B., Starling, J.M., 2016. Infectious diseases at high altitude. In: Infections of Leisure, fifth ed. American Society of Microbiology. AMS Press, Washington, DC, pp. 325–332. https://doi.org/10.1128/microbiolspec.IOL5-0006-2015.

Bhardwaj, A., Pal, M., Srivastava, M., Tulsawani, R., Sugadev, R., Misra, K., 2015. HPTLC based chemometrics of medicinal mushrooms. J. Liq. Chromatogr. Relat. Technol. 38 (14), 1392–1406.

Bok, J.W., Lermer, L., Chilton, J., Klingeman, H.G., Towers, G.N., 1999. Antitumor sterols from the mycelia of *Cordyceps sinensis*. Phytochemistry 51 (7), 891–898.

Brimson, J.M., Tencomnao, T., 2011. *Rhinacanthus nasutus* protects cultured neuronal cells against hypoxia induced cell death. Molecules 16 (8), 6322–6338.

Chakraborty, S., Chowdhury, S., Nandi, G., 2014. Review on Yarsagumba (*Cordyceps sinensis*): an exotic medicinal mushroom. Int. J. Pharmacogn. Phytochem. Res. 69 (2), 339–346.

Chang, S.-T., 1999. World production of cultivated edible and medicinal mushrooms in 1997 with emphasis on Lentinus edodes (Berk.) Sing. in China. Int. J. Med. Mushrooms 1, 291–300.

Chen, J., Zhang, W., Lu, T., Li, J., Zheng, Y., Kong, L., 2006. Morphological and genetic characterization of a cultivated *Cordyceps sinensis* fungus and its polysaccharide component possessing antioxidant property in H22 tumor-bearing mice. Life Sci. 78 (23), 2742–2748.

Chen, M., Cheung, F.W., Chan, M.H., Hui, P.K., Ip, S.P., Ling, Y.H., Che, C.T., Liu, W.K., 2012. Protective roles of *Cordyceps* on lung fibrosis in cellular and rat models. J. Ethnopharmacol. 143 (2), 448–454.

Cheung, J.K., Li, J., Cheung, A.W., Zhu, Y., Zheng, K.Y., Bi, C.W., Duan, R., Choi, R.C., Lau, D.T., Dong, T.T., Lau, B.W., 2009. Cordysinocan, a polysaccharide isolated from cultured *Cordyceps*, activates immune responses in cultured T-lymphocytes and macrophages: signaling cascade and induction of cytokines. J. Ethnopharmacol. 124 (1), 61–68.

Childs, G., Choedup, N., 2014. Indigenous management strategies and socioeconomic impacts of Yartsa Gunbu (*Ophiocordyceps sinensis*) harvesting in Nubri and Tsum, Nepal. Himalaya J. Assoc. Nepal Himalayan Stud. 34 (1), 7.

Cicerale, S., Lucas, L., Keast, R., 2010. Biological activities of phenolic compounds present in virgin olive oil. Int. J. Mol. Sci. 11 (2), 458–479.

Doetsch, P.W., Suhadolnik, R.J., Sawada, Y., Mosca, J.D., Flick, M.B., Reichenbach, N.L., Dang, A.Q., Wu, J.M., Charubala, R., Pfleiderer, W., Henderson, E.E., 1981. Core (2′-5′) oligoadenylate and the cordycepin analog: inhibitors of Epstein-Barr virus-induced transformation of human lymphocytes in the absence of interferon. Proc. Natl. Acad. Sci. U. S. A. 78 (11), 6699–6703.

Dong, C.H., Yao, Y.J., 2008. In vitro evaluation of antioxidant activities of aqueous extracts from natural and cultured mycelia of *Cordyceps sinensis*. LWT Food Sci. Technol. 41 (4), 669–677.

Ericsson, C.D., Steffen, R., Basnyat, B., Cumbo, T.A., Edelman, R., 2001. Infections at high altitude. Clin. Infect. Dis. 33 (11), 1887–1891.

Gao, B., Yang, J., Huang, J., Cui, X., Chen, S., Den, H.Y., Xiang, G.M., 2010. *Cordyceps sinensis* extract suppresses hypoxia-induced proliferation of rat pulmonary artery smooth muscle cells. Saudi Med. J. 31 (9), 974–979.

Garbyal, S.S., Aggarwal, K.K., Babu, C.R., 2004. Impact of *Cordyceps sinensis* in the rural economy of interior villages of Dharchula subdivision of Kumaon Himalayas and its implications in the society. India J. Trad. Knowledge 3 (2), 182–186.

Ghasemzadeh, A., Ghasemzadeh, N., 2011. Flavonoids and phenolic acids: role and biochemical activity in plants and human. J. Med. Plants Res. 5 (31), 6697–6703.

Gong, M.F., Xu, J.P., Chu, Z.Y., Luan, J., 2011. Effect of *Cordyceps sinensis* sporocarp on learning-memory in mice. J. Chin. Med. Mater. 34 (9), 1403–1405.

Gupta, G., Karkala, M., 2017. Yarsagumba: a miracle mushroom its history, cultivation, phytopharmacology and medicinal uses. Int. J. Herb. Med. 5 (2), 69–72.

Hackett, P.H., Roach, R.C., 2004. High altitude cerebral edema. High Alt. Med. Biol. 5 (2), 136–146.

Heleno, S.A., Barros, L., Sousa, M.J., Martins, A., Santos-Buelga, C., Ferreira, I.C., 2011. Targeted metabolites analysis in wild Boletus species. LWT-Food Sci. Technol. 44 (6), 1343–1348.

Holliday, J. and Cleaver, M., 2004. On the trail of the yak: ancient *Cordyceps* in the modern world. Online posting. June.

Holliday, J.C., Cleaver, M.P., 2008. Medicinal value of the caterpillar fungi species of the genus *Cordyceps* (Fr.) link (Ascomycetes). A review. Int. J. Med. Mushrooms 10 (3), 219–234.

Huang, L.F., Liang, Y.Z., Guo, F.Q., Zhou, Z.F., Cheng, B.M., 2003. Simultaneous separation and determination of active components in *Cordyceps sinensis* and *Cordyceps militarris* by LC/ESI-MS. J. Pharm. Biomed. Anal. 33 (5), 1155–1162.

Jayakumar, T., Chiu, C.C., Wang, S.H., Chou, D.S., Huang, Y.K., Sheu, J.R., 2014. Anti-cancer effects of CME-1, a novel polysaccharide, purified from the mycelia of *Cordyceps sinensis* against B16-F10 melanoma cells. J. Cancer Res. Therapeut. 10 (1), 43–49.

Ji, D.B., Ye, J., Li, C.L., Wang, Y.H., Zhao, J., Cai, S.Q., 2009. Antiaging effect of *Cordyceps sinensis* extract. Phytother. Res. 23 (1), 116–122.

Khan, M.S., Khan, W., Manickam, M., Kumar Tulsawani, R., 2015. Quantification of flavonoids and nucleoside by UPLC-MS in Indian *Cordyceps sinensis* and its in vitro cultures. Ind. J. Pharm. Edu. Res. 49, 353–361.

Kiho, T., Hui, J.I., Yamane, A., Ukai, S., 1993. Polysaccharides in fungi. XXXII. Hypoglycemic activity and chemical properties of a polysaccharide from the cultural mycelium of *Cordyceps sinensis*. Biol. Pharm. Bull. 16 (12), 1291–1293.

Koh, J.H., Kim, K.M., Kim, J.M., Song, J.C., Suh, H.J., 2003. Antifatigue and antistress effect of the hot-water fraction from mycelia of *Cordyceps sinensis*. Biol. Pharm. Bull. 26 (5), 691–694.

Kumar, R., Negi, P.S., Singh, B., Ilavazhagan, G., Bhargava, K., Sethy, N.K., 2011. *Cordyceps sinensis* promotes exercise endurance capacity of rats by activating skeletal muscle metabolic regulators. J. Ethnopharmacol. 136 (1), 260–266.

Li, F.H., Liu, P., Xiong, W.G., Xu, G.F., 2006. Effects of *Cordyceps sinensis* on dimethylnitrosamine-induced liver fibrosis in rats. J. Chin. Integr. Med. 4 (5), 514–517.

Li, S.P., Li, P., Dong, T.T.X., Tsim, K.W.K., 2001. Antioxidation activity of different types of natural *Cordyceps sinensis* and cultured *Cordyceps* mycelia. Phytomedicine 8 (3), 207–212.

Li, S.P., Tsim, K.W.K., 2004. The biological and pharmacological properties of *Cordyceps sinensis*, a traditional Chinese medicine, that has broad clinical applications. In: Packer, et al., (Ed.), Herbal Medicines: Molecular Basis of Biological Activity and Health. Marcel Dekker Inc., New York, p. 657

Li, X., Peng, K., Zhou, Y., Deng, F., Ma, J., 2016. Inhibitory effect of bailing capsule on hypoxia-induced proliferation of rat pulmonary arterial smooth muscle cells. Saudi Med. J. 37 (5), 498–505.

Lo, H.C., Hsieh, C., Lin, F.Y., Hsu, T.H., 2013. a systematic review of the mysterious caterpillar fungus *Ophiocordyceps sinensis* in Dong-ChongXiaCao (Dōng Chóng XiàCǎo) and related bioactive ingredients. J. Trad. Complem. Med. 3 (1), 16–32.

Mamta, Misra, K., Dhillon, G.S., Brar, S.K., Verma, M., 2013a. Antioxidant. In: Brar, S., Dhillon, G., Soccol, C. (Eds.), Biotransformation of Waste Biomass into High Value Biochemicals. Springer, New York, pp. 117–138.

Mamta, Misra, K., Brar, S.K., Verma, M., 2013b. Carotenoids in herbal medicine. In: Yamaguchi, M. (Ed.), Antioxidants in Biology and Medicine: Essentials, Advances, and Clinical Applications. Nova Science Publishers, Inc., USA, pp. 127–142

Mamta, Mehrotra, S., Kirar, V., Vats, P., Nandi, S.P., Negi, P.S., Misra, K., 2015. Phytochemical and antimicrobial activities of Himalayan *Cordyceps sinensis* (Berk.) Sacc. Ind. J. Exp. Biol. 53 (1), 36–43.

Miller, R. A., 2009. pp. 23–28 http://blog.renakrebsbach.com/wp-content/uploads/2013/06/Cordyceps_Medicinal_Mushroom.pdf, Last accessed on 06/07/2017.

Mueller, W.E., Weiler, B.E., Charubala, R., Pfleiderer, W., Leserman, L., Sobol, R.W., Suhadolnik, R.J., Schroeder, H.C., 1991. Cordycepin analogs of 2′, 5′-oligoadenylate inhibit human immunodeficiency virus infection via inhibition of reverse transcriptase. Biochemistry 30 (8), 2027–2033.

Nakamura, K., Yamaguchi, Y., Kagota, S., Kwon, Y.M., Shinozuka, K., Kunitomo, M., 2001. Inhibitory effect of *Cordyceps sinensis* on spontaneous liver metastasis of Lewis lung carcinoma and B16 melanoma cells in syngeneic mice. Jap. J. Pharmacol. 79 (3), 335–341.

Negi, C.S., Koranga, P.R., Ghinga, H.S., 2006. Yar tsa Gumba (*Cordyceps sinensis*): a call for its sustainable exploitation. Int. J. Sust. Dev. World Ecol. 13 (3), 165–172.

Neitzel, J.J., 2010. Fatty acid molecules: fundamentals and role in signaling. Nat. Educ. 3 (9), 57.

Nie, S., Cui, S.W., Xie, M., Phillips, A.O., Phillips, G.O., 2013. Bioactive polysaccharides from *Cordyceps sinensis*: isolation, structure features and bioactivities. Bioactive Carbohydr. Diet. Fiber 1 (1), 38–52.

Pal, M., Bhardwaj, A., Manickam, M., Tulsawani, R., Srivastava, M., Sugadev, R., Misra, K., 2015. Protective efficacy of the caterpillar mushroom, *Ophiocordyceps sinensis* (ascomycetes), from India in neuronal hippocampal cells against hypoxia. Int. J. Med. Mushrooms 17 (9), 829–840.

Paliwal, P., Barua, D., Bharadia, P.D., 2015. *Cordyceps sinensis* (Yarsagumba): a promising caterpillar mushroom. J. Harmon. Res. 4 (4), 337–334.

Pham-Huy, L.A., He, H., Pham-Huy, C., 2008. Free radicals, antioxidants in disease and health. Int. J. Biomed. Sci. 4 (2), 89–96.

Qian, G.M., Pan, G.F., Guo, J.Y., 2012. Anti-inflammatory and antinociceptive effects of cordymin, a peptide purified from the medicinal mushroom *Cordyceps sinensis*. Nat. Prod. Res. 26 (24), 2358–2362.

Rathi, K.R., Uppal, V., Bewal, N.M., Sen, D., Khanna, A., 2012. D-dimer in the diagnostic workup of suspected pulmonary thromboembolism at high altitude. Med. J. Arm. Forces India 68 (2), 142–144.

Rathor, R., Mishra, K.P., Pal, M., Vats, P., Kirar, V., Negi, P.S., Misra, K., 2014. Scientific validation of the Chinese caterpillar medicinal mushroom, *Ophiocordyceps sinensis* (Ascomycetes) from India: immunomodulatory and antioxidant activity. Int. J. Med. Mushrooms 16 (6), 541–553.

Ryu, E., Son, M., Lee, M., Lee, K., Cho, J.Y., Cho, S., Lee, S.K., Lee, Y.M., Cho, H., Sung, G.H., Kang, H., 2014. Cordycepin is a novel chemical suppressor of Epstein-Barr virus replication. Oncoscience 1 (12), 866.

Sharma, S., 2004. Trade of *Cordyceps sinensis* from high altitudes of the Indian Himalayas: conservation and biotechnological priorities. Curr. Sci. 86 (12), 1614–1619.

Singh, N., Pathak, R., Kathait, A.S., Rautela, D., Dubey, A., 2010. Collection of *Cordyceps sinensis* (Berk.) Sacc. in the interior villages of Chamoli district in Garhwal Himalaya (Uttarakhand) and its social impacts. J. Am. Sci. 6 (6), 120–132.

Singh, M., Tulsawani, R., Koganti, P., Chauhan, A., Manickam, M., Misra, K., 2013. *Cordyceps sinensis* increases hypoxia tolerance by inducing heme oxygenase-1 and metallothionein via Nrf2 activation in human lung epithelial cells. Biomed. Res. Int. 2013, 13.

Song, L.Q., Si-Ming, Y., Xiao-Peng, M., Li-Xia, J., 2010. The protective effects of *Cordyceps sinensis* extract on extracellular matrix accumulation of glomerular sclerosis in rats. Afr. J. Pharm. Pharmacol 4 (7), 471–478.

Spehn, E.M., Rudmann-Maurer, K., Korner, C., Maselli, D., 2012. Mountain Biodiversity and Global Change. Food and Agriculture Organization (FAO) of the U. N.

Tang, X.G., Zhang, J.H., Qin, J., Gao, X.B., Li, Q.N., Yu, J., Ding, X.H., Huang, L., 2014. Age as a risk factor for acute mountain sickness upon rapid ascent to 3700 m among young adult Chinese men. Clin. Interv. Aging 9, 1287–1294.

Wang, D., Zhang, Y., Lu, J., Wang, Y., Wang, J., Meng, Q., Lee, R.J., Teng, L., 2016. Cordycepin, a natural antineoplastic agent, induces apoptosis of breast cancer cells via caspase-dependent pathways. Nat. Prod. Commun. 11 (1), 63–68.

Wang, H.P., Liu, C.W., Chang, H.W., Tsai, J.W., Sung, Y.Z., Chang, L.C., 2013. *Cordyceps sinensis* protects against renal ischemia/reperfusion injury in rats. Mol. Biol. Rep. 40 (3), 2347–2355.

Wang, S.H., Yang, W.B., Liu, Y.C., Chiu, Y.H., Chen, C.T., Kao, P.F., Lin, C.M., 2011. A potent sphingomyelinase inhibitor from *Cordyceps* mycelia contributes its cytoprotective effect against oxidative stress in macrophages. J. Lipid Res. 52 (3), 471–479.

Wang, Z., 2017. Pharmaceutical composition for treating AIDS and preparation method thereof. U.S. Patent 9623053.

Wang, J., Liu, R., Liu, B., Yang, Y., Xie, J., Zhu, N., 2017. Systems pharmacology-based strategy to screen new adjuvant for hepatitis B vaccine from traditional Chinese medicine *Ophiocordyceps sinensis*. Sci. Rep. 7, 44788.

Wasser, S.P., 2014. Medicinal mushroom science: current perspectives, advances, evidences, and challenges. Biomed. J. 37, 345–356.

Wu, M., Guina, T., Brittnacher, M., Nguyen, H., Eng, J., Miller, S.I., 2005. The Pseudomonas aeruginosa proteome during anaerobic growth. J. Bacteriol. 187 (23), 8185–8190.

Wu, Y., Hu, N., Pan, Y., Zhou, L., Zhou, X., 2007. Isolation and characterization of a mannoglucan from edible *Cordyceps sinensis* mycelium. Carbohydr. Res. 342 (6), 870–875.

Wuthier, R.E., 1973. The role of phospholipids in biological calcification: distribution of phospholipase activity in calcifying epiphyseal cartilage. Clin. Orthop. Relat. Res. 90, 191–200.

Xiao, Z.H., Zhou, J.H., Wu, H.S., 2011. Effect of myriocin on the expression of cyclinD1 in high glucose-induced hypertrophy mesangial cells. Chin. J. Contemp. Pediatr. 13 (8), 677–679.

Yalin, W., Cuirong, S., Yuanjiang, P., 2006. Studies on isolation and structural features of a polysaccharide from the mycelium of an Chinese edible fungus (Cordyceps sinensis). Carbohydr. Polym. 63 (2), 251–256.

Yamaguchi, N., Yoshida, J., Ren, L.J., Chen, H., Miyazawa, Y., Fujii, Y., Huang, Y.X., Takamura, S., Suzuki, S., Koshimura, S., Zeng, F.D., 1990. Augmentation of various immune reactivities of tumor-bearing hosts with an extract of *Cordyceps sinensis*. Biotherapy 2 (3), 199–205.

Yan, F., Wang, B., Zhang, Y., 2014. Polysaccharides from *Cordyceps sinensis* mycelium ameliorate exhaustive swimming exercise-induced oxidative stress. Pharm. Biol. 52 (2), 157–161.

Yang, M.L., Kuo, P.C., Hwang, T.L., Wu, T.S., 2011. Anti-inflammatory principles from *Cordyceps sinensis*. J. Nat. Prod. 74 (9), 1996–2000.

Yang, Y., Ma, L., Guan, W., Wang, Y., Du, Y., Ga, Q., Ge, R.L., 2014. Differential plasma proteome analysis in patients with high-altitude pulmonary edema at the acute and recovery phases. Exp. Therap. Med. 7 (5), 1160–1166.

Yu, L., Zhao, J., Li, S.P., Fan, H., Hong, M., Wang, Y.T., Zhu, Q., 2006. Quality evaluation of *Cordyceps* through simultaneous determination of eleven nucleosides and bases by RP-HPLC. J. Sep. Sci. 29 (7), 953–958.

Yuan, M., Tang, R., Zhou, Q., Liu, K., Xiao, Z., Pouranan, V., 2013. Effect of *Cordyceps sinensis* on expressions of HIF-1α and VEGF in the kidney of rats with diabetic nephropathy. J. Central South Univ. Med. Sci. 38 (5), 448–457.

Zhang, S.S., Zhang, D.S., Zhu, T.J., Chen, X.Y., 1991. A pharmacological analysis of the amino acid components of *Cordyceps sinensis* Sacc. Acta Pharm. Sin. 26 (5), 326–330.

Zhou, X., Luo, L., Dressel, W., Shadier, G., Krumbiegel, D., Schmidtke, P., Zepp, F., Meyer, C.U., 2008. Cordycepin is an immunoregulatory active ingredient of *Cordyceps sinensis*. Am. J. Chin. Med. 36 (05), 967–980.

Zhou, X., Gong, Z., Su, Y., Lin, J., Tang, K., 2009. *Cordyceps* fungi: natural products, pharmacological functions, and developmental products. J. Pharm. Pharmacol. 61 (3), 279–291.

Zhu, J.S., Halpern, G.M., Jones, K., 1998. The scientific rediscovery of a precious ancient Chinese herbal regimen: *Cordyceps sinensis* part II. J. Alt. Complem. Med. 4 (4), 429–457.

Zong, S.Y., Han, H., Wang, B., Li, N., Dong, T.T.X., Zhang, T., Tsim, K.W., 2015. Fast simultaneous determination of 13 nucleosides and nucleobases in *Cordyceps sinensis* by UHPLC–ESI–MS/MS. Molecules 20 (12), 21816–21825.

Zou, Y.X., Chu, Z.Y., 2016. *Cordyceps sinensis* oral liquid inhibits damage induced by oxygen and glucose deprivation in SH-SY5Y cells. Altern. Ther. Health Med. 22 (2), 37–42.

Further Reading

Ghanshyam, G. and Manvitha, K., Yarsagumba: A miracle mushroom its history, cultivation, phytopharmacology, and medicinal uses. Int. J. Herbal Med. 5(2), pp. 69–72.

http://hyy.com.hk/product/cordyzen/.

CHAPTER 7

Ganoderma sp.: The Royal Mushroom for High-Altitude Ailments

Anuja Bhardwaj*, Kshipra Misra†

*Chemistry Division, Department of Biochemical Sciences (DBCS), Defence Institute of Physiology and Allied Sciences (DIPAS), Delhi, India, †Department of Biochemical Sciences (DBCS), Defence Institute of Physiology and Allied Sciences (DIPAS), Delhi, India

Abbreviations

ACC	acetyl-CoA carboxylase
ADP	adenosine diphosphate
AHR	airway hyper-reactivity
Akt	protein kinase B
AMP	Amp-activated protein
AMPK	Amp-activated protein kinase
AMS	acute mountain sickness
AP-1	activating protein-1
ASHMI	antiasthma simplified herbal medicine intervention
B. subtilis	*Bacillus subtilis*
Bax	Bcl-2 associated X protein
Bcl	B-cell lymphoma
BG	*Ganoderma lucidum* β-glucan
Bim	Bcl-2-like protein
BLA	blood lactic acid
BUN	blood urea nitrogen
BW	body weight
C. albicans	*Candida albicans*
C. glabrata	*Candida glabrata*
CaSki	Caucasian cervical epidermoid carcinoma
CAT	catalase
CD	cluster of differentiation
CECs	circulating endothelial cells
c-fos	Finkel-Biskis-Jinkins murine osteosarcoma viral (V-Fos) oncogene homolog
C-H-R	cold, hypoxia, and restraint
CICP	C-telopeptides of Type I collagen protein
c-jun	JUN proto-oncogene
c-myc	myelocytomatosis viral oncogene (V-myc) oncogene homolog
DNA	deoxyribonuclease
DPPH	2,2-diphenyl-1-picrylhydrazyl
E. coli	*Escherichia coli*

115

Management of High Altitude Pathophysiology. https://doi.org/10.1016/B978-0-12-813999-8.00007-0

EBV	Epstein-Barr virus
EBV-CA	EBV capsid antigen
EBV-EA	EBV early antigen
EF	ejection fraction
eNOS	endothelial nitric oxide synthase
EPCs	endothelial progenitor cells
ERK	extracellular signal-regulated kinase
EV	enterovirus
FAK	focal adhesion kinase
FAS	fatty acid synthase
Fas	first apoptosis signal
FSG	fasting serum glucose
FSI	fasting serum insulin
FYGL	Fudan-Yueyang-Ganoderma lucidum
G. applanatum	*Ganoderma applanatum*
G. lucidum	*Ganoderma lucidum*
GAs	ganoderic acids
GL	*Ganoderma lucidum*
GLE	*Ganoderma lucidum* extract
GLMaq	*Ganoderma lucidum* mycelium aqueous extract
GLP	*Ganoderma lucidum* peptide
GLPP	*Ganoderma lucidum* polysaccharide peptide
GL-PS	*Ganoderma lucidum* polysaccharides
GLTA	*Ganoderma lucidum* Lanosta-7,9(11),24-trien-3-one,15;26-dihydroxy
GLTB	ganoderic acid Y
GLTs	*Ganoderma lucidum* triterpenoids
GLUT	glucose transporter
GPX	glutathione peroxidase
GSH	reduced glutathione
H_2O_2	hydrogen peroxide
HA	high altitudes
HBV	hepatitis B virus
HCT	human colon carcinoma
HepG2	human hepatocellular liver carcinoma
HFMD	hand, foot, and mouth disease
HL-60	human promyelocytic leukemia
HPLC	high-performance liquid chromatography
HPLC-DAD-MS	high-performance liquid chromatography-diode array detector-mass spectrometry
HPTLC	high-performance thin layer chromatography
HSK	homoserine kinase precursor
HS-SPME-GC-MS	headspace solid-phase micro-extraction combined with gas chromatography-mass spectrometry
HUVECs	human umbilical vein endothelial cells
i.c.v.	intracerebroventricular
i.g.	intragastric
IC50	half maximal inhibitory concentration
IFN	interferon

IL	interleukin
iNO	inducible nitric oxide
IOSe	inorganic selenium
IR	insulin receptor
JNK	c-jun N-terminal kinase
LDH	lactate dehydrogenase
LDL	low-density lipoprotein
LLC	Lewis lung carcinoma
LPS	lipopolysaccharide
LVEF	left-ventricular ejection fraction
LVFS	left-ventricular fractional shortening
LZ-8	LingZhi-8
LZP	LingZhi protein
MALDI-TOF	matrix-assisted laser desorption/ionization-time of flight
MAPK	mitogen-activated protein kinase
MCAO	middle cerebral artery occlusion
MCM	mini-chromosome maintenance proteins
MDA	malondialdehyde
MMP	matrix metalloproteinase
NFO	nonfunctional overreaching
NF-κB	nuclear factor kappa B
NMR	nuclear magnetic resonance
NO	nitric oxide
NREM	nonrapid eye movement
OTS	overtraining syndrome
P. aeruginosa	*Pseudomonas aeruginosa*
p-AMPK	phospho-AMP-activated protein kinase
PARP	poly ADP ribose polymerase
PC-3	prostate cancer-3
PGE2	prostaglandin E2
PGL	polysaccharides from *Ganoderma lucidum*
PMA	phorbol-12-myristate-13-acetate
Pro-Se	organic selenium
PSP	polysaccharide peptides
PTP1B	protein tyrosine phosphatase 1B
Rad	radiation sensitive protein
REM	rapid eye movement
ROS	reactive oxygen species
SDS-PAGE	sodium dodecyl sulfate polyacrylamide gel electrophoresis
Se	selenium
SOD	superoxide dismutase
SRC	sarcoma
T2DM	type 2 diabetic mellitus
TAC	transverse aortic constriction
TCM	traditional Chinese medicine
TGF-β1	transforming growth factor-β1
Th	T helper cell
TLC	thin-layer chromatography

TLR-4	toll-like receptor
TNF	tumor necrosis factor
Trec	rectal temperature
UDP	uridine diphosphate
uPA	urokinase plasminogen activator
uPAR	urokinase-plasminogen activator receptor
UV	ultraviolet
VEGF	vascular endothelium growth factor
VOCs	volatile organic compounds
VPH	virus del papiloma humano
XTT	2,3-Bis-(2-methoxy-4-nitro-5-sulfophenyl)-2H-tetrazolium-5-carboxanilide
ZIC-HILIC	Zwitterionic hydrophilic interaction chromatographic

7.1 INTRODUCTION

Biodiversity upholds a significant role in human survival and sustenance (Sen and Samanta, 2014). Human beings rely on various components of nature—plants, fungi, and animals—for their medicinal and nutritional benefits (Wink, 2015). The diverse medicinal and therapeutic applications of different herbal products against various pharmacological targets, including cancer, brain, cardiovascular function, microbial infection, inflammation, and pain, have accelerated their acceptance worldwide (Sen and Samanta, 2014). Their therapeutic usage also is well-documented in various traditional medicine systems, such as Ayurveda, Kampo, and traditional Chinese medicine (Wink, 2015). Medicinal mushrooms have joined medicinal plants for consideration by the scientific community. Although medicinal mushrooms are used mainly today as health supplements or functional foods, they also have the potential to become real drugs of evidence-based medicine (Lindequist et al., 2014). Several medicinal mushrooms are recognized for their health benefits, including *Agaricus subrufescens*, the *Ganoderma* species, *Grifola frondosa*, *Inonotus obliquus*, *Lentinula edodes*, *Ophiocordyceps* species, and *Piptoporus betulinus*. The *Ganoderma* species and especially *Ganoderma lucidum* (*G. lucidum*) are the most widely studied and highly esteemed medicinal mushrooms of all. Numerous preclinical (in vitro and in vivo) and clinical studies corroborate its medicinal and therapeutic benefits as cited in TCM (Bishop et al., 2015). One important bioactivity of *G. lucidum* is its blood-vitalizing effect under oxygen deficit (hypoxic) environments, generated during strenuous exercise, in cancer, cardiovascular diseases, and at high altitudes (AMS) (Bishop et al., 2015; Darmananda, 1997). However, the protective role of *G. lucidum* under such hypoxia-associated clinical complications has not been reviewed critically.

This chapter reviews the major health benefits of the royal medicinal mushroom, *G. lucidum*, and the bioactive compounds that advocate its potential for alleviating hypobaric hypoxia, AMS and its associated pathophysiological conditions, and various scientific investigations in this context.

7.2 HISTORY OF *GANODERMA*

Historically, *G. lucidum* is known as "the mushroom of immortality" or "herb of spiritual potency" (Bishop et al., 2015; Hapuarachchi et al., 2015). Its chronicle record dates about 2000 years ago (Bishop et al., 2015; Wachtel-Galor et al., 2011a, b) and has been mentioned in the pharmacopeia text *Shen Nong Ben Cao Jing*, as a medicinal mushroom that promotes longevity and maintains vitality (Baby et al., 2015; Bishop et al., 2015; Hapuarachchi et al., 2015). It was a revered spiritual and therapeutic herb used by the imperial class of Chinese society (Bishop et al., 2015) whose cultural importance could be depicted from its images in paintings, carvings, furniture, carpet designs, and even women's accessories. (Bishop et al., 2015; Hapuarachchi et al., 2015). *G. lucidum* (Fig. 7.1) was used to treat chronic diseases such as arthritis, asthma, bronchitis, cancer, diabetes, hepatopathy, hypertension, insomnia, and nephritis (Baby et al., 2015). It was believed to possess therapeutic properties such as antiaging effects, enhancing vital energy, improving memory, strengthening cardiac function, and tonifying effects (Wachtel-Galor et al., 2011a, b). Its pharmacological activities have been recognized widely, as signified by its inclusion in the American Herbal Pharmacopeia and Therapeutic Compendium

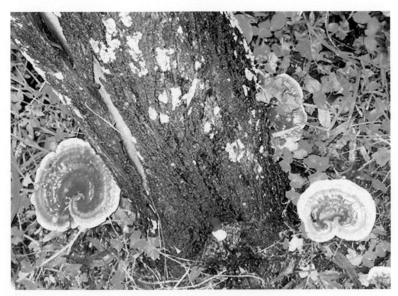

FIG. 7.1
Ganoderma lucidum fruiting body growing at the stumps of a tree.

(Chen et al., 2012a, b). Numerous *G. lucidum*-based commercial products, such as capsules, powders, tea bags, or as dietary supplements containing polysaccharides and/or triterpenoids, are marketed globally (Liu et al., 2015).

7.3 TAXONOMY AND GEOGRAPHICAL DISTRIBUTION

Ganoderma is a group of wood-degrading mushrooms with hard fruiting bodies classified as

Kingdom: Fungi, Division: Basodiomycota, Class: Homiobasidiomycetes, Order: Polyporales, Family: Ganodermataceae, Genus: *Ganoderma* (Baby et al., 2015).

Donk (1948) introduced the family Ganodermataceae (Hapuarachchi et al., 2015), which include five genera: *Ganoderma* (P. Karst 1881), *Amauroderma* (Murril 1905), *Haddowia* (Steyaert 1972), *Humphreya* (Steyaert 1972), and *Polyporopsis* (Audet 2010) (Richter et al., 2015). The genus *Ganoderma* established by Karsten is typified by *G. lucidum*(W. Curt, Fries.) as the only species (Hapuarachchi et al., 2015; Zhou et al., 2015). *Ganoderma lucidum* (*G. lucidum*) was described by Curtis (1871) based on specimen collected from Peckham, London, UK, and the addition of *"lucidum"* was sanctioned by Fries (1821) (Hapuarachchi et al., 2015; (Hennicke et al., 2016; Zhou et al., 2015). William Curtis designated and illustrated this taxon as *Boletus lucidus* in (1781), and it was named as *Polyporus lucidus* by Karsten in 1821 (Hapuarachchi et al., 2015).

Taxonomic studies report more than 300 species in genus *Ganoderma* (Baby et al., 2015). The traditional taxonomy of *Ganoderma* is based on its morphological traits and is divided into two different groups: *G. lucidum* complex (with laccate pilei) and *G. applanatum* complex (the nonlaccate species), which refer to the subgenera *Ganoderma* and *Elfvingia*, respectively (Hapuarachchi et al., 2015; Kwon et al., 2016). Environmental factors, interhybridization, morphological propensity, and variability, however, have led to the erroneous identification of *Ganoderma* species (Kwon et al., 2016). The taxonomy of the *G. lucidum* complex has long been subject to debate, and diverse opinions have been proposed regarding the validity of its members (Hapuarachchi et al., 2015; Zhou et al., 2015). Various species morphologically similar to *G. lucidum*, described from all over the world, have been included as accepted members of the *G. lucidum* complex, which is mainly characterized within the genus by laccate pilei (Zhou et al., 2015). These include *Ganoderma multipileum* (Ding Hou 1950), *Ganoderma sichuanense* (J. D. Zhao and X.Q. Zhang 1983), and *Ganoderma lingzhi* (Sheng H. Wu et al. 2012) from China; *Ganoderma resinaceum* (Boud. 1889) from Europe, and *Ganoderma oregonense* (Murrill 1908), *Ganoderma sessile* (Murrill 1902), *Ganoderma tsugae* (Murrill 1902), and *Ganoderma zonatum* (Murrill 1902) from the United States (Zhou et al., 2015).

Despite of the fact that the specimens named *G. lucidum* from Europe and East Asia are not conspecific, *lingzhi* isolates employed in research and commercially cultivated strains still are referred to as *G. lucidum* (Baby et al., 2015; Hennicke et al., 2016; Kwon et al., 2016). Several research attempts have been made based on morphological characteristics, molecular sequence data, and now secondary metabolite profiles, such as triterpenic acids composition, in attempts to reach authentic conclusions regarding the taxonomy and nomenclature of *G. lucidum* (Hennicke et al., 2016). Nevertheless, it is reasonable to categorize *Ganoderma* species as medicinal mushrooms (Baby et al., 2015).

The genus *Ganoderma* is cosmopolitan in distribution (Hapuarachchi et al., 2015). The species *G. lucidum* inhabits both tropical and temperate regions of Asia (India, China, Korea), Europe, and North America, but is tolerant to hot and humid conditions (Hapuarachchi et al., 2015; Nadu, 2014). The basidiocarp (fruiting body) of this fungus grows from decaying wood logs, stumps, roots, or trunks of deciduous trees, especially those of the *Quercus, Acer, Alnus, Betula, Pyrus*, and *Magnolia* genus (Gurung et al., 2013; Nadu, 2014), and less frequently from conifers (Nadu, 2014). It usually grows as a facultative parasite but can thrive saprophytically (Hapuarachchi et al., 2015). This annual mushroom is rare in the wild but now is being commercially cultivated. The methods most widely used for its commercial production include wood logs, short wood segments, tree stumps, sawdust bags, and bottle procedures (Gurung et al., 2013).

7.4   METABOLITE PROFILE OF *G. LUCIDUM*

Metabolite composition substantially influences the quality and health-promoting properties of herbs (Tugizimana et al., 2013). Several research publications signify the abundance and variety of *G. lucidum* in biological actions conferred by either primary metabolites such as fatty acids, nucleotides, polysaccharides, and proteins/peptides, or secondary metabolites, such as alkaloids, polyphenols, triterpenoids, steroids, and tannins (Grienke et al., 2015). Metabolomics have been a valuable tool for high-throughput screening of bioactive principles that could be used as therapeutic agents with high selectivity, unique modes of action, and acceptable toxicological profiles (Tugizimana et al., 2013). The metabolite analysis of the basidiocarp, mycelia, and spores of *G. lucidum* has identified more than 400 different compounds (Chen et al., 2012a, b; Sanodiya et al., 2009). Triterpenoids and polysaccharides are the two main pharmacologically active compounds in *G. lucidum* (Bishop et al., 2015; Chen et al., 2012a, b; Siwulski et al., 2015). Some of the major bioactive compounds and reported activities are found in Table 7.1.

Table 7.1 Some of the Major Bioactive Compounds in *Ganoderma lucidum* With Reported Activities[a]

Compound Class	Bioactive Compound	Reported Activity
Nucleoside and derivatives	Adenosine	Analgesic and antiplatelet aggregation activities
	5′-Deoxy-5′methylsulphinyl adenosine	Antiplatelet aggregation activity
Polyphenols	Tannic acid	Antioxidant, antihyperglycemic, hypolipedemic activities and renal protective effects
Polysaccharides	β-D-glucans	Immunomodulation, anticancer, antitumor, antiviral activity, antioxidant, antihyperglycemic, hypolipedemic activity
	β-Glucan (BG)	Radioprotective activity
	GLP	Antiviral activity
	PGY, GL-PP, F3	Immunomodulation
	GLIS	Immunomodulation and antitumor activity
	Gl-PS	Antihypoxia activity
	SeGLP-2B-1	Antiproliferative activity
	FYGL-n	Antidiabetic activity
	F31	Antidiabetic activity
	GL-PS	Radioprotective activity
Proteins and peptides	LZ-8	immunomodulation, mitogenic activity
	Ganodermin	Antifungal activity
	LZP-1, LZP-2, LZP-3	Mitogenic activity
	GLP	Antioxidant activity
	Enzyme α-galactocidase	α-galactocidase actvity
	Lignolytic enzyme	Anti-HIV activity
	Se-GL	Antioxidant activity
Peptidoglycan	Ganoderan A, B, and C	Hypoglycemic activity
Triterpenes	*Ganoderma* total sterol [GS (1)]	Neuroprotective activity
	GLhw, GLlw, GLMe-1-8	Activity against Influenza virus
	Ganodermadiol	Activity against Herpes Simplex virus
	Methyl GA-A, methyl GA-B, GA-S1, and GA-TQ	Neuro-protective and antifatigue activity
	Ganodermanontriol	Antimetastatic activity, anti-HIV activity
	Ganoderic acids (GA-A, GA-DM, GA-F, GA-H, GA-Me, GA-T)	Antimetastatic activity
	Ganoderic acids (GA-AM1, GA-B, GA-F, GA-K, GA-Me, GA-T, GA-D)	Antitumor activity
	Ganoderic acids (GA-B, -C1, -α, -H), Ganoderiol (A, B, F), Ganolucidic acid A, Ganodermanondiol, Ganodermanontriol, Lucidenic acid O, lucidenic lactone	Anti-HIV activity
	Ganoderic acid-R, -S, Ganosporeric acid A	Hepatoprotective activity
	Ganoderic acid-C, -D	Antiallergic activity
Volatile organic compounds	2-Naphthyl esters of nonadecanoic acid and *cis*-9-nonadecenoic acid	Antitumor activity
	Oleic acid and cyclooctasulfur	Antihistamine release activity

[a]*Boh et al. (2007), Cheng and Sliva (2015), Sanodiya et al. (2009), and Siwulski et al. (2015).*

7.4.1 Amino Acids, Peptides, and Proteins

G. lucidum has been analyzed for numerous amino acids by a variety of analytical techniques, including as amino acid analyzer, capillary electrophoresis, and Cu-Sephadex G-25 column chromatography. These amino acids include glycine, L-alanine, L-isoleucine, L-leucine, L-methionine, L-phenylalanine, L-proline, L-serine, L-tyrosine, L-tryptophan, L-threonine, and valine. (Huie and Di, 2004; Kim et al., 2009; Mau et al., 2001).

Bioactive proteins with therapeutic effects also have been isolated from *G. lucidum*. Ling Zhi-8 (LZ-8) was the first protein isolated from the mycelia of *G. lucidum* (Siwulski et al., 2015; Trigos and Suárez Medellín, 2011). It is a 12 kDa polypeptide of 110 amino-acid residues that possess immunomodulatory, mitogenic (Siwulski et al., 2015; Trigos and Suárez Medellín, 2011; Wachtel-Galor et al., 2011a, b), and immunosuppressive activities (Boh et al., 2007; Wachtel-Galor et al., 2011a, b). Few investigations have shown that the immunomodulatory protein LZ-8 could enhance the efficacy of DNA vaccine by activating dendritic cells and promoting innate and adaptive immune responses via toll-like receptor 4 (TLR-4) in vivo, thereby highlighting its potential use as an immunoadjuvant for cancer therapy (Lin et al., 2011; Paliya et al., 2014). Another important protein isolated from *G. lucidum* fruiting body is Ganodermin (Boh et al., 2007; Paliya et al., 2014). This 15-kDa protein has demonstrated antifungal activity against *Botrytis cinerea*, *Fusarium oxysporum*, and *Physalospora piricola* (Boh et al., 2007; Paterson, 2006; Siwulski et al., 2015). Three more bioactive proteins viz., LZP-1, LZP-2, LZP-3 with mitogenic activity were discovered from *G. lucidum* fruiting body and the spores (Huie and Di, 2004; Paterson, 2006). With the aid of HPLC, SDS-PAGE and MALDI TOF/TOF, Ansor et al. found four antihypertensive-related proteins from *G. lucidum* mycelia, which are proposed to lower blood pressure level through the inhibition of angiotensin converting enzyme activity. These were cystathionine α-synthase-like protein, DEAD/DEAH box helicase-like protein, paxillin-like protein, and α/β hydrolase-like protein (Ansor et al., 2013). A proteome analysis of *G. lucidum* by Sethy et al. in 2017 led to the identification of seven proteins, among which two were enzymes. These proteins included DNA replication licensing factor MCM7 (involved in eukaryotic genome replication), myosin (muscle contraction and mobility processes), malate dehydrogenase (human metabolic process such as citric acid cycle and glucogenesis), kinesins (mitosis, meiosis, transport of cellular cargo and in axonal transport), Rad50 (DNA repair protein), UDP-*N*-acetylglucosamine pyrophosphorylase (amino-sugars metabolism), and Hsk1-interacting molecule 1 protein (binding catalytic activity) (Sethy et al., 2017). Proteins as enzymes of *G. lucidum* origin also have been isolated and characterized (Huie and Di, 2004; Paliya et al., 2014; Paterson, 2006). For example, α-galactocidase (Paliya et al., 2014), laccase isoenzymes

(Nahata, 2013; Paliya et al., 2014; Trigos and Suárez Medellín, 2011), a ligninolytic enzyme (Paliya et al., 2014), proteinase A inhibitor (Boh et al., 2007), and ribonuclease (Wang et al., 2007) have been reported from the fruiting body.

G. lucidum is also a natural source of low molecular weight peptide (e.g., GLP) (Paliya et al., 2014; Trigos and Suárez Medellín, 2011; Wachtel-Galor et al., 2011a, b), polysaccharide-protein complexes (e.g., GLPP) (Paliya et al., 2014; Trigos and Suárez Medellín, 2011), glycoprotein (Se-GL and lectins such as GLL-F, GLL-M) (Paliya et al., 2014; Trigos and Suárez Medellín, 2011; Wachtel-Galor et al., 2011a, b), proteoglycans (e.g., GLIS; ganoderan) (Paliya et al., 2014; Siwulski et al., 2015), and glycans (such as ganoderans A, B, and C) (Paliya et al., 2014; Siwulski et al., 2015; Trigos and Suárez Medellín, 2011) with established biological activities (Table 7.1).

7.4.2   Nucleobases, Nucleosides, and Nucleotides

Scientific investigations employing analytical techniques, including UV-spectrophotometry, thin layer chromatography (TLC), high-performance thin-layer chromatography (HPTLC), high-performance liquid chromatography-diode array detector-mass spectrometry (HPLC-DAD-MS), NMR-based metabolomics, and zwitterionic hydrophilic interaction chromatographic (ZIC-HILIC) have revealed the presence of nucleosides, nucleotides, and nucleobases in *G. lucidum* and its related species. These studies have detected nucleosides, such as adenosine, cytidine, guanosine, inosine, thymidine, uridine, 2′-deoxyuridine, and 2′-deoxyadenosine adenine, and nucleobases, such as adenine, cytosine, guanine, hypoxanthine, thymine, uracil, and xanthine, from different parts of *G. lucidum*. Nucleotides 5′-adenosine monophosphate, 5′-cytosine monophosphate, 5′-guanosine monophosphate, 5′-inosine monophosphate, 5′-uridine monophosphate, and 5′-xanthosine monophosphate, also have been identified (Chen et al., 2012a, b; Gao et al., 2012; Kasahara and Hikino, 1987; Nahata, 2013; Sanodiya et al., 2009; Siwulski et al., 2015; Wasser, 2005; Weng and Yen, 2010).

Kasahara et al. isolated adenosine from *G. lucidum* through bioguided fractionation. The isolated adenosine demonstrated numerous activities of pharmaceutical importance (Table 7.1). In another study, adenosine and its derivative 5′-deoxy-5′-methylsulphinyl adenosine from *G. lucidum*, also were detected (Sanodiya et al., 2009; Siwulski et al., 2015; Wasser, 2005). The content of nucleosides adenine and adenosine have been used for routine quality control by Khan et al., 2015. They quantified adenine (0.16%) and adenosine (0.14%) in *G. lucidum* fruiting body extract of Indian origin by HPLC.

7.4.3 Polyphenolic Compounds

Polyphenols, mainly the flavonoids and phenolics, are the most important and widely occurring groups of secondary metabolites (Jing et al., 2015; Saeed et al., 2012; Wang and Xu, 2014). Several studies have been conducted to evaluate the antioxidant potential of *G. lucidum* in terms of total phenolic content and flavonoids content either in whole or crude extracts (Kirar et al., 2015) or extracts from fruiting body (Ćilerdžić et al., 2014), mycelium (Saltarelli et al., 2009), and spores (Heleno et al., 2013). Some of the phenolic compounds detected in *G. lucidum* are gallic acid, hesperidine, rutin, Kaempferol, chlorogenic acid, vanillin, cinnamic acid, and genistein, as well as phenolic acids (*p*-hydroxybenzoic, *p*-coumaric acids, and cinnamic acid) (Bhardwaj et al., 2015; Heleno et al., 2012; Liu et al., 2008; Sheikh et al., 2014; Yildiz et al., 2015). Reis et al. (2015) reported induction of autophagy in human gastric adenocarcinoma cells by a *G. lucidum* methanolic extract that contained substantial amounts of phenolic compounds, such as *p*-hydroxybenzoic acid ($123 \pm 9\,\mu g/g$ extract), *p*-coumaric acid ($80 \pm 6\,\mu g/g$ extract), and cinnamic acid ($59 \pm 6\,\mu g/g$ extract).

The major flavonoids detected included morin, myricetin, quercetin, and rutin (Saltarelli et al., 2015). A publication in 2016 reported isolation of a novel tannic acid from the fruiting body aqueous extract of *G. lucidum*, which was extracted by TLC and confirmed by HPLC (Elhussainy et al., 2016). Examples of polyphenolic compounds with reported activities are found in Table 7.1.

7.4.4 Polysaccharides

The occurrence of >200 polysaccharides isolated from *G. lucidum* fruiting body, mycelium, and spores have been reported (Ferreira et al., 2015; Trigos and Suárez Medellín, 2011). Polysaccharides of *G. lucidum* are mostly high molecular weight (4×10^5 to 1×10^6 Daltons) heteropolymers containing sugar molecules, such as glucose, arabinose, fucose, galactose, mannose, and xylose, combined in different conformations and joined together by 1–3 or 1–6 glycosidic linkages. Structural analysis has revealed that the polysaccharides of *G. lucidum* include (1.3), (1.6)-α/β-glucans, glycoproteins, and water soluble heteropolysaccharides, with (1.3), (1.6)-β-D-glucans being the most biologically active entities (Bishop et al., 2015; Siwulski et al., 2015; Trigos and Suárez Medellín, 2011). The polysaccharides of *G. lucidum* have been reported to exhibit various bioactivities, including immunomodulation, antihepatotoxicity, free radical scavenging, influence on cell cycle, inhibition of leukemic cells differentiation into monocytes/macrophages, inhibition of platelet aggregation, and inhibition of the interaction between virus and cell membranes with an increase in the production of IL-2 (Trigos and Suárez Medellín, 2011).

7.4.5   Triterpenes

Triterpenes are the most abundant metabolites besides polysaccharides (Baby et al., 2015; Bishop et al., 2015; Grienke et al., 2015; Taofiq et al., 2017). Triterpenes identified and isolated from *G. lucidum* are the ganoderic acids (A, AM1, B, C1, C2, D, DM, F, G, H, K, Me, Mk, S, T, TR, Y), ganoderenic acids (A, B, D), ganoderols (A, B), ganoderiol F, ganodermanontriol, ganoderal (A, Me), ganoderates (D, G), and lucidenic acids. These triterpenes are highly oxygenated lanostane-tetracyclic triterpenes composed of six isoprene units and most contain 30 or 27 carbon atoms with few having 24 carbon atoms (Siwulski et al., 2015; Taofiq et al., 2017; Trigos and Suárez Medellín, 2011). These have been considered important contributors to antitumor, antiinflammatory, antioxidant, antihepatitis, antimalarial, hypoglycemic, antimicrobial, and anti-HIV properties of *G. lucidum* (Baby et al., 2015; Bishop et al., 2015; Grienke et al., 2015; Taofiq et al., 2017). Among triterpenes, the ganoderic acids (containing four cyclic and two linear isoprene units) are most abundant (Baby et al., 2015; Bishop et al., 2015). The characteristic bitterness of *G. lucidum* owes to its triterpene content, which in turn depends on the strain, cultivation conditions, and the extraction process employed (Baby et al., 2015). An exhaustive review of triterpenes characterized from *Ganoderma* species are detailed elsewhere (Baby et al., 2015; Xia et al., 2014).

7.4.6   Other Bioactive Compounds

Numerous other bioactive compounds of medicinal significance have been identified and/or isolated from *G. lucidum*. These are volatile organic compounds (VOCs), vitamins (C and E and β-carotene) (Huie and Di, 2004; Siwulski et al., 2015; Wachtel-Galor et al., 2011a, b), and elements such as calcium, copper, iron, magnesium, manganese, phosphorous, potassium, silica, sulfur, sodium, and zinc. Heavy metals such as lead, cadmium, and mercury also were detected in trace amounts (Sanodiya et al., 2009; Wachtel-Galor et al., 2011a, b).

Chromatographic techniques have enabled detection of transanethol, *R*-(−)-linalool, *S*-(+)-carvone, and α-bisabolol as the major volatiles from the hydro-distillates and solvent extracts of *G. lucidum* fruiting body (Ziegenbein et al., 2006). Fifty-eight volatile compounds in *G. lucidum* mycelium using headspace solid-phase microextraction combined with gas chromatography-mass spectrometry (HS-SPME-GC-MS) were identified by Chen et al. (2010a, b). The volatile flavor compounds—1-octen-3-ol, ethanol, hexanal, 1-hexanol, sesquirosefuran, 3-octanol, and 3-octanone—were present in considerable amounts (Chen et al., 2010a, b). HS-SPME-GC-MS also has proved an efficient technique in detection of volatile aroma compounds, including acids, alcohols, aldehydes, phenols, L-alanine, D-alanine, 3-Methyl, 2-butanamine,

2-propanamine, and identification of 1-octen-3-ol and 3-methylbutanal in a *G. lucidum* specimen from Turkey (Taskin et al., 2013).

G. lucidum also contains sterols, oleic acid, cyclooctasulfur, ergosterol peroxide (5,8-epidioxy-ergosta-6,22E-dien-3-ol), and the cerebrosides (4E′,8E)-*N*-ᴅ-2′-hydroxystearoyl-1-*O*-ᴅ-glucopyranosyl-9-methyl-4-8-sphingadienine and (4E,8E)-*N*-ᴅ-2′-hydroxypamitoyl-1-*O*-ᴅ-glucopyranosyl-9-methyl-4-8-sphingadienine (Paterson, 2006; Sanodiya et al., 2009; Siwulski et al., 2015). Alkaloids, such as choline and betaine, tetracosanoic acid, stearic acid, palmitic acid, ergosta-7, 22-dien-3-ol, nonadecanoic acid, behenic acid, tetracosane, hentriacontane, ergosterol, and β-sitosterol and the lipid pyrophosphatidic acid also have been identified from spores (Sanodiya et al., 2009).

7.5 PHARMACOLOGICAL EFFECTS OF *G. LUCIDUM*

Development of hypobaric hypoxia is usual for people who live or visit to high altitudes (HA). Its various pathophysiological conditions include sleep apnea (systemic hypoxia), cancer, and obesity (cellular hypoxia). Hypoxia also is considered a potent inducer of inflammation and as an immunosuppressive agent (Caris et al., 2017). Many at-risk travelers, being naive to the health risks of HA, develop serious altitude-related illness because of underlying diseases, such as cardiovascular disorders, respiratory disorders, gastrointestinal conditions, chronic kidney diseases, diabetes mellitus, obesity, and neurological disorders that might be intensified on exposure to HA (Mieske et al., 2009). In this section, we review the health benefits of *G. lucidum* against hypobaric hypoxia and the preexisting clinical complications that often are exaggerated by HA exposure.

7.5.1 Cardiovascular Disorders

Research pertaining to *G. lucidum* suggests that triterpenes and polysaccharides are two major bioactive components that have effects on the cardiovascular system. Triterpenes such as ganoderic acids have demonstrated hypoglycaemic and hypotensive effects, inhibitory effects on enzymes (α-glycosidase and angiotensin-converting enzyme), inhibition of cholesterol synthesis, and are supposed to suppress sympathetic efferent activity. While the polysaccharide molecules, including beta-ᴅ-glucans, heteropolysaccharides, and glycoproteins identified in *G. lucidum*, have hypoglycaemic effects (Klupp et al., 2015).

Sugita (2014) mentioned in a review article that β-ᴅ-glucan derived from the purified extract of East Java Indonesian *G. lucidum* mycelia had a remarkable clinical effect after 8 months treatment with β-glucan dosage (3×200 mg β-glucan per day) in five cardiovascular diseases patients. A significant improvement of the ejection fraction (EF) was observed from 22% to 44% in comparison to the standard drug therapy, where the improvement of EF was not significant (from 13% as baseline to 22%) was observed (Sugita, 2014).

In another study, a transverse aortic constriction (TAC) mouse model of pressure overload-induced cardiomyopathy and heart failure was employed to examine the role of orally administered *Ganoderma* spore oil (for 14 days). The TAC model induced an initial compensatory cardiac remodeling that enhanced cardiac contractility. Gradually, the response to chronic overload leads to cardiac dilatation and heart failure within the model animal. Control groups were administered vegetable oil and an antihypertensive β2-adrenergic receptor antagonist. Transthoracic echocardiography was performed to measure left-ventricular ejection fraction (LVEF), left-ventricular fractional shortening (LVFS), left-ventricular end diastolic diameter, and cardiac output. Treatment with the oil in TAC mice normalized ejection fraction and corrected the fractional shortening generated by this model. Additionally, as assessed by left-ventricular end diastolic diameter, a reduction in left-ventricular hypertrophy also was observed. Treatment with *Ganoderma* spore oil decreased level of circular Foxo3, which inhibits tumor cell cycle progression and promotes cardiac senescence (Xie et al., 2016).

Several studies also highlight the antiatherosclerotic effect of *G. lucidum*. Atherosclerosis has been considered to be the hallmark of cardiovascular disease, which involves endothelial dysfunction, inflammation, and oxidative stress, and stimulates cytokines and other biomarkers. A study was designed by Handayani et al., 2016, to evaluate the effects of polysaccharide peptides (PSP) of *G. lucidum* on circulating endothelial cells (CECs), endothelial progenitor cells (EPCs), and nitric oxide (NO) as the indicators of endothelial vascular injury, and TNF-alpha and IL-6 as inflammatory markers in 34 patients with stable angina and 37 high-risk patients. The patients were administered PSP at a concentration of 750 mg/day in divided dose for 3 months as adjuvant therapy to their previous medications. Results indicated significant reduction in CEC and EPC levels in both stable angina and high-risk patients. A decrease in TNF-alpha and IL-6 level, followed by a reduced NO level, also was observed after PSP administration. A considerable reduction in total cholesterol level (205.49 ± 48.49 mg/dL to 182.11 ± 73.81 mg/dL) and LDL (126.17 ± 38.87 mg/dL to 116.17 ± 54.16 mg/dL) in patients with stable angina was noted. A similar reduction pattern was observed in high-risk patients, where the level of total cholesterol and LDL cholesterol was reduced after 3 months treatment. Overall, the study highlighted that PSP by virtue of its potent protective vascular effect and antilipid in stable angina pectoris could be used as an adjuvant therapy for the treatment of atherosclerotic cardiovascular diseases (Handayani et al., 2016). Another study led by Sargowo et al. in 2015 also emphasized the antiinflammatory and antioxidant effects of *G. lucidum* PSP in atherosclerosis (Sargowo et al., 2015). Lasukova et al., 2015, demonstrated that the antioxidant activity of *G. lucidum* imparts cardioprotective properties to it. The cardioprotective effects of *G. lucidum* extract (400 mg/kg for 15 days) were investigated in isolated and perfused rat heart subjected to prior global ischemia (45 min) and reperfusion (30 min). The results revealed that administration of the extract reduced necrotic death of cardiomyocytes and decreased reperfusion contracture.

7.5.2 Respiratory Disorders

A survey of literature provides sufficient review articles that emphasize the protective efficacy of *G. lucidum* against respiratory diseases such as bronchitis, pneumonia, and asthma (Bishop et al., 2015; Wasser, 2005). In a comparative study, Min-Chang et al. (2014) demonstrated significant differences between inorganic selenium (IOSe) and organic selenium (Pro-Se) in preventing asthma in ovalbumin-induced asthmatic mice. Following the oral administration of Pro-Se, the endogenous antioxidant enzyme level was restored, and there was a reduction in levels of tumor necrosis factor alpha (TNF-α) and interleukin 1 beta (IL-1β) because of the decrease in NF-κB expression in the asthmatic mice (Min-Chang et al., 2014). A significant study was by antiasthma simplified herbal medicine intervention (ASHMI). It is a traditional Chinese medicine (TCM) herbal asthma formula consisting of three herbs: Ling-Zhi (*G. lucidum*), Ku-Shen (*Sophora flavescens*), and Gan-Cao (*Glycyrrhiza uralensis*). This formulation is reported to decrease airway hyperreactivity (AHR) and eosinophilic inflammation via down-regulation of Th2 response in murine models. In clinical trials, ASHMI treatment significantly improved lung function and reduced symptoms of asthma to a similar extent as standard asthma therapy (López-Expósito et al., 2015). The protective effects of polysaccharides from (PGL) *G. lucidum* on bleomycin-induced pulmonary fibrosis in rats has been studied. Treatment with PGL (100–300 mg/kg for 28 days) resulted in significant decrease in the inflammatory cell infiltration, collagen deposition, and pulmonary index in rats with this condition. Additionally, the levels of catalase, glutathione, glutathione peroxidase, and superoxide dismutase were increased, and the amount of malondialdehyde and hydroxyproline in the lung were decreased. Therefore, PGL demonstrated a protective role in pulmonary fibrosis possibly by improving lung antioxidant status (Chen et al., 2016).

7.5.3 Neurological Disorders

G. lucidum has been shown to have both antioxidative and antiinflammatory effects, and noticeably decreases both the infarct area and neuronal apoptosis of the ischemic cortex. Zhang et al. (2014a, b) led a study that investigated the protective effects of intragastric administration of *G. lucidum* in cerebral ischemia/reperfusion injury in rats. Pretreatment with *G. lucidum* for 3 and 7 days lessened neuronal loss in the hippocampus. The malondialdehyde content in the hippocampus and serum and the levels of TNF-α and IL-8 in the hippocampus were reduced. On the other hand, the activity of superoxide dismutase in the hippocampus and serum was increased. This study illustrated protective role of *G. lucidum* against cerebral ischemia/reperfusion injury through its antioxidative and antiinflammatory properties (Zhang et al., 2014b).

G. lucidum also possesses tranquilizing action, and its use for the treatment of anxiety, insomnia, and palpitation is well documented in TCM. Cui et al. (2012), considering this action of *G. lucidum*, investigated the influence on the sleep of freely moving rats and the potential mechanism of *G. lucidum* fruiting body extract (GLE). Three-day treatment of GLE considerably augmented total sleep time and nonrapid eye movement (NREM) sleep time at a dose of 80 mg/kg without affecting slow-wave sleep or REM sleep. Serum, hypothalamus, and dorsal raphe nucleus TNF-α levels were significantly increased concomitantly. An inhibitory effect of TNF-α antibody (2.5 μg/rat) intracerebroventricular injection on the hypnotic effect of GLE (80 mg/kg, i.g.) also was observed. Coadministration of GLE (40 mg/kg, i.g.) and TNF-α (12.5 ng/rat, i.c.v.), both at ineffective doses, revealed an additive hypnotic effect. The authors suggested the hypnotic effects of GLE in freely moving rats apparently was related to the modulation of cytokines such as TNF-α (Cui et al., 2012).

Ke et al. (2017), demonstrated that *G. lucidum* polysaccharides (GLP), because of their immuneregulatory and antitumor activities, improved neurological deficits, cerebral infarct volume, and brain edema rate, indicating the protective role of GLP in middle cerebral artery occlusion (MCAO) in rats. GLP treatment also increased the cerebral vascular density in the ischemic area, as well as angiogenesis after cerebral ischemia. These authors proposed that the underlying mechanism for this effect was enhanced activation of AMP-activated protein (AMP) kinase (AMPK) and eNOS (via their phosphorylation) following treatment with GLP, which in turn stimulated angiogenesis and neuron regeneration (Ke et al., 2017). In another study pertaining to protective efficacy of *G. lucidum* against hydrogen peroxide (H_2O_2)-induced neuronal ischemia/reperfusion injury, *G. lucidum* polysaccharides remarkably suppressed H_2O_2-induced apoptosis by diminishing the expression of caspase-3, Bax, and Bim and increasing Bcl-2 expression. These findings confirmed that *G. lucidum* polysaccharides have significant neuroprotective effects, but postulated another mechanism of *G. lucidum* polysaccharides against neuronal ischemia/reperfusion injury. They proposed that the polysaccharides derived from *G. lucidum* exhibited the neuroprotective effects by regulating the expression of apoptosis-associated proteins and inhibited oxidative stress-induced neuronal apoptosis (Sun et al., 2017).

7.5.4 Diabetes Mellitus

Diabetes mellitus inducing a leading cause of morbidity is widespread worldwide. Teng et al. (2012) were probably the first to propose the underlying mechanism regarding the antidiabetic effect of *G. lucidum*. In their study, the antidiabetic effect of a proteoglycan extract of *G. lucidum* fruiting bodies (FYGL; Fudan-Yueyang-*G. lucidum*) was evaluated using streptozotocin-induced type 2 diabetic mellitus (T2DM) rat model and compared with those of metformin and rosiglitazone. A dose- and time-dependent increase in insulin concentration and decrease in fasting plasma glucose in FYGL-treated T2DM rats was observed. The protein tyrosine phosphatase 1B (PTP1B) expression level and activity were decreased, while the tyrosine phosphorylation level of the insulin receptor (IR) β-subunit was increased in the skeletal muscles. Considering the plasma biochemistry indexes relative to T2DM accompanied metabolic disorders, FYGL significantly reduced the levels of free fatty acids, triglycerides, total cholesterol, and low-density lipoprotein-cholesterol, and raised the level of high density lipoprotein-cholesterol (Teng et al., 2012). Later in 2015, Pan et al., isolated a protein tyrosine phosphatase 1B (PTP1B) inhibitor, named FYGL-n (72.9 kDa), from *G. lucidum* fruiting bodies and characterized for its structure and bioactivity. FYGL-n was a hyperbranched heteropolysaccharide bonded with protein via both serine and threonine residues by O-type glycoside. The hyperbranched chain structure of FYGL-n was suggested to play an important role for its bioactivities of PTP1B inhibition and antihyperglycemic activity (Pan et al., 2015).

In another study, F31, a 15.9 kDa β-heteropolysaccharide demonstrated the antidiabetic potential in type 2 diabetic mice model. Results showed that F31 significantly lowered fasting serum glucose (FSG) ($P < 0.05$), fasting serum insulin (FSI), and epididymal fat/BW ratio ($P < 0.01$). In the liver, F31 decreased the mRNA levels of hepatic glucose regulatory enzymes and upregulated the ratio of phospho-AMP-activated protein kinase (p-AMPK)/AMPK. In epididymal fat tissue, F31 increased the mRNA levels of GLUT4 but reduced fatty acid synthase (FAS), acetyl-CoA carboxylase (ACC1) and resistin levels. Immunohistochemistry studies revealed that F31 raised the protein levels of GLUT4 and decreased resistin level (Xiao et al., 2017).

7.5.5 Radioprotective Effect

G. lucidum has demonstrated a protective role against both γ-radiation and UV radiation. The radioprotective effect of *G. lucidum* against UV radiation could be depicted from the work carried by Zeng et al. (2017). This group evaluated the efficacy of *G. lucidum* polysaccharides GL-PS-associated inhibition of ultraviolet B (UVB)-induced photoaging in human fibroblasts in vitro. Primary human skin fibroblasts were cultured, and a fibroblast photoaging model was built through exposure to UVB. Cells exposed to UVB and treated with 10, 20, and 40 µg/mL GL-PS demonstrated increased cell viability, decreased aged cells, increased C-telopeptides of Type I collagen (CICP) protein expression. Furthermore, decreased matrix metalloproteinase (MMP)-1 protein expression and decreased cellular ROS levels compared with UVB exposed/GL-PS untreated cells were observed. The findings suggested that the underlying mechanism for antiphotoaging activity of GL-PS against UVB was elimination of UVB-induced ROS (Zeng et al., 2017). The radioprotective effect of *G. lucidum* against gamma (γ) radiation was illustrated by Pillai et al. (2014). They studied the radioprotective action of a *G. lucidum* β-glucan (BG) on gamma radiation exposure (0, 1, 2, and 4 Gy) of human lymphocytes using comet assay. The results suggested that BG exhibited substantial radioprotective activity with DNA repairing ability and antioxidant activity as the plausible mechanism (Pillai et al., 2014). A similar mechanism was reported for the radioprotective effect exhibited by the total triterpenes of *G. lucidum* in mouse splenic lymphocytes in vitro. Total triterpenes successfully reduced (even at a low concentration of 25 µg/mL) the formation of intracellular ROS and enhanced endogenous antioxidant enzyme activity in splenic lymphocytes following irradiation (Smina et al., 2011a). Also studied were the effects of total triterpenes on γ-radiation-induced DNA strand breaks in pBR 322 plasmid DNA in vitro, human peripheral blood lymphocytes ex vivo, and in mice bone marrow cells in vivo. The results indicated the significant effectiveness of *Ganoderma* triterpenes in protecting DNA and membrane damage following irradiation. The findings further supported the potential use of *Ganoderma* triterpenes in radiotherapy (Smina et al., 2011b). The same group confirmed protective efficacy of total triterpenes isolated from *G. lucidum* followed by a whole-body exposure to γ-radiation when administered for 14 days (Smina et al., 2016).

7.5.6 Protection Against Infections

The harsh environmental conditions prevailing at HA often alter the host immune response and, consequently, the susceptibility to infection increases (Basnyat and Starling, 2016). Various bioactive compounds such as polysaccharides, triterpenoids, lectins, and proteins, with potential antimicrobial activity have been reported in *Ganoderma* species (Shikongo et al., 2013). This section illustrates some of the major antimicrobial activities, including antibacterial, antifungal, and antiviral potential exhibited by *G. lucidum*.

7.5.6.1 Antibacterial Activity

Various solvent extracts of *G. lucidum* have demonstrated antibacterial activity against *Escherichia coli*, *Staphylococcus aureus*, *Klebsiella pneumoniae*, *Bacillus subtilis*, *Salmonella typhi*, and *Pseudomonas aeruginosa* (Quereshi et al., 2010). The polysaccharides derived from *G. lucidum* have been found effective against harmful food-spoiling microorganisms, including *Bacillus cereus, B. subtilis, E. coli, Aspergillus niger,* and *Rhizopus nigricans,* and plant pathogens such as *Erwinia carotovora, Penicillium digitatum,* and *B. cinerea* (Bai et al., 2008). *G. lucidum* extract showed higher activity against *S. aureus* and *B. cereus* than the antibiotics ampicillin and streptomycin in a study by Heleno et al. (2013). In a recent study, Vazirian et al. (2014) observed that addition of purified triterpenoids and steroids from the crude fractions (hexane and chloroform) of *G. lucidum* exhibited antimicrobial activity against the yeast, *C. albicans;* Gram-positive (*S. aureus, B. subtilis*) and Gram-negative (*P. aeruginosa, E. coli*) bacteria (Vazirian et al., 2014). Extract from *G. lucidum* also has demonstrated bacteriostatic activity against *Helicobacter pylori,* a causative agent for the formation of gastric ulcers and gastric cancer (Siwulski et al., 2015).

7.5.6.2 *Antifungal Activity*

A significant study considering the antifungal activity of *G. lucidum* is the inhibitory effect of the protein, ganodermin isolated from *G. lucidum* mycelium against *B. cinerea, F. oxysporum,* and *P. piricola* (Wang et al., 2005). *G. lucidum* extract has demonstrated antifungal action against *Trichoderma viride,* which produces allergic symptoms, especially in immune-compromised individuals (Heleno et al., 2012; Siwulski et al., 2015). Some studies have looked at the activity of *G. lucidum* against the planktonic forms of *Candida,* resulting in the evaluation of *G. lucidum*-formulated toothpastes (Adwan et al., 2012; Nayak et al., 2010). In another study, the sesquiterpenoid extracts of *G. lucidum* were found highly effective against *C. albicans,* as compared to standard antibiotics (Shekhar et al., 2010). Moreover, the antifungal effects of ethanolic and methanolic extracts of *G. lucidum* against *C. glabrata* and *C. albicans* also have been examined, with higher inhibitory activity observed against *C. glabrata* than *C. albicans* (Celik et al., 2014).

Candida species are the most common opportunistic pathogenic fungi that constitute the normal microbiota of an individual's mucosal oral cavity, gastrointestinal tract, and vagina (Adwan et al., 2012). These ubiquitous fungi are known to cause superficial to invasive infections, mainly in immune-compromised patients (Guinea, 2014). In the light of these facts, we evaluated the aqueous extracts and the methanolic and ethyl acetate fractions of *G. lucidum* mycelium and the fruiting body against two *Candida* species: *C. albicans* and *C. glabrata* biofilms using XTT [2,3-Bis-(2-methoxy-4-nitro-5-sulfophenyl)-2H-tetrazolium-5-carboxanilide] reduction assay. The extracts reduced the adhesion, biofilm formation, and mature biofilm of the two test pathogens., The aqueous extract of *G. lucidum* mycelium, however, was most effective among all the extracts, especially against *C. albicans,* probably because of the presence of substantial ascorbic acid content, polyphenolic compounds, and volatile organic compounds in it (Bhardwaj et al., 2017).

7.5.6.3   *Antiviral Activity*

Antiviral activity of G. *lucidum* is owed to the triterpenes and polysaccharides both, but in some cases the water extracts have also demonstrated antiviral activity. For instance, G. *lucidum* water extract inhibited proliferation of HPV (human papilloma virus) transformed cells (Hernández-Márquez et al., 2014). Another study found that water soluble substances (GLhw and GLlw) and methanol soluble substances (GLMe-1-8) isolated from fruiting bodies inhibited replication of influenza-A virus (Siwulski et al., 2015).

G. *lucidum* polysaccharides have demonstrated a direct action toward hepatitis B virus (HBV) by inhibiting DNA polymerase. Ganodermadiol exhibited activity against herpes simplex virus type 1 (Bisko and Mitropolskaya, 2003). Ganoderic acids isolated from fruiting bodies of G. *lucidum* exhibited antiviral activity against HIV and Epstein-Barr virus (Eo et al., 2000).

Two G. *lucidum* triterpenoids (GLTs); Lanosta-7,9(11),24-trien-3-one,15;26-dihydroxy (GLTA), and Ganoderic acid Y (GLTB) have been studied for their antiviral activity against EV71 infection. Enterovirus 71 (EV71) is a major causative agent for hand, foot, and mouth disease (HFMD), and fatal neurological and systemic complications in children. The results suggested that GLTA and GLTB prevent EV71 infection by blocking the viral adsorption to the cells. Molecular docking studies also revealed that GLTA and GLTB probably bound to the viral capsid protein block uncoating of EV71 (Zhang et al., 2014a). G. *lucidum* triterpenoides, ganoderic acid A, ganoderic acid B, ganoderol B, ganodermanontriol, and ganodermanondiol inhibited the activation of Epstein-Barr virus (EBV) antigens, EBV early antigen (EBV-EA), and EBV capsid antigen (EBV-CA) by inhibiting the activity of telomerase (Zheng and Chen, 2017).

7.5.7   Protection Against Cancer

A literature survey provides ample scientific evidence for the potent anticancerous activity of G. *lucidum* and have identified polysaccharides and triterpenoids as the major anticancer bioactive agents. Scientific studies suggest that G. *lucidum* triterpenoids exhibit anticancer activity by virtue of their antiangiogenic, antiinflammatory, antimetastatic, and antioxidative ability to arrest the cell cycle and direct cytotoxic effects, in addition to its immunomodulation (Boh, 2013; Kao et al., 2013). G. *lucidum* polysaccharides display their anticancer action through antiangiogenic activity, antioxidative activity, and immunomodulatory action (Kao et al., 2013). The contributing bioactivities that confer anticancer activities to G. *lucidum* are detailed in this section.

7.5.7.1   *Antitumor Activity*

Tumors can be benign or malignant, with the malignant form commonly referred to as cancer (Lodish et al., 2000). In tumors, hypoxia is one of the major forces that drive metastasis and decreases the susceptibility to undergo apoptosis (Gupta and Massagué, 2006). *G. lucidum* has been reported to exhibit both antitumor and anticancer activities (Cheng and Sliva, 2015; Kao et al., 2013). Numerous mechanisms have been proposed by various researchers for the antitumor activity of *G. lucidum* (Cheng and Sliva, 2015; Kao et al., 2013; Trigos and Suárez Medellín, 2011; Xu et al., 2011) as illustrated in Fig. 7.2. Some of the major antitumor bioactivities of *G. lucidum* are discussed as later.

G. lucidum polysaccharide (GLP) exhibited cytotoxic and apoptotic effects by triggering intracellular calcium release and involving death receptor pathway through upregulation of caspase-3, -8, and *Fas* (Liang et al., 2014a). It also has demonstrated induction of cell apoptosis by activating mitochondrial and mitogen-activated protein kinase (MAPK) pathways via upregulation of Bax/Bcl-2, caspase-3, -9, and poly (ADP-ribose) polymerase (PARP) levels in HCT-116 cells (Liang et al., 2014b). Another study suggested that the polysaccharide obtained from *G. lucidum* exhibits antitumor activity in HL-60 acute myeloid leukemia cells by blocking the extracellular signal-regulated kinase/MAPK signaling pathway and simultaneously activated p38 and JNK MAPK pathways, and therefore, regulated their downstream genes and proteins,

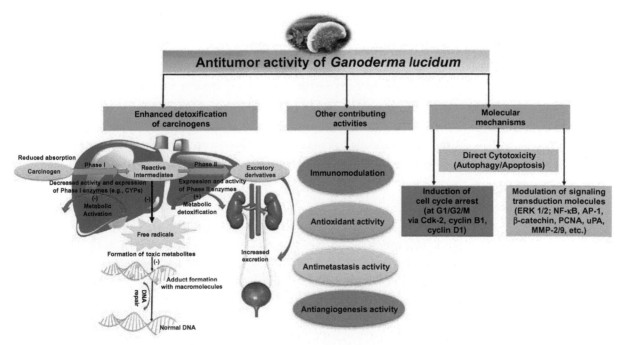

FIG. 7.2

The various mechanisms by which *Ganoderma lucidum* exhibits antitumor activity.

namely p53, c-myc, c-fos, c-jun, Bcl-2, Bax, cleaved caspase-3, and cyclin D1. As a result, cycle arrest and apoptosis of HL-60 cells were induced. Therefore, GLP exerted antitumor activity via MAPK pathways in HL-60 acute leukemia cells (Yang et al., 2016).

An earlier study highlighted that by activation of macrophages and consequent release of cytokines such as IFN-γ and tumor necrosis factor-α (TNF-α), and nitric oxide along with interleukins (IL-1β and IL-6) and other mediators and activated nuclear factor kappa B (NF-κB), Ganopoly, a polysaccharide rich aqueous extract of *G. lucidum*, demonstrated antitumor activity against sarcoma-180 in mice. It also demonstrated cytotoxicity in vitro, with marked apoptotic effects observed in human Caucasian cervical epidermoid carcinoma (CaSki), human hepatocellular liver carcinoma (Hep G2), and human colon carcinoma (HCT 116) cell lines (Gao et al., 2005b). Ganopoly also has displayed enhanced immune response in patients with advanced-stage cancer (Gao et al., 2005a).

Polysaccharide-protein or polysaccharide-peptide complexes, such as *G. lucidum* polysaccharide peptide (GLPP), also have been ascribed with antitumor effects because of their antiproliferative and antiangiogenic activity, both in vitro and in vivo (Cao and Lin, 2004). The proteoglycan, from *G. lucidum*, GLIS, exhibited an effective antitumor effect by increasing both humoral and cellular immune activities (Zhang et al., 2010). *G. lucidum* polysaccharide-peptide conjugate also displayed antitumor potential by the virtue of its antiproliferative effect in human umbilical vein endothelial cells (HUVECs) and a murine sarcoma 180 model, and antiapoptotic action by reducing vascular endothelium growth factor (VEGF) expression in human lung cancer cells (Cao and Lin, 2004; Li et al., 2008).

Several studies accentuate triterpenoids, especially the ganoderic acids (GAs), as the bioactive constituents of *G. lucidum* with potential antitumor activity in addition to the polysaccharides. The cytotoxic and antiproliferative effects of many GAs against tumor cells have been demonstrated in numerous investigations (Xu et al., 2010). GA-Me, a lanostane triterpenoid purified from *G. lucidum* mycelia by increasing expressions of IL-2 and Interferon-γ (IFN-γ) and upregulating expression of NF-κB effectively inhibited both tumor growth and lung metastasis of Lewis lung carcinoma in C57BL/6 mice (Wang et al., 2007), and tumor invasion through downregulating matrix metalloproteinase 2/9 (MMP2/9) gene expression. It was found that GA-Me depressed the same viability of tumor cells at much lower concentrations than against normal cells (Chen et al., 2008). It might be an effective potential therapeutic drug for prevention and treatment of tumors in clinic. GA-D was found to inhibit the proliferation, induce cell cycle arrest at G2/M phase, and trigger cell apoptosis (Yue et al., 2008), while GA-F, GA-K, GA-B, and GA-AM1 exhibited cytotoxicity with half maximal inhibitory concentration (IC50) values of 15–20 μM in HeLa human cervical carcinoma cell line (Yue et al., 2010).

Inhibition of DNA polymerase and posttranslational modification of oncoproteins also has been reported to contribute to the antitumor activity of *G. lucidum* (Sanodiya et al., 2009).

7.5.7.2 Antiangiogenic Activity

Angiogenesis is an essential step in tumor transition and antiangiogenic therapy could be critical in cancer therapy (Sanodiya et al., 2009; Xu et al., 2011). Antiangiogenic effects of *G. lucidum* have been demonstrated by the ethanolic extract of its fruiting body through inducible nitric oxide (iNO) production inhibition (Song et al., 2004; Yuen, 2007). *G. lucidum* is reported to inhibit the early event in angiogenesis in prostate cancer-3 (PC-3) cells by modulation of MAPK and Akt signaling pathways, which caused downregulation of VEGF and TGF-β1 from PC-3 cells (Stanley et al., 2005; Yuen, 2007).

G. lucidum polysaccharides peptide (GLPP) isolated from its fruiting body had antiangiogenic effect by direct inhibition on human umbilical vein endothelial cells (HUVECs) proliferation in vitro and had an indirect effect on PG cell (a human lung carcinoma cell line) proliferation (Cao and Lin, 2004; Weng and Yen, 2010; Xu et al., 2011; Yuen, 2007). GLPP also was found to possess proapoptotic action, mediated by the reduction of Bcl-2 expression, increase of Bax expression and downregulation of VEGF secretion on HUVECs (Cao and Lin, 2004; Xu et al., 2011; Yuen, 2007).

Triterpenoid derived from *G. lucidum* also exhibits antiangiogenic activities. A reduction of primary splenic tumor size as well as suppression of secondary metastasis in intrasplenic implant mice with LLC cells was noted as an inhibitory effect of *G. lucidum* triterpenoids fraction against Matrigel-induced angiogenesis (Kimura et al., 2002; Xu et al., 2011; Yuen, 2007). Further studies identified Ganoderic acid F as the active constituent for the tumor-induced antiangiogenic activity of the *G. lucidum* triterpenoids fraction (Kimura et al., 2002; Weng and Yen, 2010).

7.5.7.3 Antimetastatic Activity

Cancer metastasis is the spread of malignant cells from a primary tumor to distant sites, a process common in the late stages of cancer (Cheng and Sliva, 2015; Geiger and Peeper, 2009). Oral administration of standardized *G. lucidum* extract (GLE), containing 6% chemically characterized triterpenes and 13.5% polysaccharides, suppressed breast-to-lung cancer metastases in vivo through the downregulation of genes responsible for cell invasiveness (Adamec et al., 2009; Loganathan et al., 2014). Another study demonstrated inhibition of migration and adhesion of highly metastatic breast cancer cells by a ganoderiol

A-enriched G. *lucidum* extract via suppression of the focal adhesion kinase (FAK)-SRC-paxillin cascade pathway (Cheng and Sliva, 2015).

G. *lucidum* triterpenes also regulate several key proteins involved in cancer metastasis (Kao et al., 2013). GA-A and GA-H suppressed invasive behavior of breast cancer cells by inhibiting activating protein-1 (AP-1) and NF-κB signaling and suppressed secretion of urokinase plasminogen activator (uPA) (Jiang et al., 2008). Lucidenic acid B also exhibited a similar effect in hepatoma cells and inhibited cell invasion by suppressing the phosphorylation of extracellular signal-regulated kinase (ERK1/2) (Weng et al., 2007b). Lucidenic acids -A, -B, -C, and -N were found to possess potential antiinvasive activity in hepatoma cells by suppressing phorbol-12-myristate-13-acetate (PMA)-induced MMP-9 activity (Weng et al., 2007a). GA-Me effectively inhibited the invasive behavior of highly metastatic lung and breast cancer cells via inhibition of matrix metalloproteinase 2/9 (MMP2/9) gene expression (Chen et al., 2008) that degrade the extracellular matrix and can promote cancer metastasis (Kao et al., 2013). Another ganoderic acid, GA-T also inhibited colon cancer cell migration, invasion, and adhesion in vitro but through p53-dependent inhibition of MMP expression (Chen et al., 2010a, b). Nevertheless, Ganodermanontriol also has exerted its antiinvasiveness effect in breast cancer cells by inhibiting secretion of urokinase-plasminogen activator (uPA) and expression of its receptor (uPAR) (Jiang et al., 2011).

7.5.7.4   *Antiinflammatory Activity*

Inflammation has been regarded as the etiology of about 20% of cancers (Kao et al., 2013). It is an innate mechanism mediated via various mediators that are secreted by activated macrophages during the pathogenesis of many diseases, including carcinogenesis, chronic pulmonary inflammatory disease, Crohn's disease, rheumatoid arthritis, osteoarthritis, sepsis, and ulcerative colitis. Activated macrophages secrete a number of different inflammatory mediators, including tumor necrosis factor-α (TNF-α), interleukin-6 (IL-6), reactive oxygen species (ROS), prostaglandin E2 (PGE2), and nitric oxide (NO) (Dudhgaonkar et al., 2009). Chronic overexpression of these inflammatory cytokines could promote carcinogenesis (Kao et al., 2013).

G. *lucidum* polysaccharides (GLP) have shown significant antiinflammatory activity in acute and chronic inflammation in in vivo models (Joseph et al., 2011). An extract of G. *lucidum* polysaccharides (EORP) attenuated lipopolysaccharide (LPS)-induced expression of adhesion molecule and monocyte adherence by ERK phosphorylation suppression and NF-κB activation, both in vitro and in vivo (Liu et al., 2008). A triterpene-enriched ethanol extract from G. *lucidum* on exposure to proinflammatory stimuli also exhibited antiinflammatory activity in human colon carcinoma cells (Hong and Jung, 2004). Whereas, another ethanol extract from the mycelium of G. *lucidum* showed antiinflammatory activity against carrageenan-induced (acute) and formalin-induced (chronic) inflammation in mice and phorbol ester-induced mouse skin inflammation (Lakshmi et al., 2006). *G. lucidum* triterpene extract showed antiinflammatory effects in macrophages, which were mediated by the inhibition of NF-κB and AP-1 signaling pathways (Dudhgaonkar et al., 2009), and also has prevented colitis-associated carcinogenesis in mice (Sliva et al., 2012).

7.5.7.5 Antioxidative Activity

Oxidative stress is another major contributor to increased cancer risk. It could be countered by antioxidative enzymes and repair mechanisms. Excess oxidative stress, however, can overwhelm the innate protective systems leading to a variety of physiological disorders including cancer (Kao et al., 2013). Under such circumstances, antioxidants are suggested to defend the cancer-inducing oxidative damage (Sun et al., 2002). The triterpenes extracted from *G. lucidum* exhibit antioxidative activity in vitro by directly scavenging free radicals, increasing the activities of antioxidant enzymes in blood and tissues, and reducing radiation-induced oxidative DNA damage in mice splenocytes (Smina et al., 2011a, b). In another study, polysaccharides from *G. lucidum* mycelium demonstrated antioxidative activity by reducing ROS-induced oxidative damage and suppressing aberrant crypt foci formation in rat colon (Kao et al., 2013). Inhibition of iron-induced lipid peroxidation in rat brain homogenates, inactivation of hydroxyl radicals and superoxide anions, and reduction of UV-induced DNA strand breakage in differentiated HL-60 (human promyelocytic leukemia) cells by an amino-polysaccharide *G. lucidum* fraction (G009) also has been reported (Lee et al., 2001).

7.5.7.6 Cytotoxic and Cytostatic Activity

Deregulation of the cell cycle often triggers aberrant cell reproduction, leading to cancer formation when the cell cycle controls are bypassed (Cheng and Sliva, 2015). Anticancer drugs targeting cell cycle phases have either cytotoxic or cytostatic action. Cytotoxic drugs end up in apoptosis and autophagy; cytostatic drugs halt the rapid proliferation of cancer cells via cell cycle arrest at a particular phase (Villasana et al., 2010). The triterpenes isolated from *G. lucidum* (GTS) are reported to regulate the expression of 14 proteins in human cervical carcinoma cells, which have critical roles in cell proliferation, cell cycle, oxidative stress, and apoptosis (Yue et al., 2008). A study by Thyagarajan et al. (2010) signified that *G. lucidum* triterpene extract (GLT) could suppress proliferation (human colon cancer cells) as well as inhibit tumor growth (in a xenograft model) by triggering cell cycle arrest at the G0/G1 phase and stimulating programmed cell death Type II-autophagy (Thyagarajan et al., 2010). It also was found that GA-A induced apoptosis in lymphoma cells through caspase-3 and -9, and enhanced HLA class II-mediated antigen presentation and CD4+ T-cell recognition (Radwan et al., 2015).

7.5.7.7 *Immunomodulatory Effect*

G. lucidum owe its potent immunomodulatory action to the presence of poly-saccharides (β-D-glucans), proteins (Ling Zhi-8), and triterpenoids as its major active constituents (Sanodiya et al., 2009; Siwulski et al., 2015). *G. lucidum* polysaccharides have been demonstrated to modulate and improve immune functions in both human studies and animal models (Bishop et al., 2015; Gao et al., 2003; Wang et al., 2007). Tsai et al. (2012) established the relation-ship between the structural features of *G. lucidum* polysaccharides (GLPS) and their immunomodulating mechanisms. Two different bioactive components (GLPS-SF1 and GLPS-SF2) were isolated from GLPS, which exhibited distinct bioactivities. The GLPS-SF1 was glycopeptides, whereas GLPS-SF2 was acidic oligosaccharides. GLPS-SF1 demonstrated different immunostimulating prop-erties in human monocytes, NK cells, and T lymphocytes as well as induced cytokine expression via TLR4-dependent signaling in mouse macrophages. On the other hand, GLPS-SF2 exhibited immunomodulating activities specific on NK cells and T lymphocytes by inducing the IL-2 cytokine expression and cell proliferation (Tsai et al., 2012).

Furthermore, the polysaccharides from this fungus are reported to enhance the major histocompatibility complex (MHC) expression in a melanoma cell line with subsequent improvement in antigen presentation, thus promoting viral and cancer immunity (Bishop et al., 2015; Sun et al., 2011). Another study reports that the administration of *G. lucidum* extracts resulted in increased secre-tion of cytokines from immune cells, which leads to increased cellular activity and survival of immune cells of innate (macrophages) and adaptive immunity (lymphocytes) (Bach et al., 2009; Bishop et al., 2015; Mantovani and Sica, 2010). The various mechanisms as reported by various researchers (Sanodiya et al., 2009; Xu et al., 2011), by which *G. lucidum* exhibits immunomodulatory effect, are depicted in Fig. 7.3.

7.5.8 Endurance-Promoting Effects of *G. lucidum*

The medicinal mushroom, *G. lucidum*, has been used in traditional medicine systems of the Orient with the belief that it promotes vigor and provides vitality (Cheng and Sliva, 2015; Paterson, 2006; Sanodiya et al., 2009; Wachtel-Galor et al., 2011a, b).

Ganoderma spores oil of Changbai mountain origin was assessed for its hypoxia tolerance effect in mice. It was observed that as compared with the control group, the time of hypoxia tolerance, and time and frequency of buccal respi-ration, among mice receiving high doses were increased significantly (Lan and Xu, 2009). In another study, the protective efficacy of *G. lucidum* polysaccha-rides (GL-PS) were evaluated against exhaustive exercise-induced oxidative stress in skeletal muscle tissues of mice. GL-PS was administered for 28 days

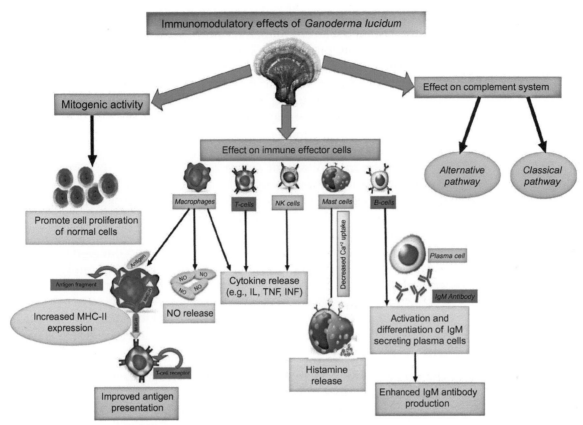

FIG. 7.3

The various mechanisms by which *Ganoderma lucidum* exhibits immunomodulating effects.

and subsequently, the mice were subjected to exhaustive swimming exercise. Then measurements were taken of biochemical parameters such as superoxide dismutase (SOD), glutathione peroxidase (GPX), catalase (CAT) activities, and malondialdehyde (MDA) levels in mice skeletal muscle. The findings showed an increase in antioxidant enzyme activities and a decrease in MDA levels after GL-PS supplementation (Zhonghui et al., 2014). That same year, another group investigated the antihypoxia and antifatigue effects of *G. lucidum* polysaccharides (Gl-PS). The findings of that study suggested antihypoxia and antifatigue effects of Gl-PS as indicated by prolonged survival and exhaustive swimming times, decreased blood lactic acid (BLA) and urea nitrogen (BUN) contents, along with increased liver and muscle glycogen contents after Gl-PS administration (Luo et al., 2014).

Rossi et al. (2014) investigated the endurance-promoting effect of *G. lucidum* capsules given as a food supplement for 3 months (two capsules per day) to seven healthy male athlete volunteers (30–40 years old). The effects of fungal supplements on the level of physical fitness was monitored and compared just before and after physical exertion by measuring the testosterone/cortisol ratio in the saliva and oxidative stress (DPPH free radical scavenging activity). A decrease of more than 30% in the testosterone/cortisol ratio after racing compared to before racing was considered to be a risk factor for nonfunctional overreaching (NFO) or overtraining syndrome (OTS). The results indicated that after 3 months of supplementation, the testosterone/cortisol ratio changed in a statistically significant manner, thereby protecting the athletes from NFO and OTS. Furthermore, antioxidant activity measurement by DPPH assay revealed an increased scavenging capacity of free radicals in the serum after the race, thereby protecting the athletes from oxidative stress (Rossi et al., 2014). Pawar Vinod and Shivakumar (2011) evaluated the adaptogenic effect of *G. lucidum* fruiting bodies methanolic extract against swimming endurance followed by post swimming antifatigue and motor coordination and hypoxic stress tolerance test in mice. Oral administration (100, 300, and 500 mg/kg/day) of the extract showed a significant dose-dependent increment in the parameters stated earlier. The 100 mg/kg dose was most effective in increasing the hypoxia tolerance time. Concomitant treatment with the extract at concentrations of 300 and 500 mg/kg demonstrated significant increase (Pawar Vinod and Shivakumar, 2011).

The studies mentioned above endorse the adaptogenic, antihypoxic, antifatigue, and endurance-promoting activities of *G. lucidum*, but neither describes these protective efficacies of the fungus under hypobaric hypoxic conditions. Considering that fact, we designed a study to evaluate the adaptogenic activity of *G. lucidum* mycelium aqueous extract (GLMaq) using cold, hypoxia, and restraint (C-H-R) animal model. The C-H-R model is the only passive multiple stress animal model among other stress models in which the experimental animals in restraint are exposed to cold (5°C) and hypoxia (428 mmHg) equivalent to an altitude of 4572 m (Sharma et al., 2012). This model assesses endurance promotion of an adaptogen by thermoregulation under stressful conditions (CHR) via rectal temperature (T_{rec}) as an indirect indicator of physical and mental performance capacity, the metabolic status of the animal, and hypothalamic and higher brain function (Ramachandran et al., 1990). The C-H-R model also known as animal decompression chamber is an ideal stress model to study adaptogenic potential of any herbal preparation against the multifactorial pathophysiology that is associated with high altitude.

GLMaq was administered orally at three doses (50, 100, 150 mg/kg B.W.). After C-H-R exposure, the rats were taken out of the chamber as soon as they acquired T_{rec} of 23°C. The time taken (minutes) to reach T_{rec} 23°C during exposure was used as a measure of endurance. The dose of 100 mg/kg body weight (B.W.) provided maximum resistance (93.55%) to CHR-induced hypothermia by delaying fall in $T_{rec} = 23$°C and was found comparable with other reported potent adaptogenic agents. Therefore, it was regarded as the dose with highest adaptogenic potential. However, a 50 mg/kg B.W. dose showed significant results as observed from hematological and biochemical [catalase (CAT), lactate dehydrogenase (LDH), lipid peroxidation (MDA), reduced glutathione (GSH), and superoxide dismutase (SOD)] parameters in comparison to 100 mg/kg B.W. and 150 mg/kg B.W. doses. Overall, GLMaq imparted protective adaptogenic efficacy against C-H-R exposure, suggesting it could be used as a protective herbal remedy against high altitude-induced pathologies (Bhardwaj and Misra, 2018; Forthcoming).

7.6 CONCLUSIONS

The species complex around the medicinal fungus *G. lucidum* (Ganodermataceae) has been revered in traditional medicines as well as in modern applications such as functional food or nutraceuticals. Current research pertaining to the various pharmacological and therapeutic applications strongly support its potential role in ameliorating the ill-effects imparted by the adversities of high altitudes, making it of therapeutic importance in managing the pathophysiology at high altitudes.

Acknowledgment

The authors thank the Director, Defense Institute of Physiology and Allied Sciences, Delhi, India, for his constant support and encouragement.

References

Adamec, J., Jannasch, A., Dudhgaonkar, S., Jedinak, A., Sedlak, M., Sliva, D., 2009. Development of a new method for improved identification and relative quantification of unknown metabolites in complex samples: determination of a triterpenoid metabolic fingerprint for the in situ characterization of Ganoderma bioactive compounds. J. Sep. Sci. 32 (23–24), 4052–4058.

Adwan, G., Salameh, Y., Adwan, K., Barakat, A., 2012. Assessment of antifungal activity of herbal and conventional toothpastes against clinical isolates of Candida albicans. Asian Pacific J. Trop. Biomed. 2 (5), 375–379.

Ansor, N.M., Abdullah, N., Aminudin, N., 2013. Antiangiotensin converting enzyme (ACE) proteins from mycelia of Ganoderma lucidum (Curtis) P. Karst. BMC Complement. Altern. Med. 13, 256.

Baby, S., Johnson, A.J., Govindan, B., 2015. Secondary metabolites from Ganoderma. Phytochemistry 114, 66–101.

Bach, J.P., Deuster, O., Balzer-Geldsetzer, M., Meyer, B., Dodel, R., Bacher, M., 2009. The role of macrophage inhibitory factor in tumorigenesis and central nervous system tumors. Cancer 115 (10), 2031–2040.

Bai, D., Chang, N.T., Li, D.H., Liu, J.X., You, X.Y., 2008. Antiblastic activity of Ganoderma lucidum polysaccharides. Acta Agric. Bor. Sin., 23, S1.

Basnyat, B., Starling, J.M., 2016. Infectious diseases at high altitude. In: Schlossberg, D. (Ed.), Infections of Leisure, fifth ed. American Society of Microbiology, Washington, DC, (pp. 325–332).

Bhardwaj, A. and Misra, K. Therapeutic medicinal mushroom (Ganoderma lucidum): a review of bioactive compounds and their applications. In: Goyal, M. R. and Chauhan, D. N. (Eds.), Plant and Marine Based Phytochemicals for Human Health: Attributes, Potential and Uses. Apple Academic Press Inc.; CRC Press (Taylor & Francis Group). Forthcoming June 15, 2018.

Bhardwaj, A., Pal, M., Srivastava, M., Tulsawani, R., Sugadev, R., Misra, K., 2015. HPTLC based chemometrics of medicinal mushrooms. J. Liq. Chromatogr. Relat. Technol. 38 (14), 1392–1406.

Bhardwaj, A., Gupta, P., Kumar, N., Mishra, J., Kumar, A., Misra, K., 2017. Lingzhi or Reishi medicinal mushroom, Ganoderma lucidum (Agaricomycetes), inhibits Candida biofilms: a Metabolomic approach. Int. J. Med. Mushrooms 19 (8), 685–696.

Bishop, K.S., Kao, C.H., Xu, Y., Glucina, M.P., Paterson, R.R.M., Ferguson, L.R., 2015. From 2000 years of Ganoderma lucidum to recent developments in nutraceuticals. Phytochemistry 114, 56–65.

Bisko, N.A., Mitropolskaya, N.Y., 2003. Some biologically active substances from medicinal mushroom Ganoderma lucidum (W. Curt.: Fr.) P. Karst. (Aphyllophoromycetideae). Int. J. Med. Mushrooms 5 (3), 301–305.

Boh, B., 2013. Ganoderma lucidum: a potential for biotechnological production of anticancer and immunomodulatory drugs. Recent Pat. Anti-Cancer Drug Discov. 8 (3), 255–287.

Boh, B., Berovic, M., Zhang, J., Zhi-Bin, L., 2007. Ganoderma lucidum and its pharmaceutically active compounds. Biotechnol. Annu. Rev. 13, 265–301.

Cao, Q.Z., Lin, Z.B., 2004. Antitumor and anti-angiogenic activity of Ganoderma lucidum polysaccharides peptide. Acta Pharmacol. Sin. 25, 833–838.

Caris, A.V., Ysis, W., de Aquino Lemos, V., Bottura, R., dos Santos, R.V.T., 2017. Nutrition and exercise can attenuate inflammatory and psychobiological changes in hypoxia? Asian Pacific J. Trop. Biomed. 7 (1), 86–90.

Celik, G.Y., Onbasli, D., Altinsoy, B., Alli, H., 2014. In vitro antimicrobial and antioxidant properties of Ganoderma lucidum extracts grown in Turkey. Eur. J. Med. Plants 4 (6), 709–722.

Chen, N.H., Liu, J.W., Zhong, J.J., 2008. Ganoderic acid me inhibits tumor invasion through down-regulating matrix metalloproteinases 2/9 gene expression. J. Pharmacol. Sci. 108 (2), 212–216.

Chen, N.H., Liu, J.W., Zhong, J.J., 2010a. Ganoderic acid T inhibits tumor invasion in vitro and in vivo through inhibition of MMP expression. Pharmacol. Rep. 62 (1), 150–163.

Chen, Z.J., Yang, Z.D., Gu, Z.X., 2010b. Determination of volatile flavor compounds in Ganoderma lucidum by HS-SPME-GC-MS. Food Res Dev 31 (2), 132–135.

Chen, S., Xu, J., Liu, C., Zhu, Y., Nelson, D.R., Zhou, S., Li, C., Wang, L., Guo, X., Sun, Y., Luo, H., 2012a. Genome sequence of the model medicinal mushroom Ganoderma lucidum. Nat. Commun. 3, 1–9.

Chen, Y., Bicker, W., Wu, J., Xie, M., Lindner, W., 2012b. Simultaneous determination of 16 nucleosides and nucleobases by hydrophilic interaction chromatography and its application to the quality evaluation of Ganoderma. J. Agric. Food Chem. 60 (17), 4243–4252.

Chen, J., Shi, Y., He, L., Hao, H., Wang, B., Zheng, Y., Hu, C., 2016. Protective roles of polysaccharides from Ganoderma lucidum on bleomycin-induced pulmonary fibrosis in rats. Int. J. Biol. Macromol. 92, 278–281.

Cheng, S., Sliva, D., 2015. Ganoderma lucidum for cancer treatment: we are close but still not there. Integr. Cancer Therap. 14 (3), 249–257.

Ćilerdžić, J., Vukojević, J., Stajić, M., Stanojković, T., Glamočlija, J., 2014. Biological activity of Ganoderma lucidum basidiocarps cultivated on alternative and commercial substrate. J. Ethnopharmacol. 155 (1), 312–319.

Cui, X.Y., Cui, S.Y., Zhang, J., Wang, Z.J., Yu, B., Sheng, Z.F., Zhang, X.Q., Zhang, Y.H., 2012. Extract of Ganoderma lucidum prolongs sleep time in rats. J. Ethnopharmacol. 139 (3), 796–800.

Darmananda, S., 1997. Reduction of Mountain Sickness With Chinese Herbs. Institute for Traditional Medicine, Portland, Oregon.

Dudhgaonkar, S., Thyagarajan, A., Sliva, D., 2009. Suppression of the inflammatory response by triterpenes isolated from the mushroom Ganoderma lucidum. Int. Immunopharmacol. 9 (11), 1272–1280.

Elhussainy, E.M., Elzawawy, N.A., Shorbagy, S.H., 2016. A novel tannic acid from Ganoderma lucidum fruiting bodies extract ameliorates early diabetic nephropathy in streptozotocin induced diabetic rats. Int. J. Pharm. Sci. Res. 7 (1), 62.

Eo, S.K., Kim, Y.S., Lee, C.K., Han, S.S., 2000. Possible mode of antiviral activity of acidic protein bound polysaccharide isolated from Ganoderma lucidum on herpes simplex viruses. J. Ethnopharmacol. 72 (3), 475–481.

Ferreira, I.C., Heleno, S.A., Reis, F.S., Stojkovic, D., Queiroz, M.J.R., Vasconcelos, M.H., Sokovic, M., 2015. Chemical features of Ganoderma polysaccharides with antioxidant, antitumor, and antimicrobial activities. Phytochemistry 114, 38–55.

Gao, Y., Zhou, S., Jiang, W., Huang, M., Dai, X., 2003. Effects of Ganopoly® (A ganoderma lucidum polysaccharide extract) on the immune functions in advanced-stage cancer patients. Immunol. Investig. 32 (3), 201–215.

Gao, Y., Gao, H., Chan, E., Tang, W., Xu, A., Yang, H., Huang, M., Lan, J., Li, X., Duan, W., Xu, C., 2005a. Antitumor activity and underlying mechanisms of ganopoly, the refined polysaccharides extracted from Ganoderma lucidum, in mice. Immunol. Investig. 34 (2), 171–198.

Gao, Y., Tang, W., Dai, X., Gao, H., Chen, G., Ye, J., Chan, E., Koh, H.L., Li, X., Zhou, S., 2005b. Effects of water-soluble Ganoderma lucidum polysaccharides on the immune functions of patients with advanced lung cancer. J. Med. Food 8 (2), 159–168.

Gao, P., Hirano, T., Chen, Z., Yasuhara, T., Nakata, Y., Sugimoto, A., 2012. Isolation and identification of C-19 fatty acids with anti-tumor activity from the spores of Ganoderma lucidum (reishi mushroom). Fitoterapia 83 (3), 490–499.

Geiger, T.R., Peeper, D.S., 2009. Metastasis mechanisms. Biochim. Biophys. Acta Rev. Cancer 1796 (2), 293–308.

Grienke, U., Kaserer, T., Pfluger, F., Mair, C.E., Langer, T., Schuster, D., Rollinger, J.M., 2015. Accessing biological actions of Ganoderma secondary metabolites by in silico profiling. Phytochemistry 114, 114–124.

Guinea, J., 2014. Global trends in the distribution of Candida species causing candidemia. Clin. Microbiol. Infect. 20 (s6), 5–10.

Gupta, G.P., Massagué, J., 2006. Cancer metastasis: building a framework. Cell 127 (4), 679–695.

Gurung, O.K., Budathoki, U., Parajuli, G., 2013. Effect of different substrates on the production of Ganoderma lucidum (Curt.: Fr.) Karst. Nature 10 (1), 191–198.

Handayani, O., Siwi, K., Widya, A., Ubaidillah, N., Vitriyaturidda, V., Failasufi, M., Ramadhan, F., Wulandari, H., Waranugraha, Y., Hayuningputri, D., Sargowo, D., 2016. PS 16-11 Ganoderma lucidum polysaccharide peptides: a potent protective endothelial vascular and antilipid in atherosclerosis. J. Hypertension 34, e468.

Hapuarachchi, K.K., Wen, T.C., Deng, C.Y., Kang, J.C., Hyde, K.D., 2015. Mycosphere essays 1: taxonomic confusion in the Ganoderma lucidum species complex. Mycosphere 6 (5), 542–559.

Heleno, S.A., Barros, L., Martins, A., Queiroz, M.J.R., Santos-Buelga, C., Ferreira, I.C., 2012. Fruiting body, spores and in vitro produced mycelium of Ganoderma lucidum from Northeast Portugal: a comparative study of the antioxidant potential of phenolic and polysaccharidic extracts. Food Res. Int. 46 (1), 135–140.

Heleno, S.A., Ferreira, I.C., Esteves, A.P., Ćirić, A., Glamočlija, J., Martins, A., Soković, M., Queiroz, M.J.R., 2013. Antimicrobial and demelanizing activity of Ganoderma lucidum extract, p-hydroxybenzoic and cinnamic acids and their synthetic acetylated glucuronide methyl esters. Food Chem. Toxicol. 58, 95–100.

Hennicke, F., Cheikh-Ali, Z., Liebisch, T., Maciá-Vicente, J.G., Bode, H.B., Piepenbring, M., 2016. Distinguishing commercially grown Ganoderma lucidum from Ganoderma lingzhi from Europe and East Asia on the basis of morphology, molecular phylogeny, and triterpenic acid profiles. Phytochemistry 127, 29–37.

Hernández-Márquez, E., Lagunas-Martínez, A., Bermudez-Morales, V.H., Burgete-García, A.I., León-Rivera, I., Montiel-Arcos, E., García-Villa, E., Gariglio, P., Madrid-Marina, V.V., Ondarza-Vidaurreta, R.N., 2014. Inhibitory activity of Lingzhi or Reishi medicinal mushroom, Ganoderma lucidum (higher basidiomycetes) on transformed cells by human papillomavirus. Int. J. Med. Mushrooms 16 (2), 179–187.

Hong, S.G., Jung, H.S., 2004. Phylogenetic analysis of Ganoderma based on nearly complete mitochondrial small-subunit ribosomal DNA sequences. Mycologia 96 (4), 742–755.

Huie, C.W., Di, X., 2004. Chromatographic and electrophoretic methods for Lingzhi pharmacologically active components. J. Chromatogr. B 812 (1), 241–257.

Jiang, J., Grieb, B., Thyagarajan, A., Sliva, D., 2008. Ganoderic acids suppress growth and invasive behavior of breast cancer cells by modulating AP-1 and NF-κB signaling. Int. J. Oncol. 21 (5), 577–584.

Jiang, J., Jedinak, A., Sliva, D., 2011. Ganodermanontriol (GDNT) exerts its effect on growth and invasiveness of breast cancer cells through the down-regulation of CDC20 and uPA. Biochem. Biophys. Res. Commun. 415 (2), 325–329.

Jing, L., Ma, H., Fan, P., Gao, R., Jia, Z., 2015. Antioxidant potential, total phenolic, and total flavonoid contents of Rhododendron anthopogonoides and its protective effect on hypoxia-induced injury in PC12 cells. BMC Complement. Altern. Med. 1 (15), 1–12.

Joseph, S., Sabulal, B., George, V., Antony, K., Janardhanan, K., 2011. Antitumor and anti-inflammatory activities of polysaccharides isolated from Ganoderma lucidum. Acta Pharm. 61 (3), 335–342.

Kao, C., Jesuthasan, A.C., Bishop, K.S., Glucina, M.P., Ferguson, L.R., 2013. Anti-cancer activities of Ganoderma lucidum: active ingredients and pathways. Funct. Foods Health Dis. 3 (2), 48–65.

Kasahara, Y., Hikino, H., 1987. Central actions of adenosine, a nucleotide of Ganoderma lucidum. Phytother. Res. 1 (4), 173–176.

Ke, Y., Chen, Y., Duan, J., Shao, Y., 2017. Ganoderma lucidum polysaccharides improves cerebral infarction by regulating AMPK/eNOS signaling. Int. J. Clin. Exp. Med. 10 (11), 15286–15293.

Khan, M.S., Parveen, R., Mishra, K., Tulsawani, R., Ahmad, S., 2015. Determination of nucleosides in Cordyceps sinensis and Ganoderma lucidum by high performance liquid chromatography method. J. Pharm. Bioallied Sci. 7 (4), 264–266.

Kim, M.Y., Lee, S.J., Ahn, J.K., Kim, E.H., Kim, M.J., Kim, S.L., Moon, H.I., Ro, H.M., Kang, E.Y., Seo, S.H., Song, H.K., 2009. Comparison of free amino acid, carbohydrates concentrations in Korean edible and medicinal mushrooms. Food Chem. 113 (2), 386–393.

Kimura, Y., Taniguchi, M., Baba, K., 2002. Antitumor and antimetastatic effects on liver of triterpenoid fractions of Ganoderma lucidum: mechanism of action and isolation of an active substance. Anticancer Res. 22 (6A), 3309–3318.

Kirar, V., Mehrotra, S., Negi, P.S., Nandi, S.P., Misra, K., 2015. HPTLC fingerprinting, antioxidant potential, and antimicrobial efficacy of Indian Himalayan lingzhi: Ganoderma lucidum. Int. J. Pharm. Sci. Res. 6 (10), 4259.

Klupp, N.L., Chang, D., Hawke, F., Kiat, H., Cao, H., Grant, S.J., Bensoussan, A., 2015. Ganoderma lucidum mushroom for the treatment of cardiovascular risk factors. Cochrane Database Syst. Rev. (2), CD007259.

Kwon, O., Park, Y.J., Kim, H.I., Kong, W.S., Cho, J.H., Lee, C.S., 2016. Taxonomic position and species identity of the cultivated Yeongji Ganoderma lucidum in Korea. Mycobiology 44 (1), 1–6.

Lakshmi, B., Ajith, T.A., Jose, N., Janardhanan, K.K., 2006. Antimutagenic activity of methanolic extract of Ganoderma lucidum and its effect on hepatic damage caused by benzo [a] pyrene. J. Ethnopharmacol. 107 (2), 297–303.

Lan, Y., Xu, Q.S., 2009. Effects of the Ganoderma spores oil in Changbai mountain on hypoxia-tolerance of mice [J]. J. Med. Sci. Yanbian Univ. 4, 009.

Lasukova, T.V., Maslov, L.N., Arbuzov, A.G., Burkova, V.N., Inisheva, L.I., 2015. Cardioprotective activity of Ganoderma lucidum extract during total ischemia and reperfusion of isolated heart. Bull. Exp. Biol. Med. 158 (6), 739–741.

Lee, J.M., Kwon, H., Jeong, H., Lee, J.W., Lee, S.Y., Baek, S.J., Surh, Y.J., 2001. Inhibition of lipid peroxidation and oxidative DNA damage by Ganoderma lucidum. Phytother. Res. 15 (3), 245–249.

Li, L., Lei, L.S., Yu, C.L., 2008. Changes of serum interferon-gamma levels in mice bearing S-180 tumor and the interventional effect of immunomodulators. Nan fang yi ke da xue xue bao. J. South. Med. Univ. 28 (1), 65–68.

Liang, Z., Guo, Y.T., Yi, Y.J., Wang, R.C., Hu, Q.L., Xiong, X.Y., 2014a. Ganoderma lucidum polysaccharides target a Fas/caspase dependent pathway to induce apoptosis in human colon cancer cells. Asian Pac. J. Cancer Prev. 15 (9), 3981–3986.

Liang, Z., Yi, Y., Guo, Y., Wang, R., Hu, Q., Xiong, X., 2014b. Chemical characterization and anti-tumor activities of polysaccharide extracted from Ganoderma lucidum. Int. J. Mol. Sci. 15 (5), 9103–9116.

Lin, C.C., Yu, Y.L., Shih, C.C., Liu, K.J., Ou, K.L., Hong, L.Z., Chen, J.D., Chu, C.L., 2011. A novel adjuvant Ling Zhi-8 enhances the efficacy of DNA cancer vaccine by activating dendritic cells. Cancer Immunol. Immunother. 60 (7), 1019–1027.

Lindequist, U., Kim, H.W., Tiralongo, E., Van Griensven, L., 2014. Medicinal mushrooms. Evid. Based Complement. Alternat. Med 2014, 806180.

Liu, S.Y., Wang, Y., He, R.R., Qu, G.X., Qiu, F., 2008. Chemical constituents of Ganoderma lucidum (Leys. ex Fr.) Karst [J]. J. Shenyang Pharm. Univ. 3, 004.

Liu, Y.H., Lin, Y.S., Lin, K.L., Lu, Y.L., Chen, C.H., Chien, M.Y., Shang, H.F., Lin, S.Y., Hou, W.C., 2015. Effects of hot-water extracts from Ganoderma lucidum. Bot. Stud. 56 (1), 1–10.

Lodish, H., Berk, A., Zipursky, S.L., Matsudaira, P., Baltimore, D., Darnell, J., 2000. Tumor cells and the onset of cancer. In: Freeman, W.H. (Ed.), Molecular Cell Biology. fourth ed. W. H. Freeman and Company, New York.

Loganathan, J., Jiang, J., Smith, A., Jedinak, A., Thyagarajan-Sahu, A., Sandusky, G.E., Nakshatri, H., Sliva, D., 2014. The mushroom Ganoderma lucidum suppresses breast-to-lung cancer metastasis through the inhibition of pro-invasive genes. Int. J. Oncol. 44 (6), 2009–2015.

López-Expósito, I., Srivastava, K.D., Birmingham, N., Castillo, A., Miller, R.L., Li, X.M., 2015. Maternal ASHMI-therapy prevents airway inflammation and modulates pulmonary innate immune responses in young offspring mice. Ann. Allergy Asthma Immunol. 114 (1), 43.

Luo, L., Cai, L.M., Hu, X.J., 2014. Evaluation of the anti-hypoxia and antifatigue effects of Ganoderma lucidum polysaccharides. Appl. Mech. Mater. 522, 303–306. Trans Tech Publications.

Mantovani, A., Sica, A., 2010. Macrophages, innate immunity, and cancer: balance, tolerance, and diversity. Curr. Opin. Immunol. 22 (2), 231–237.

Mau, J.L., Lin, H.C., Chen, C.C., 2001. Nonvolatile components of several medicinal mushrooms. Food Res. Int. 34 (6), 521–526.

Mieske, K., Flaherty, G., O'brien, T., 2009. Journeys to high altitude: risks and recommendations for travelers with preexisting medical conditions. J. Travel Med. 17 (1), 48–62.

Min-Chang, G., Wei-Hong, T., Zhen, X., Jie, S., 2014. Effects of selenium-enriched protein from Ganoderma lucidum on the levels of IL-1 β and TNF-α, oxidative stress, and NF-κ B activation in ovalbumin-induced asthmatic mice. Evid. Based Complement. Alternat. Med. 2014, 182817

Nadu, T., 2014. Evaluation of bioactive potential of basidiocarp extracts of Ganoderma lucidum. Int. J. 3 (1), 36–46.

Nahata, A., 2013. Ganoderma lucidum: a potent medicinal mushroom with numerous health benefits. Pharm. Anal. Acta 4 (10), 1000e159.

Nayak, A., Nayak, R.N., Bhat, K., 2010. Antifungal activity of a toothpaste containing Ganoderma lucidum against Candida albicans: an in vitro study. Journal of International Oral Health 2 (2), 51–57.

Paliya, B.S., Verma, S., Chaudhary, H.S., 2014. Major bioactive metabolites of the medicinal mushroom: Ganoderma lucidum. Int. J. Pharm. Res. 6 (1), 12–24.

Pan, D., Wang, L., Chen, C., Hu, B., Zhou, P., 2015. Isolation and characterization of a hyperbranched proteoglycan from Ganoderma lucidum for anti-diabetes. Carbohyd. Polym. 117, 106–114.

Paterson, R.R.M., 2006. Ganoderma: a therapeutic fungal biofactory. Phytochemistry 67 (18), 1985–2001.

Pawar Vinod, S., Shivakumar, H., 2011. Adaptogegic (antistress) activity of methanolic extract of Ganoderma lucidum against physical and hypoxic stress in mice. Pharmacology 2, 989–995.

Pillai, T.G., Maurya, D.K., Salvi, V.P., Janardhanan, K.K., Nair, C.K., 2014. Fungal beta glucan protects radiation induced DNA damage in human lymphocytes. Ann. Transl. Med. 2 (2), 13.

Quereshi, S., Pandey, A., Sandhu, S., 2010. Evaluation of antibacterial activity of different Ganoderma lucidum extracts. Department of Biological Sciences, R. D. University, Jabalpur. Centre for Scientific Research and Development, People's Group Bhanpur, Bhopal, India 462037 (M.P.). People's J. Sci. Res. 3 (1), 9–13.

Radwan, F.F., Hossain, A., God, J.M., Leaphart, N., Elvington, M., Nagarkatti, M., Tomlinson, S., Haque, A., 2015. Reduction of myeloid-derived suppressor cells and lymphoma growth by a natural triterpenoid. J. Cell. Biochem. 116 (1), 102–114.

Ramachandran, U., Divekar, H.M., Grover, S.K., Srivastava, K.K., 1990. New experimental model for the evaluation of adaptogenic products. J. Ethnopharmacol. 29 (3), 275–281.

Reis, F.S., Lima, R.T., Morales, P., Ferreira, I.C., Vasconcelos, M.H., 2015. Methanolic extract of Ganoderma lucidum induces autophagy of AGS human gastric tumor cells. Molecules 20 (10), 17872–17882.

Richter, C., Wittstein, K., Kirk, P.M., Stadler, M., 2015. An assessment of the taxonomy and chemotaxonomy of Ganoderma. Fungal Divers. 71 (1), 1–15.

Rossi, P., Buonocore, D., Altobelli, E., Brandalise, F., Cesaroni, V., Iozzi, D., Savino, E., Marzatico, F., 2014. Improving training condition assessment in endurance cyclists: effects of Ganoderma lucidum and Ophiocordyceps sinensis dietary supplementation. Evid. Based Complem. Alternat. Med. 2014, 979613.

Saeed, N., Khan, M.R., Shabbir, M., 2012. Antioxidant activity, total phenolic and total flavonoid contents of whole plant extracts Torilis leptophylla L. BMC Complem. Alternat. Med. 12 (1), 221.

Saltarelli, R., Ceccaroli, P., Iotti, M., Zambonelli, A., Buffalini, M., Casadei, L., Vallorani, L., Stocchi, V., 2009. Biochemical characterisation and antioxidant activity of mycelium of Ganoderma lucidum from Central Italy. Food Chem. 116 (1), 143–151.

Saltarelli, R., Ceccaroli, P., Buffalini, M., Vallorani, L., Casadei, L., Zambonelli, A., Iotti, M., Badalyan, S., Stocchi, V., 2015. Biochemical characterization and antioxidant and antiproliferative activities of different Ganoderma collections. J. Mol. Microbiol. Biotechnol. 25 (1), 16–25.

Sanodiya, B.S., Thakur, G.S., Baghel, R.K., Prasad, G.B.K.S., Bisen, P.S., 2009. Ganoderma lucidum: a potent pharmacological macrofungus. Curr. Pharm. Biotechnol. 10 (8), 717–742.

Sargowo, D., Prasetya, I., Ashriyah, R., Setyawati, I., Andri, W.T., Heriansyah, T., 2015. Anti inflammation and antioxidant effect of active agent polysaccharide peptide (Ganoderma Lucidum) in preventing atherosclerotic diseases. Biomed. Pharmacol. J. 8 (1), 27–33.

Sen, T., Samanta, S.K., 2014. Medicinal plants, human health, and biodiversity: a broad review. In: - Mukherjee, J. (Ed.), Biotechnological Applications of Biodiversity. Springer, Berlin Heidelberg, pp. 59–110.

Sethy, N.K., Bhardwaj, A., Singh, V.K., Sharma, R.K., Deswal, R., Bhargava, K., Mishra, K., 2017. Characterization of Ganoderma lucidum: phytochemical and proteomic approach. J. Proteins Proteom. 8 (1), 25–33.

Sharma, P., Kirar, V., Meena, D.K., Suryakumar, G., Misra, K., 2012. Adaptogenic activity of Valeriana wallichii using cold, hypoxia and restraint multiple stress animal model. Biomed. Aging Pathol. 2 (4), 198–205.

Sheikh, I.A., Vyas, D., Ganaie, M.A., Dehariya, K., Singh, V., 2014. HPLC determination of phenolics and free radical scavenging activity of ethanolic extracts of two polypore mushrooms. Int. J. Pharm. Pharm. Sci. 6 (2), 679–684.

Shekhar, R.B., Bapat, G., Jitendra, G.V., Sandhya, A.G., Hiralal, B.S., 2010. Antimicrobial activity of terpenoid extracts from Ganoderma samples. Int. J. Pharm. Life Sci. 1 (4), 234–240.

Shikongo, L.T., Chimwamurombe, P.M., Lotfy, H.R., 2013. Antimicrobial screening of crude extracts from the indigenous Ganoderma lucidum mushrooms in Namibia. Afr. J. Microbiol. Res. 7 (40), 4812–4816.

Siwulski, M., Sobieralski, K., Golak-Siwulska, I., Sokół, S., Sękara, A., 2015. Ganoderma lucidum (Curt.: Fr.) Karst.–health-promoting properties. A review. Herba Polon. 61 (3), 105–118.

Sliva, D., Loganathan, J., Jiang, J., Jedinak, A., Lamb, J.G., Terry, C., Baldridge, L.A., Adamec, J., Sandusky, G.E., Dudhgaonkar, S., 2012. Mushroom Ganoderma lucidum prevents colitis-associated carcinogenesis in mice. PLoS One 7.

Smina, T.P., De, S., Devasagayam, T.P.A., Adhikari, S., Janardhanan, K.K., 2011. Ganoderma lucidum total triterpenes prevent radiation-induced DNA damage and apoptosis in splenic lymphocytes in vitro. Mut. Res./Genet. Toxicol. Environ. Mutagen. 726 (2), 188–194.

Smina, T.P., Mathew, J., Janardhanan, K.K., Devasagayam, T.P.A., 2011b. Antioxidant activity and toxicity profile of total triterpenes isolated from Ganoderma lucidum (Fr.) P. Karst occurring in South India. Environ. Toxicol. Pharmacol. 32 (3), 438–446.

Smina, T.P., Joseph, J., Janardhanan, K.K., 2016. Ganoderma lucidum total triterpenes prevent γ-radiation induced oxidative stress in Swiss albino mice in vivo. Redox Rep. 21 (6), 254–261.

Song, Y.S., Kim, S.H., Sa, J.H., Jin, C., Lim, C.J., Park, E.H., 2004. Anti-angiogenic and inhibitory activity on inducible nitric oxide production of the mushroom Ganoderma lucidum. J. Ethnopharmacol. 90 (1), 17–20.

Stanley, G., Harvey, K., Slivova, V., Jiang, J., Sliva, D., 2005. Ganoderma lucidum suppresses angiogenesis through the inhibition of secretion of VEGF and TGF-β1 from prostate cancer cells. Biochem. Biophys. Res. Commun. 330 (1), 46–52.

Sugita, P., 2014. Abstract 138. Circulation Res. 115 (Suppl_1), A138.

Sun, J., Chu, Y.F., Wu, X., Liu, R.H., 2002. Antioxidant and antiproliferative activities of common fruits. J. Agricult. Food Chem. 50 (25), 7449–7454.

Sun, L.X., Lin, Z.B., Li, X.J., Li, M., Lu, J., Duan, X.S., Ge, Z.H., Song, Y.X., Xing, E.H., Li, W.D., 2011. Promoting effects of Ganoderma lucidum polysaccharides on B16F10 cells to activate lymphocytes. Basic Clin. Pharmacol. Toxicol. 108 (3), 149–154.

Sun, X.Z., Liao, Y., Li, W., Guo, L.M., 2017. Neuroprotective effects of Ganoderma lucidum polysaccharides against oxidative stress-induced neuronal apoptosis. Neural Regen. Res. 12 (6), 953–958.

Taofiq, O., Heleno, S.A., Calhelha, R.C., Alves, M.J., Barros, L., González-Paramás, A.M., Barreiro, M.F., Ferreira, I.C., 2017. The potential of Ganoderma lucidum extracts as bioactive ingredients in topical formulations, beyond its nutritional benefits. Food Chem. Toxicol. 108, 139–147.

Taskin, H., Kafkas, E., Çakiroglu, Ö., Büyükalaca, S., 2013. Determination of volatile aroma compounds of Ganoderma lucidum by gas chromatography mass spectrometry (HS-GC/MS). Afr. J. Tradit. Complem. Altern. Med. 10 (2), 353–355.

Teng, B.S., Wang, C.D., Zhang, D., Wu, J.S., Pan, D., Pan, L.F., Yang, H.J., Zhou, P., 2012. Hypoglycemic effect and mechanism of a proteoglycan from Ganoderma lucidum on streptozotocin-induced type 2 diabetic rats. Eur. Rev. Med. Pharmacol. Sci. 16 (2), 166–175.

Thyagarajan, A., Jedinak, A., Nguyen, H., Terry, C., Baldridge, L.A., Jiang, J., Sliva, D., 2010. Triterpenes from Ganoderma lucidum induce autophagy in colon cancer through the inhibition of p38 mitogen-activated kinase (p38 MAPK). Nutr. Cancer 62 (5), 630–640.

Trigos, Á., Suárez Medellín, J., 2011. Biologically active metabolites of the genus Ganoderma: three decades of myco-chemistry research. Rev. Mexicana Micol. 34, 63–83.

Tsai, C.C., Yang, F.L., Huang, Z.Y., Chen, C.S., Yang, Y.L., Hua, K.F., Li, J., Chen, S.T., Wu, S.H., 2012. Oligosaccharide and peptidoglycan of Ganoderma lucidum activate the immune response in human mononuclear cells. J. Agricult. Food Chem. 60 (11), 2830–2837.

Tugizimana, F., Piater, L., Dubery, I., 2013. Plant metabolomics: a new frontier in phytochemical analysis. S. Afr. J. Sci. 109 (5–6), 01–11.

Vazirian, M., Faramarzi, M.A., Ebrahimi, S.E.S., Reza, H., Esfahani, M., Samadi, N., Hosseini, S.A., Asghari, A., Manayi, A., Mousazadeh, A., Asef, M.R., 2014. Antimicrobial effect of the lingzhi or reishi medicinal mushroom, Ganoderma lucidum (higher basidiomycetes) and its main compounds. Int. J. Med. Mushrooms 16 (1), 77–84.

Villasana, M., Ochoa, G., Aguilar, S., 2010. Modeling and optimization of combined cytostatic and cytotoxic cancer chemotherapy. Artif. Intell. Med. 50 (3), 163–173.

Wachtel-Galor, S., Yuen, J.W.M., Buswell, J.A., Benzie, I.F.F., 2011a. Ganoderma lucidum (Lingzhi; Reishi): a medicinal mushroom. In: IFF, B., Wachtel-Galor, S. (Eds.), Herbal Medicine: Biomolecular and Clinical Aspects, second ed. CRC Press, Boca Raton (FL), pp. 53–76.

Wachtel-Galor, S., Yuen, J., Buswell, J.A., Benzie, I.F., 2011b. Ganoderma lucidum (Lingzhi or Reishi). Chapter 9. In: Benzie, I.F.F., Wachtel-Galor, S. (Eds.), Herbal Medicine: Biomolecular and Clinical Aspects, second ed. CRC Press/Taylor & Francis, Boca Raton (FL), pp. 175–197.

Wang, Y., Xu, B., 2014. Distribution of antioxidant activities and total phenolic contents in acetone, ethanol, water and hot water extracts from 20 edible mushrooms via sequential extraction. Aust. J. Nutr. Food Sci. 2 (1), 1–5.

Wang, M., Lamers, R.J.A., Korthout, H.A., van Nesselrooij, J.H., Witkamp, R.F., van der Heijden, R., Voshol, P.J., Havekes, L.M., Verpoorte, R., van der Greef, J., 2005. Metabolomics in the context of systems biology: bridging traditional Chinese medicine and molecular pharmacology. Phytother. Res. 19 (3), 173–182.

Wang, G., Zhao, J., Liu, J., Huang, Y., Zhong, J.J., Tang, W., 2007. Enhancement of IL-2 and IFN-γ expression and NK cells activity involved in the anti-tumor effect of ganoderic acid Me in vivo. Int. Immunopharmacol. 7 (6), 864–870.

Wasser, S.P., 2005. Reishi or Ling Zhi (Ganoderma lucidum). Encyclopedia Diet. Suppl. 1, 603–622.

Weng, C.J., Yen, G.C., 2010. The in vitro and in vivo experimental evidences disclose the chemopreventive effects of Ganoderma lucidum on cancer invasion and metastasis. Clin. Exp. Metastasis 27 (5), 361–369.

Weng, C.J., Chau, C.F., Chen, K.D., Chen, D.H., Yen, G.C., 2007a. The anti-invasive effect of lucidenic acids isolated from a new Ganoderma lucidum strain. Mol. Nutr. Food Res. 51 (12), 1472–1477.

Weng, C.J., Chau, C.F., Hsieh, Y.S., Yang, S.F., Yen, G.C., 2007b. Lucidenic acid inhibits PMA-induced invasion of human hepatoma cells through inactivating MAPK/ERK signal transduction pathway and reducing binding activities of NF-κB and AP-1. Carcinogenesis 29 (1), 147–156.

Wink, M., 2015. Modes of action of herbal medicines and plant secondary metabolites. Medicines 2 (3), 251–286.

Xia, Q., Zhang, H., Sun, X., Zhao, H., Wu, L., Zhu, D., Yang, G., Shao, Y., Zhang, X., Mao, X., Zhang, L., 2014. A comprehensive review of the structure elucidation and biological activity of triterpenoids from Ganoderma spp. Molecules 19 (11), 17478–17535.

Xiao, C., Wu, Q., Zhang, J., Xie, Y., Cai, W., Tan, J., 2017. Antidiabetic activity of Ganoderma lucidum polysaccharides F31 downregulated hepatic glucose regulatory enzymes in diabetic mice. J. Ethnopharmacol. 196, 47–57.

Xie, Y.Z., Yang, F., Tan, W., Li, X., Jiao, C., Huang, R., Yang, B.B., 2016. The anti-cancer components of Ganoderma lucidum possesses cardiovascular protective effect by regulating circular RNA expression. Oncoscience 3 (7–8), 203–207.

Xu, J.W., Zhao, W., Zhong, J.J., 2010. Biotechnological production and application of ganoderic acids. Appl. Microbiol. Biotechnol. 87 (2), 457–466.

Xu, Z., Chen, X., Zhong, Z., Chen, L., Wang, Y., 2011. Ganoderma lucidum polysaccharides: immunomodulation and potential anti-tumor activities. Am. J. Chin. Med. 39 (01), 15–27.

Yang, G., Yang, L., Zhuang, Y., Qian, X., Shen, Y., 2016. Ganoderma lucidum polysaccharide exerts anti-tumor activity via MAPK pathways in HL-60 acute leukemia cells. J. Receptors Signal Transduct. 36 (1), 6–13.

Yildiz, O., Can, Z., Laghari, A.Q., Şahin, H., Malkoç, M., 2015. Wild edible mushrooms as a natural source of phenolics and antioxidants. J. Food Biochem. 39 (2), 148–154.

Yue, Q.X., Cao, Z.W., Guan, S.H., Liu, X.H., Tao, L., Wu, W.Y., Li, Y.X., Yang, P.Y., Liu, X., Guo, D.A., 2008. Proteomics characterization of the cytotoxicity mechanism of ganoderic acid D and computer-automated estimation of the possible drug target network. Mol. Cell. Proteomics 7 (5), 949–961.

Yue, Q.X., Song, X.Y., Ma, C., Feng, L.X., Guan, S.H., Wu, W.Y., Yang, M., Jiang, B.H., Liu, X., Cui, Y.J., Guo, D.A., 2010. Effects of triterpenes from Ganoderma lucidum on protein expression profile of HeLa cells. Phytomedicine 17 (8), 606–613.

Yuen, W.M.J. (Ed.), 2007. Chemopreventive effects of Ganoderma lucidum on human uroepithelial cell carcinoma. PhD Thesis; The Hong Kong Polytechnic University; pages 282; PolyU Library Call No.: [THS] LG51 .H577P HTI 2007 Yuen; http://hdl.handle.net/10397/2771.

Zeng, Q., Zhou, F., Lei, L., Chen, J., Lu, J., Zhou, J., Cao, K., Gao, L., Xia, F., Ding, S., Huang, L., 2017. Ganoderma lucidum polysaccharides protect fibroblasts against UVB-induced photoaging. Mol. Med. Rep. 15 (1), 111–116.

Zhang, J., Tang, Q., Zhou, C., Jia, W., Da Silva, L., Nguyen, L.D., Reutter, W. and Fan, H., 2010. GLIS, a bioactive proteoglycan fraction from Ganoderma lucidum, displays anti-tumour activity by increasing both humoral and cellular immune response. Life Sci., 87(19), pp.628–637.

Zhang, W., Tao, J., Yang, X., Yang, Z., Zhang, L., Liu, H., Wu, K., Wu, J., 2014a. Antiviral effects of two Ganoderma lucidum triterpenoids against enterovirus 71 infection. Biochem. Biophys. Res. Commun. 449 (3), 307–312.

Zhang, W., Zhang, Q., Deng, W., Li, Y., Xing, G., Shi, X., Du, Y., 2014b. Neuroprotective effect of pretreatment with Ganoderma lucidum in cerebral ischemia/reperfusion injury in rat hippocampus. Neural Regen. Res. 9 (15), 1446–1452.

Zheng, D.S., Chen, L.S., 2017. Triterpenoids from Ganoderma lucidum inhibit the activation of EBV antigens as telomerase inhibitors. Exp. Therapeut. Med. 14 (4), 3273–3278.

Zhonghui, Z., Xiaowei, Z., Fang, F., 2014. Ganoderma lucidum polysaccharides supplementation attenuates exercise-induced oxidative stress in skeletal muscle of mice. Saudi J. Biol. Sci. 21 (2), 119–123.

Zhou, L.W., Cao, Y., Wu, S.H., Vlasák, J., Li, D.W., Li, M.J., Dai, Y.C., 2015. Global diversity of the Ganoderma lucidum complex (Ganodermataceae, Polyporales) inferred from morphology and multilocus phylogeny. Phytochemistry 114, 7–15.

Ziegenbein, F.C., Hanssen, H.P., König, W.A., 2006. Secondary metabolites from Ganoderma lucidum and Spongiporus leucomallellus. Phytochemistry 67 (2), 202–211.

Further Reading

Smina, T.P., Maurya, D.K., Devasagayam, T.P.A., Janardhanan, K.K., 2015. Protection of radiation induced DNA and membrane damages by total triterpenes isolated from Ganoderma lucidum (Fr.) P. Karst. Chem. Biol. Interact. 233, 1–7.

CHAPTER 8

Curcuma sp.: The Nature's Souvenir for High-Altitude Illness

Jigni Mishra*, Anuja Bhardwaj*, Kshipra Misra‡

*Chemistry Division, Department of Biochemical Sciences (DBCS), Defence Institute of Physiology and Allied Sciences (DIPAS), Delhi, India, ‡Department of Biochemical Sciences (DBCS), Defence Institute of Physiology and Allied Sciences (DIPAS), Delhi, India

Abbreviations

μg	microgram
ABTS	2,2′-azino-bis(3-ethylbenzthiazoline-6-sulfonic acid)
AFP	alpha-feto protein
ALT	alanine aminotransferase
AST	aspartate aminotransferase
Bcl	B-cell lymphoma
CD	cluster of differentiation
COX-2	cyclooxygenase-2
DEN	diethyl nitrosamine
DIPAS	Defence Institute of Physiology and Allied Sciences
DMPD	*N,N*-dimethyl-*p*-phenylenediamine dihydrochloride
DNA	deoxyribonucleic acid
DPPH	1,1-diphenyl-2-picryl-hydrazyl
HACE	high altitude cerebral edema
HAPE	high altitude pulmonary edema
HAT	histone acetyltransferases
HBV	hepatitis B virus
HIF-1α	hypoxia inducible factor-1α
HIV	human immunodeficiency virus
HO	heme oxygenase
HSV	herpes simplex virus
IFN	interferon
IgG	immunoglobulin G
ILs	interleukins
INF	interferons
iNOS	inducible nitric oxide synthase
LPS	lipopolysachharide
LTR	long terminal repeat

Management of High Altitude Pathophysiology. https://doi.org/10.1016/B978-0-12-813999-8.00008-2

MDA	malondialdehyde
MIC	minimal inhibitory concentration
ml	millilitre
MMP-9	matrix metalloproteinase-9
NFκ-B	nuclear factor kappa-B
NK	natural killer cell
PEPT 1	peptide transporter 1
PKB	protein kinase B
PKC	protein kinase C
ROS	reactive oxygen species
sp.	species
TNF-α	tumor necrosis factor alpha
VEGF	vascular endothelial growth factor

8.1 INTRODUCTION

Kingdom *Plantae* includes numerous herbal sources that are abundant in medicinally important phytoconstituents. One such pharmacologically important genus of *Plantae* is *Curcuma*, belonging to the family *Zingiberaceae*. Plants belonging to *Curcuma* genus abound in several medicinal values, being antiallergic, anticancer, antidiabetic, anti-inflammatory, antivenom, cardioprotective, digestive stimulant, hepatoprotective, hypolipidemic, and neuroprotective (Chaturvedi et al., 2014). Numerous *Curcuma* species are found predominantly to be native to Asia; however, some species have been exported from their native, tropical countries (India, Indonesia, Myanmar, Thailand) and naturalized in temperate regions (Canada, Mexico, and the United States) (Wohlmuth, 2008), for exploration of their curative values.

This chapter presents a compilation of the chemical composition, occurrence, and potential health benefits of the most essential *Curcuma* sp. distributed across different countries. Bioefficacy of different *Curcuma* sp. is accredited to the various curcuminoids in them. This chapter also aims at elaborating the efficacy of curcuminoids in alleviating a number of health complications, with a special focus on curcumin isolated from *Curcuma longa*.

8.2 GENERAL DESCRIPTION OF *CURCUMA* SPECIES

All the plants included in *Curcuma* sp. are perennial, rhizomatous herbs. The leaves are simple and distichous. Flowers are arranged in inflorescences that arise terminally from the lateral shoot or leafy shoot. Fruits are capsular. Multiple parts of the plants, including flowers, leaves, and rhizomes possess medicinal properties. The rhizomes typically are aromatic with certain species having a characteristic peculiar smell (Apavatjrut et al., 1999). Chemical composition of almost all the *Curcuma* sp. investigated so far customarily comprise curcuminoids and volatile oils.

Curcuma sp. grow natively in tropical climates, requiring substantial humidity for growth, with most of the growth and cultivation of *Curcuma* sp. happens in India, Southeast Asia, and tropical regions of Australia.

> The taxonomical classification of *Curcuma* is Kingdom—Plantae
> Class—Liliopsida
> Subclass—Commelinids
> Order—Zingiberales
> Family—Zingiberaceae
> Genus—*Curcuma*

8.3 DIFFERENT SPECIES OF *CURCUMA* GENUS

The entire *Curcuma* genus comprises more than a hundred different species, distributed across different continents. Thorough research pertaining to pharmacological properties, however, is limited to only a few selected species (Fig. 8.1). Some of these *Curcuma* species and their reported bioactivities are described later.

8.3.1 *Curcuma alismatifolia*

This species of *Curcuma* is commonly known as "Siam tulip." It grows mostly in Bangladesh, Cambodia, Laos, and Thailand. Large, pink bracts are a characteristic feature of this plant. The major classes of biomolecules present are alkaloids, flavonoids, and gums. Leaves of *Curcuma alismatifolia* have been reported to exhibit anti-inflammatory, antioxidant, and wound-healing properties (Akter et al., 2008; Hasan et al., 2009).

8.3.2 *Curcuma amada*

Curcuma amada grows abundantly in India, Indonesia, Malaysia, Northern Australia, and Thailand. The rhizomes are sympodially branched, with circularly arranged scaly rings at the nodes. The rhizomes have a unique raw mango-like flavor, hence its name "mango ginger." *C. amada* contains several classes of biomolecules, such as curcuminoids (curcumin, bisdemethoxy curcumin, and dimethoxy curcumin)|; phenolics (caffeic acid, coumaric acid, ferulic acid, gallic acid, gentistic acid, and syringic acid), and terpenoids (amadaldehyde, amadannulen, and difurocumenonol). The volatile oil of *C. amada* contains azulenogenic oil, car-3-ene, *cis*-ocimene, curcumene, phytosterol, and turmerone. It exhibits antiallergic, analgesic, antiinflammatory, antimicrobial, antioxidant, antipyretic, and laxative properties (Gupta et al., 1999; Jain and Mishra, 1964; Jatoi et al., 2007; Policegoudra et al., 2010; Policegoudra et al., 2011).

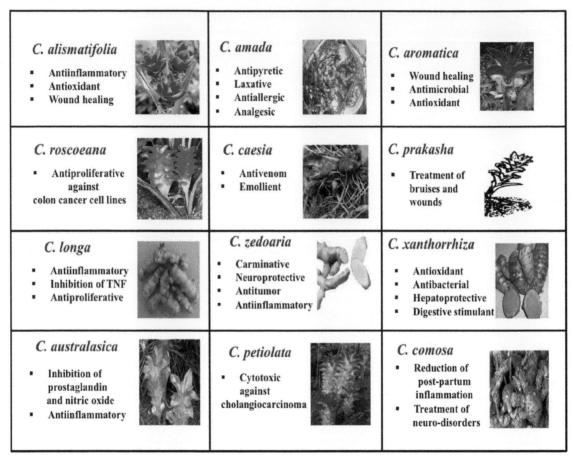

FIG. 8.1
Selected species of *Curcuma* genus and their major bioactivities.

8.3.3   *Curcuma aromatica*

Commonly known as "wild turmeric," *Curcuma aromatica* is widespread in eastern Himalayan regions and Western Ghats of India. The rhizomes, which are its most important parts, remain dormant in winters, while in early spring inflorescence arises from its base. The main active constituents are borneol, camphor, curcumene, curzerenone, and zingiberine. Biological activities of this species pertain largely to antibacterial and antifungal effects (Chattopadhyay et al., 2004; Rachana and Venugopalan, 2014; Santhanam and Nagarajan, 1990).

8.3.4   *Curcuma australasica*

Curcuma australasica is the only species of *Curcuma* genus that is native to the Australian subcontinent. Rhizomes and leaves arising from a pseudostem are known to have medicinal properties. This plant is rich in furanodien-6-one, ketone, sesquiterpene, and zederone. *C. australasica* has been proven to be effective as a contraceptive and in inhibiting nitric oxide synthesis and prostaglandin production (Rajkumari and Sanatombi, 2018; Wohlmuth, 2008).

8.3.5 *Curcuma caesia*

This species of *Curcuma* is mainly localized to Central India, Northeastern India, and West Bengal. Owing to the bluish-black or grayish-black color of its rhizome, *Curcuma caesia* is colloquially termed "kali haldi" or "black turmeric." Essential oil extracted from its rhizomes is rich in borneol, ar-curcumene, ar-turmerone, camphor, curcuminoids, elemane, guinane, β-ocimene, and γ-curcumene. The rhizome oil generally is used as a rubefacient to treat wounds or bites from scorpions and snakes. Local people use paste from the rhizomes to treat various skin diseases (Sarangthem and Haokip, 2010; Rajkumari and Sanatombi, 2018).

8.3.6 *Curcuma comosa*

Curcuma comosa grows mainly in the Southeast Asian countries of India, Indonesia, Malaysia, and Thailand. It has been widely used in Thailand to reduce post-partum bleeding and inflammation (Piyachaturawat et al., 1995). Glucosides present in *C. comosa* exhibit choleretic activity (Suksamran et al., 1997). Rhizome extracts of this plant contain diarylheptanoids, which improve functioning of peptide transporter 1 (PEPT1), and accentuating intestinal absorption of certain drugs (Su et al., 2013; Kawami et al., 2017). The rhizome extracts also exert protective effect against neurodegenerative diseases related to microglial activation (Thampithak et al., 2009).

8.3.7 *Curcuma longa*

C. longa is by far the most explored species of *Curcuma*. The medicinal properties of *C. longa* are above those of its genus counterparts. *C. longa* is predominantly found in India and Southeast Asia. Several phytoconstituents are present in *C. longa*, but the most prominent are curcuminoids, elemenone, germacrone, isolongifolol, turmerone, and zingiberine (Osman et al., 2017). Rhizomes are typically yellow to orange in color. *C. longa* has been widely used in treating inflammatory disorders (Jurenka, 2009). The antioxidant potential of curcuminoids from *C. longa* has been extensively reported (Akram et al., 2010; Jatoi et al., 2007). *C. longa* extracts also have been found to inhibit tumor necrosis factor-alpha (TNF-α) (Yue et al., 2010) and contributes toward antitumor activity against colorectal cancer by inducing apoptosis (Shakibaei et al., 2015). Other pharmacological activities of curcumin, including the most bioactive component of *C. longa*, are elaborated later.

8.3.8 *Curcuma petiolata*

Curcuma petiolata occurs natively in Malaysia and Thailand, where it is more widespread. It is called "jewel of Thailand" because of its dominant use as a medicinal and an ornamental plant by the Thai people. Its rhizome oil consists of curcumol, 2-methyl-5-pentanol and 1H-pyrrol-1-amine, 2-(4-methoxyphenyl)-*n,n*,5-trimethyl. The rhizome oil has been proven to be antioxidant in nature (Thakam and Saewan, 2012). Labdanes from the rhizomes have demonstrated cytotoxic effects against cholangiocarcinoma cell lines (Jittra Suthiwong et al., 2014).

8.3.9 *Curcuma prakasha*

Curcuma prakasha, named in honor of ethnobotanist, Dr. Ved Prakash, has been reported to grow solely in the Garo hills of Meghalaya. Paste from the rhizomes of *C. prakasha* is often used by local people as a treatment for bruises and wounds (Tripathi, 2001).

8.3.10 *Curcuma roscoeana*

This species of *Curcuma* is native to Burma, India, and Malaysia. It is popularly called "orange ginger" because of its characteristic bright-orange colored flowers (Kuehny et al., 2002). It is frequently used in Burma as an ornamental plant, and so is also known as "the pride of Burma." *Curcuma roscoeana* flowers and rhizomes are used locally to treat skin diseases and white spots (Apavatjrut et al., 1999).

8.3.11 *Curcuma xanthorrhiza*

Curcuma xanthorrhiza is widely distributed in the Java island of Indonesia, Malaysia, and Thailand and sparsely occurs in China and India. Its common name in Indonesia is "Java ginger." *C. xanthorrhiza* possesses many different phytoconstituents that give rise to a number of therapeutic activities. For instance, the curcuminoids in *C. xanthorrhiza* elicit antioxidant effects against the autooxidation of linoleic acid (Masuda et al., 1992). Xanthorrhizol has been reported to show antibacterial activities against *Lactobacillus* sp., *Porphyromonas gingivalis*, and *Streptococcus* sp. Xanthorrhizol also acts as a remedial agent for gastrointestinal and constipation-related disorders. Other medicinally important constituents of this plant are curcumene, curzerenone, and turmerone (Hwang et al., 2000). *C. xanthorrhiza* rhizome extracts also have been used in traditional medicines as analgesic and anti-inflammatory agents (Ozaki, 1990).

8.3.12 *Curcuma zedoaria*

Curcuma zedoaria typically grows in India, Indonesia, and Taiwan. This species also is called "white turmeric" because of the white color at the interior of its rhizome. The rhizome oil has been reported to contain antioxidants such as curzerene and epicurzerene (Mau et al., 2003). The tuber extracts of *Curcuma zedoaria* act as carminative, digestive stimulant, and exhibit antimicrobial action against *Aspergillus niger*, *Bacillus subtilis*, *Candida albicans*, and *Klebsiella pneumonia* (Wilson et al., 2005). It also contains curcemenol and sesquiterpene, which possess anti-inflammatory, antitumor, hepatoprotective and neuroprotective effects. These compounds also aid in decreasing lipopolysachharide (LPS)-induced nitric oxide production, thus reducing the level of proinflammatory cytokines (Lo et al., 2015). Curcuzedoalide, found in the rhizomes of *C. zedoaria*, displays antiproliferative action against human gastric cancer cell lines by inciting apoptosis (Jung et al., 2018).

8.4 MEDICINAL IMPLICATIONS OF *CURCUMA* SPECIES

Plant-derived bioactive compounds are preferred as therapeutic agents because they are lack side effects and are nontoxic in action. Both primary and secondary metabolites synthesized by plants demonstrate appreciable bioactivities, some of which are found in traditional medicinal systems and some of which are being explored in current healthcare research. The most beneficial *Curcuma* sp. distributed across several countries, along with their promising bioactivities, are found in Fig. 8.1.

C. longa is the predominant of the species found in the Indian subcontinent and is known as "Indian turmeric." It is revered for its numerous healing properties, as well as for its beauty and culinary purposes. Turmeric has been mentioned in Ayurveda, Siddha, and Unani systems of medicine. It is used as a food preservative and spice in Indian households. Traditional Indian medicine has witnessed the application of turmeric to treat allergies, biliary disorders, diabetes, fever, and high blood pressure (Chattopadhyay et al., 2004).

The nutritional composition of *C. longa* comprises several classes of biomolecules such as carbohydrates, fats, and proteins (Chattopadhyay et al., 2004) are depicted in Fig. 8.2.

Most of the bioactivities in *C. longa* are ascribed to the abundant presence of curcuminoids. The most vital curcuminoids that have been studied for pharmacological relevance are, curcumin and its derivatives, bisdemethoxy curcumin and demethoxy curcumin. The structures of these three curcuminoids are given in Fig. 8.3. Curcuminoids serve as the marker compounds for overall *Curcuma* sp. The curcuminoids are polyphenolic in nature and are responsible for the

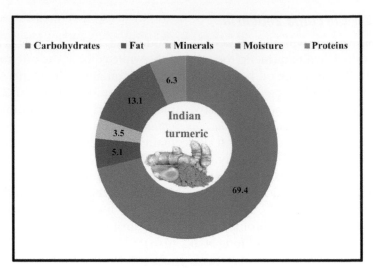

FIG. 8.2

Nutritional composition of Indian turmeric, *Curcuma longa*.

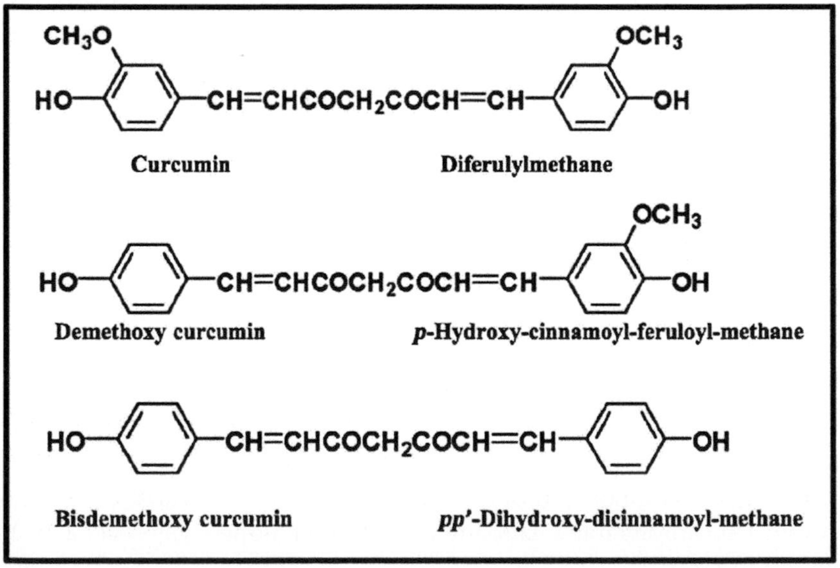

CH₃O and structures...

Curcumin — Diferulylmethane

Demethoxy curcumin — *p*-Hydroxy-cinnamoyl-feruloyl-methane

Bisdemethoxy curcumin — *pp'*-Dihydroxy-dicinnamoyl-methane

FIG. 8.3

Structures of three main curcuminoids—curcumin, demethoxy curcumin, and bisdemethoxy curcumin.

yellowish to orange hue of *Curcuma* rhizomes. Curcumin could exist in both enol and keto forms, with the enol form being more stable (Akram et al., 2010).

For several centuries, from ancient scriptures to modern scientific research, curcumin has been mentioned as an antiinflammatory and antioxidant compound. Curcumin imparts curative action to all the vital organs of human physiology. Recent research has considered the anticancer and apoptotic potential of curcumin against a number of tumor cell lines. Curcumin also has been explored for its ameliorative properties in treating ailments arising from hypobaric hypoxia. Some of the promising medicinal effects of curcumin are described later.

8.4.1 Antioxidant Activity

The antioxidant properties and free radical scavenging potential of curcumin has been widely reported in several studies. Free radical-mediated peroxidation of membrane lipids and oxidative damage of deoxyribonucleic acid (DNA) and proteins are known to aggravate physiological dysfunctions in cases of atherosclerosis, cancer, and neurodegenerative diseases by elevating oxidative stress. Curcumin has been found to substantially downregulate such stress, providing defensive action against oxidative species and their aftereffects (Menon and Sudheer, 2007). The antioxidant effects of curcumin in scavenging free radicals for e.g. 2,2'-azino-bis(3-ethylbenzthiazoline-6-sulfonic acid (ABTS), *N,N*-dimethyl-*p*-phenylenediamine dihydrochloride (DMPD), 1,1-diphenyl-2-picryl-hydrazyl (DPPH), and superoxide anions has been demonstrated (Ak and Gülçin, 2008).

Endothelial heme oxygenase-1 (HO-1) is a stress protein that breaks down heme to biliverdin and carbon monoxide under oxidative stress conditions in vascular endothelial cells. It has been demonstrated that curcumin was able to sufficiently upregulate the activity of HO-1 protein, resulting in improved cytoprotective action to the endothelial cells subjected to oxidative stress (Motterlini et al., 2000).

8.4.2 Antiinflammatory Activity

The anti-inflammatory potential of curcumin and its molecular mechanisms have been studied by a number of researchers. Curcumin is known to reduce inflammation levels by modulating the levels of enzymes such as cyclooxygenase-2 (COX-2) and inducible nitric oxide synthase (iNOS). Proinflammatory cytokines, such as interleukins (ILs), matrix metalloproteinase-9 (MMP-9), and tumor necrosis factor alpha (TNF-α), also are known to be downregulated by curcumin. The role of curcumin in managing inflammation also has been observed to happen via suppression of transcription factors such as nuclear factor kappa-B (NKκ-B) (Abe et al., 1999; Goel et al., 2008; Jurenka, 2009; Kant et al., 2014; Surh et al., 2001).

8.4.3 Antitumor Activity

Curcumin wielded antitumor activity in human leukemia cell lines HL-60 via induction of apoptosis. Apoptosis caused by curcumin has been verified by a decreased level of B-cell lymphoma-2 (Bcl-2) protein and distorted cell morphology as observed under flow cytometer (Kuo et al., 1996). The progress of carcinogenesis was truncated in macrophages activated with curcumin, accompanied by an inhibition of nitric oxide production (Brouet and Ohshima, 1995).

Apart from exerting antitumor effect as a standalone prophylactic, curcumin also has been confirmed to potentiate the efficacy of some commonly used antitumor drugs in situ. For example, the efficiency of gemcitabine, which is a highly regarded drug for treatment of pancreatic cancer, was enhanced when administered in combination with curcumin. This combination treatment successfully curbed angiogenesis and cell proliferation. NKκ-B and its regulated gene products also were reduced (Sandur et al., 2007). Similarly, nanoformulations comprising curcumin enhanced the ameliorative potential of paclitaxel in treating ovarian adenocarcinoma (Ganta et al., 2010). Doxorubicin, co-encapsulated with curcumin, aided in overcoming multidrug resistance to chemotherapeutic drugs (Wang et al., 2015). This kind of coencapsulated drug also was beneficial in appropriate delivery of antitumor drugs to target sites.

8.4.4 Hepatoprotective Activity

Curcumin is capable of exerting appreciable hepatoprotective action; providing defense against various forms of hepatotoxicity. In a previous study, administration of curcumin along with cisplatin showed a marked increase in liver enzyme levels, namely alanine aminotransferase (ALT) and aspartate aminotransferase (AST), and led to the improvement of liver histopathology in hepatocytes of male Wistar rats exposed to oxidative stress (Palipoch et al., 2014). Diethyl nitrosamine (DEN)-induced hepatic tumorigenesis in male albino rats has been found to be cured upon treatment with curcumin, which brought down the serum levels of alpha-feto protein (AFP), ALT, AST, IL-2, IL-6, and malondialdehyde (MDA) (Kadasa et al., 2015). The protective effect of curcumin also has been demonstrated in paracetamol-induced acute liver damage in mice. Here, curcumin normalized the abnormal levels of liver enzymes and reversed centrizonal necrosis and fat deposition in hepatocytes (Girish et al. 2009).

8.4.5 Cardioprotective Activity

The protective role of curcumin against cardiovascular diseases could largely be accredited to its anti-inflammatory activities, which prevent atrial and ventricular arrhythmias. Inhibition of histone acetyltransferases (HAT) by curcumin brings about a restorative action in cardiac cells after hypertrophy and heart failure. Curcumin also plays a major role in maintaining calcium ion homeostasis during adverse cardiovascular conditions (Wongcharoen et al., 2012). It aids in modulating the amounts of HAT, mitogen activated protein kinase (MAPK), protein kinase B (PKB), PKC, and other similar signaling factors that are essential for cardiovascular health (Kapakos et al., 2012). Dyslipidemia, an elevated level of serum lipids, is another risk associated with cardiovascular diseases. Curcumin contributes toward the lowering of these higher serum lipid levels, thus establishing its cardioprotective role.

8.4.6 Immune Enhancing Activity

Certain research has shed light on the immune enhancing potency of curcumin. In addition to the suppressing proinflammatory cytokines such as IL, interferons (IFN), and TNF-α, curcumin was capable of elevating CD8$^+$ T cell population in skin melanoma cancer cells, B16F10 (Lu et al., 2016). A similar increase in CD8$^+$ T cell population was brought about by relatively low doses of curcumin in Lewis lung tumor cell lines-3LL, as studied in vivo mice models (Chidambaram and Krishnasamy, 2014). Curcumin has also proven to accentuate antibody response in rat models, wherein immunoglobulin G (IgG) and natural killer (NK) cell levels were significantly enhanced (South et al., 1997).

8.4.7 Antimicrobial Activity

Literature cites that curcumin has demonstrated appreciable antimicrobial activities against bacteria, fungi, and viruses. In one study, curcumin was able to inhibit 10 different strains of *Staphylococcus aureus* with minimal inhibitory concentration (MIC) values ranging from 125 to 250 μg/mL. When curcumin was combined with commercial antibiotics such as ampicillin, ciprofloxacin, and norfloxacin, the MIC values were markedly reduced suggesting that a combination treatment comprising antibiotics and curcumin was a better approach to treat *S. aureus* infections (Mun et al., 2013). Curcumin also has been reported to inhibit the growth of several other bacterial pathogens, including *B. subtilis*, *Enterococcus faecalis*, *Escherichia coli*, *K. pneumoniae*, and *Pseudomonas aeruginosa*. The MIC values in this case were appreciably low and ranged from 129 to 253 μg/mL (Gunes et al., 2016). Curcumin also has been observed to reduce bacterial contamination by *Bacillus sereus*, *Listeria monocytogenes*, and *Salmonella typhimurium* in dairy products (Hosny et al., 2011).

Curcumin has been proven to inhibit the growth of 19 different strains of *Helicobacter pylori* and reduce gastric and duodenal diseases caused by *H. pylori* (Mahady et al., 2002). Histopathological examinations have verified that curcumin at doses of 5 to 50 µg/mL almost completely removed *H. pylori* infection (De et al., 2009), suggesting that curcumin has substantial chemopreventive potency to treat colon and gastric cancer conditions.

Certain studies also have reported the antibacterial mechanism of action exhibited by curcumin. For instance, suppression of cytokinetic Z-ring formation in *B. subtilis* by curcumin has been adjudged to be the major cause of its inhibition. Inhibition of NKκ-B, and the resultant decrease in IL-8 levels reduced inflammation in gastric cells, are largely responsible for controlling *H. pylori* infection (Zorofchian Moghadamtousi et al., 2014).

Curcumin has been reported to impart strong antiviral action against several pathogenic viral strains, including Hepatitis B virus (HBV), Human immunodeficiency virus-1 (HIV-1), Herpes simplex virus (HSV), and influenza virus. Introduction of curcumin suppressed p53 levels in HBV strains, consequently impeding the replication of HBV. Curcumin had direct inhibitory effects on the activity of long terminal repeat (LTR) regions of HIV-1, in turn, completely halting its transcription. Additionally, curcumin also has been reported to interfere with the functioning of HIV-1 integrases and proteases, in a way controlling the viral growth. Replications of influenza strains H1N1 and H6N1 have been found to be retarded on curcumin supplementation, wherein inhibition of hemagglutinin interaction by curcumin was the causative antiviral mechanism (Kutluay et al., 2008; Zorofchian Moghadamtousi et al., 2014; Zandi et al., 2010).

8.5   CURCUMIN AS A THERAPEUTIC AGAINST HYPOXIA

Hypoxia is an adverse health condition that is triggered by an individual's exposure to a circumstance where the prevailing partial pressure of oxygen is drastically low. When exposed to hypoxia, the negative physiological repercussions might range from moderate (dyspepsia, localized edema, headache, and tachycardia) to extremely severe (acute mountain sickness, cerebral edema, and pulmonary edema) clinical conditions. In this section, the role of curcumin in abating hypoxia caused by various reasons, such as exposure to high altitudes, obesity, and tumor, is discussed.

Hypobaric hypoxia resulting from exposure to high altitudes can cause several health problems, including high altitude cerebral edema (HACE), high-altitude pulmonary edema (HAPE), and increased risk of heart attack (Himadri et al., 2010; Priyanka et al., 2014; Sagi et al., 2014). The Defense Institute of Physiology and Allied Sciences (DIPAS), based in Delhi, India, is actively oriented toward treating physiological problems that are frequently faced by soldiers posted at high altitudes. The following observations pertaining to high altitude-hypobaric hypoxia are outcomes of experimental work that have been carried out by the researchers in DIPAS.

In one study, rats were exposed to a simulated hypobaric hypoxic environment, mimicking a situation similar to altitudes as high as 25,000 ft. The rats experienced a great degree of oxidative stress and inflammation that arose because of HACE and associated transvascular leakage in brain cells. It was reported that introduction of curcumin appreciably lowered reactive oxygen species (ROS) levels and lessened NFκ-B levels, thus restoring cerebral transvascular leakage (Himadri et al., 2010).

In another study, supplementation of curcumin helped to maintain the integrity of tight junction proteins and ion-channel expression in cerebral vasculature in rats enduring HACE (Sarada et al., 2015). In a similar experiment, rats exposed to hypobaric hypoxia in a simulated environment underwent pulmonary transvascular leakage and an increase in oxidative stress markers. Curcumin inhibited excessive fluid efflux from lungs and reduced the amount of hypoxia inducible factor-1α (HIF-1α), all of which establish its therapeutic value against HAPE (Sagi et al., 2014).

Research pertaining to hypoxia arising from obesity is gaining worldwide attention because of its overtly negative effects on adipocytes. In a recent study, curcumin exercised its protective effects on human adipocytes 3T3-L1, by reducing HIF-1α, lipid, and protein oxidation, and ROS levels. An improvement in mitochondrial biogenesis in obese adipocytes was a vital curative effect, which substantiated the action of curcumin in overcoming obesity-induced hypoxia (Priyanka et al., 2014).

In hepatocellular carcinoma cells, HepG2, curcumin influenced a vast decrease in levels of HIF-1α and its associated vascular endothelial growth factor (VEGF). Additionally, curcumin was observed to block angiogenesis in vascular endothelial cells (Bae et al., 2006).

8.6 CONCLUSION

This chapter clearly concludes that curcumin, the main bioactive compound of the *Curcuma* species plays a major therapeutic role in the management of a plethora of diseases, including high-altitude induced hypoxia, thereby establishing its role as a worthy prophylactic agent.

Acknowledgments

The authors would like to express their gratitude to the director, DIPAS, for his constant support and encouragement. One of the authors, Miss Jigni Mishra thanks the Defense Research and Development Organization, India, for her doctoral fellowship.

References

Abe, Y., Hashimoto, S.H.U., Horie, T., 1999. Curcumin inhibition of inflammatory cytokine production by human peripheral blood monocytes and alveolar macrophages. Pharmacol. Res. 39 (1), 41–47.

Ak, T., Gülçin, İ., 2008. Antioxidant and radical scavenging properties of curcumin. Chem. Biol. Interact. 174 (1), 27–37.

Akram, M., Shahab-Uddin, A.A., Usmanghani, K., Hannan, A., Mohiuddin, E., Asif, M., 2010. *Curcuma longa* and curcumin: a review article. Roman. J. Biol. Plant Biol. 55 (2), 65–70.

Akter, R., Hasan, S.R., Siddiqua, S.A., Majumder, M.M., Hossain, M.M., Alam, M.A., Haque, S., Ghani, A., 2008. Evaluation of analgesic and antioxidant potential of the leaves of *Curcuma alismatifolia* Gagnep. Stamford J. Pharm. Sci. 1 (1), 3–9.

Apavatjrut, P., Anuntalabhochai, S., Sirirugsa, P., Alisi, C., 1999. Molecular markers in the identification of some early flowering Curcuma L. (Zingiberaceae) species. Ann. Bot. 84 (4), 529–534.

Bae, M.K., Kim, S.H., Jeong, J.W., Lee, Y.M., Kim, H.S., Kim, S.R., Yun, I., Bae, S.K., Kim, K.W., 2006. Curcumin inhibits hypoxia-induced angiogenesis via down-regulation of HIF-1. Oncol. Rep. 15 (6), 1557–1562.

Brouet, I., Ohshima, H., 1995. Curcumin, an anti-tumor promoter and anti-inflammatory agent, inhibits induction of nitric oxide synthase in activated macrophages. Biochem. Biophys. Res. Commun. 206 (2), 533–540.

Chattopadhyay, I., Biswas, K., Bandyopadhyay, U., Banerjee, R.K., 2004. Turmeric and curcumin: biological actions and medicinal applications. Curr. Sci. 87 (1), 44–53.

Chaturvedi, P., Soundar, S., Parekh, K., Lokhande, S., Chowdhary, A., 2014. Media optimization in immobilized culture to enhance the content of Kaempferol in *Tylophora indica* (Asclepeadaceae) and curcumin in *Curcuma longa* (Zingiberaceae). IOSR J. Pharm. Biol. Sci. 9 (2), 86–90.

Chidambaram, M., Krishnasamy, K., 2014. Nanoparticulate drug delivery system to overcome the limitations of conventional curcumin in the treatment of various cancers: a review. Drug Deliv. Lett. 4 (2), 116–127.

De, R., Kundu, P., Swarnakar, S., Ramamurthy, T., Chowdhury, A., Nair, G.B., Mukhopadhyay, A.K., 2009. Antimicrobial activity of curcumin against *Helicobacter pylori* isolates from India and during infections in mice. Antimicrob. Agents Chemother. 53 (4), 1592–1597.

Ganta, S., Devalapally, H., Amiji, M., 2010. Curcumin enhances oral bioavailability and anti-tumor therapeutic efficacy of paclitaxel upon administration in nanoemulsion formulation. J. Pharm. Sci. 99 (11), 4630–4641.

Girish, C., Koner, B.C., Jayanthi, S., Ramachandra Rao, K., Rajesh, B., Pradhan, S.C., 2009. Hepatoprotective activity of picroliv, curcumin, and ellagic acid compared to silymarin on paracetamol induced liver toxicity in mice. Fundam. Clin. Pharmacol. 23 (6), 735–745.

Goel, A., Kunnumakkara, A.B., Aggarwal, B.B., 2008. Curcumin as "curecumin": from kitchen to clinic. Biochem. Pharmacol. 75 (4), 787–809.

Gunes, H., Gulen, D., Mutlu, R., Gumus, A., Tas, T., Topkaya, A.E., 2016. Antibacterial effects of curcumin: an *in vitro* minimum inhibitory concentration study. Toxicol. Ind. Health 32 (2), 246–250.

Gupta, A.P., Gupta, M.M., Kumar, S., 1999. Simultaneous determination of curcuminoids in *Curcuma* samples using high performance thin layer chromatography. J. Liq. Chromatogr. Relat. Technol. 22 (10), 1561–1569.

Hasan, S.R., Hossain, M.M., Akter, R., Jamila, M., Mazumder, M.E.H., Rahman, S., 2009. DPPH free radical scavenging activity of some Bangladeshi medicinal plants. J. Med. Plants Res. 3 (11), 875–879.

Himadri, P., Kumari, S.S., Chitharanjan, M., Dhananjay, S., 2010. Role of oxidative stress and inflammation in hypoxia-induced cerebral edema: a molecular approach. High Alt. Med. Biol. 11 (3), 231–244.

Hosny, I.M., El Kholy, W.I., Murad, H.A. and El Dairouty, R.K., 2011. Antimicrobial activity of cur-cumin upon pathogenic microorganisms during manufacture and storage of a novel style cheese "Karishcum." J. Am. Sci., 7(5), pp.611–618.

Hwang, J.K., Shim, J.S., Pyun, Y.R., 2000. Antibacterial activity of xanthorrhizol from *Curcuma xanthorrhiza* against oral pathogens. Fitoterapia 71 (3), 321–323.

Jain, M.K., Mishra, R.K., 1964. Chemical examination of *Curcuma amada* Roxb. Indian J. Chem. 2 (1), 39.

Jatoi, S.A., Kikuchi, A., Gilani, S.A., Watanabe, K.N., 2007. Phytochemical, pharmacological, and ethnobotanical studies in mango ginger (*Curcuma amada* Roxb.; Zingiberaceae). Phytother. Res. 21 (6), 507–516.

Jung, E.B., Trinh, T.A., Lee, T.K., Yamabe, N., Kang, K.S., Song, J.H., Choi, S., Lee, S., Jang, T.S., Kim, K.H. and Hwang, G.S., 2018. Curcuzedoalide contributes to the cytotoxicity of *Curcuma zedoaria* rhizomes against human gastric cancer AGS cells through induction of apoptosis. J. Ethnopharmacol., 213, pp.48–55. https://doi.org/10.1016/j.jep.2017.10.025

Jurenka, J.S., 2009. Anti-inflammatory properties of curcumin, a major constituent of *Curcuma longa*: a review of preclinical and clinical research. Altern. Med. Rev. 14 (2), 141–153.

Kadasa, N.M., Abdallah, H., Afifi, M., Gowayed, S., 2015. Hepatoprotective effects of curcumin against diethyl nitrosamine induced hepatotoxicity in albino rats. Asian Pac. J. Cancer Prev. 16 (1), 103–108.

Kant, V., Gopal, A., Pathak, N.N., Kumar, P., Tandan, S.K., Kumar, D., 2014. Antioxidant and anti-inflammatory potential of curcumin accelerated the cutaneous wound healing in streptozotocin-induced diabetic rats. Int. Immunopharmacol. 20 (2), 322–330.

Kapakos, G., Youreva, V., Srivastava, A.K., 2012. Cardiovascular protection by curcumin: molecular aspects. Indian J. Biochem. Biophys. 49, 306–315.

Kawami, M., Yamada, Y., Toshimori, F., Issarachot, O., Junyaprasert, V.B., Yumoto, R., Takano, M., 2017. Effect of *Curcuma comosa* extracts on the functions of peptide transporter and P-glycoprotein in intestinal epithelial cells. Die Pharm 72 (2), 123–127.

Kuehny, J.S., Sarmiento, M.J., Branch, P.C., 2002. Cultural studies in ornamental ginger. In: Janick, J., Whipkey, A. (Eds.), Trends in New Crops and New Uses. ASHS Press, VA, pp. 477–482.

Kuo, M.L., Huang, T.S., Lin, J.K., 1996. Curcumin, an antioxidant and anti-tumor promoter, induces apoptosis in human leukemia cells. Biochim. Biophys. Acta Mol. Basis Dis. 1317 (2), 95–100.

Kutluay, S.B., Doroghazi, J., Roemer, M.E., Triezenberg, S.J., 2008. Curcumin inhibits herpes sim-plex virus immediate-early gene expression by a mechanism independent of p300/CBP histone acetyltransferase activity. Virology 373 (2), 239–247.

Lo, J.Y., Kamarudin, M.N.A., Hamdi, O.A.A., Awang, K., Kadir, H.A., 2015. Curcumenol isolated from *Curcuma zedoaria* suppresses Akt-mediated NF-κB activation and p38 MAPK signaling pathway in LPS-stimulated BV-2 microglial cells. Food Funct. 6 (11), 3550–3559.

Lu, Y., Miao, L., Wang, Y., Xu, Z., Zhao, Y., Shen, Y., Xiang, G., Huang, L., 2016. Curcumin micelles remodel tumor microenvironment and enhance vaccine activity in an advanced melanoma model. Mol. Ther. 24 (2), 364–374.

Mahady, G.B., Pendland, S.L., Yun, G., Lu, Z.Z., 2002. Turmeric (*Curcuma longa*) and curcumin inhibit the growth of *Helicobacter pylori*, a group 1 carcinogen. Anticancer Res. 22 (6C), 4179–4181.

Masuda, T., Isobe, J., Jitoe, A., Nakatani, N., 1992. Antioxidative curcuminoids from rhizomes of *Curcuma xanthorrhiza*. Phytochemistry 31 (10), 3645–3647.

Mau, J.L., Lai, E.Y., Wang, N.P., Chen, C.C., Chang, C.H., Chyau, C.C., 2003. Composition and anti-oxidant activity of the essential oil from *Curcuma zedoaria*. Food Chem. 82 (4), 583–591.

Menon, V.P., Sudheer, A.R, 2007. Antioxidant and anti-inflammatory properties of curcumin. In: Aggarwal, B.B., Surh, Y.J., Shishodia, S (Eds.), The Molecular Targets and Therapeutic Uses of Curcumin in Health and Disease. Springer, MA, pp. 105–125.

Motterlini, R., Foresti, R., Bassi, R., Green, C.J., 2000. Curcumin, an antioxidant and anti-inflammatory agent, induces heme oxygenase-1 and protects endothelial cells against oxidative stress. Free Radic. Biol. Med. 28 (8), 1303–1312.

Mun, S.H., Joung, D.K., Kim, Y.S., Kang, O.H., Kim, S.B., Seo, Y.S., Kim, Y.C., Lee, D.S., Shin, D.W., Kweon, K.T., Kwon, D.Y., 2013. Synergistic antibacterial effect of curcumin against methicillin-resistant *Staphylococcus aureus*. Phytomedicine 20 (8–9), 714–718.

Osman, A.H., El-Far, A.H., Sadek, K.M., Abo-Ghanema, I.I., Abdel-Latif, M.A., 2017. Immunity, antioxidant status, and performance of broiler chickens fed turmeric (*Curcuma Longa*) rhizome powder. Alex. J. Vet. Sci. 54 (2), 19–28.

Ozaki, Y., 1990. Anti-inflammatory effect of *Curcuma xanthorrhiza* ROXB and its active principles. Chem. Pharm. Bull. 38 (4), 1045–1048.

Palipoch, S., Punsawad, C., Koomhin, P., Suwannalert, P., 2014. Hepatoprotective effect of curcumin and alpha-tocopherol against cisplatin-induced oxidative stress. BMC Complem. Altern. Med. 14 (1), 111.

Piyachaturawat, P., Ercharuporn, S., Suksamrarn, A., 1995. Uterotrophic effect of *Curcuma comosa* in rats. Int. J. Pharmacogn. 33 (4), 334–338.

Policegoudra, R.S., Rehna, K., Rao, L.J., Aradhya, S.M., 2010. Antimicrobial, antioxidant, cytotoxicity and platelet aggregation inhibitory activity of a novel molecule isolated and characterized from mango ginger (Curcuma amada Roxb.) rhizome. J. Biosci. 35 (2), 231–240.

Policegoudra, R.S., Aradhya, S.M., Singh, L., 2011. Mango ginger (*Curcuma amada* Roxb.): a promising spice for phytochemicals and biological activities. J. Biosci. 36 (4), 739–748.

Priyanka, A., Anusree, S.S., Nisha, V.M., Raghu, K.G., 2014. Curcumin improves hypoxia-induced dysfunctions in 3T3-L1 adipocytes by protecting mitochondria and down regulating inflammation. Biofactors 40 (5), 513–523.

Rachana, S., Venugopalan, P., 2014. Antioxidant and bactericidal activity of wild turmeric extracts. J. Pharmacogn. Phytochem. 2 (6), 89–94.

Rajkumari, S. and Sanatombi, K., 2018. Nutritional value, phytochemical composition, and biological activities of edible *Curcuma* species: a review. Int. J. Food Prop., pp.1–20. https://doi.org/10.1080/10942912.2017.1387556.

Sagi, S.S.K., Matthew, T., Patir, H., 2014. Prophylactic administration of curcumin abates the incidence of hypobaric hypoxia induced pulmonary edema in rats: a molecular approach. J. Pulm. Respir. Med. 4 (1), 1–12.

Sandur, S.K., Ichikawa, H., Pandey, M.K., Kunnumakkara, A.B., Sung, B., Sethi, G., Aggarwal, B.B., 2007. Role of pro-oxidants and antioxidants in the anti-inflammatory and apoptotic effects of curcumin (diferuloylmethane). Free Radic. Biol. Med. 43 (4), 568–580.

Santhanam, G., Nagarajan, S., 1990. Wound healing activity of *Curcuma aromatica* and *Piper betle*. Fitoterapia 61 (5), 458–459.

Sarada, S.K.S., Titto, M., Himadri, P., Saumya, S., Vijayalakshmi, V., 2015. Curcumin prophylaxis mitigates the incidence of hypobaric hypoxia-induced altered ion channels expression and impaired tight junction proteins integrity in rat brain. J. Neuroinflamm. 12 (1), 113.

Sarangthem, K., Haokip, M.J., 2010. Bioactive components in *Curcuma caesia* Roxb. grown in Manipur. Bioscan 5, 113–115.

Shakibaei, M., Kraehe, P., Popper, B., Shayan, P., Goel, A., Buhrmann, C., 2015. Curcumin potentiates antitumor activity of 5-fluorouracil in a 3D alginate tumor microenvironment of colorectal cancer. BMC Cancer 15 (1), 250.

South, E.H., Exon, J.H., Hendrix, K., 1997. Dietary curcumin enhances antibody response in rats. Immunopharmacol. Immunotoxicol. 19 (1), 105–119.

Su, J., Sripanidkulchai, K., Hu, Y., Chaiittianan, R., Sripanidkulchai, B., 2013. Increased in situ intestinal absorption of phytoestrogenic diarylheptanoids from *Curcuma comosa* in nanoemulsions. AAPS PharmSciTech. 14 (3), 1055–1062.

Suksamran, A., Eiamon, S., Piyachaturawat, P., Byrnes, L.T., 1997. A phloracetaphenone glucoside with choleretic activity from *Curcuma comosa*. Phytochemistry 145 (1), 103–105.

Surh, Y.J., Chun, K.S., Cha, H.H., Han, S.S., Keum, Y.S., Park, K.K., Lee, S.S., 2001. Molecular mechanisms underlying chemopreventive activities of anti-inflammatory phytochemicals: down-regulation of COX-2 and iNOS through suppression of NF-κB activation. Mut. Res./Fundam. Mol. Mech. Mutagen. 480, 243–268.

Suthiwong, J., Pitchuanchom, S., Wattanawongdon, W., Hahnvajanawong, C., Yenjai, C., 2014. Isolation, docking, and cytotoxicity evaluation against cholangiocarcinoma of labdanes from *Curcuma petiolata*. Asian J. Chem. 26 (14), 4286.

Thakam, A., Saewan, N., 2012. Chemical composition of essential oil and antioxidant activities of *Curcuma petiolata* Roxb. Rhizomes. In: Tunkasiri, T. (Ed.), Advanced Materials Research. In: 506, Trans Tech Publications, Switzerland, pp. 393–396.

Thampithak, A., Jaisin, Y., Meesarapee, B., Chongthammakun, S., Piyachaturawat, P., Govitrapong, P., Supavilai, P., Sanvarinda, Y., 2009. Transcriptional regulation of iNOS and COX-2 by a novel compound from *Curcuma comosa* in lipopolysaccharide-induced microglial activation. Neurosci. Lett. 462 (2), 171–175.

Tripathi, S., 2001. *Curcuma prakasha* sp. nov.(Zingiberaceae) from Northeastern India. Nord. J. Bot. 21 (5), 549–550.

Wang, J., Ma, W., Tu, P., 2015. Synergistically improved anti-tumor efficacy by co-delivery doxorubicin and curcumin polymeric micelles. Macromol. Biosci. 15 (9), 1252–1261.

Wilson, B., Abraham, G., Manju, V.S., Mathew, M., Vimala, B., Sundaresan, S., Nambisan, B., 2005. Antimicrobial activity of *Curcuma zedoaria* and *Curcuma malabarica* tubers. J. Ethnopharmacol. 99 (1), 147–151.

Wohlmuth, H. 2008, "Phytochemistry and pharmacology of plants from the ginger family, Zingiberaceae," PhD Thesis, Southern Cross University, Lismore, NSW.

Wongcharoen, W., Jai-Aue, S., Phrommintikul, A., Nawarawong, W., Woragidpoonpol, S., Tepsuwan, T., Sukonthasarn, A., Apaijai, N., Chattipakorn, N., 2012. Effects of curcuminoids on frequency of acute myocardial infarction after coronary artery bypass grafting. Am. J. Cardiol. 110 (1), 40–44.

Yue, G.G., Chan, B.C., Hon, P.M., Lee, M.Y., Fung, K.P., Leung, P.C., Lau, C.B., 2010. Evaluation of in vitro anti-proliferative and immunomodulatory activities of compounds isolated from *Curcuma longa*. Food Chem. Toxicol. 48 (8–9), 2011–2020.

Zandi, K., Ramedani, E., Mohammadi, K., Tajbakhsh, S., Deilami, I., Rastian, Z., Fouladvand, M., Yousefi, F., Farshadpour, F., 2010. Evaluation of antiviral activities of curcumin derivatives against HSV-1 in Vero cell line. Nat. Prod. Commun. 5 (12), 1935–1938.

Zorofchian Moghadamtousi, S., Abdul Kadir, H., Hassandarvish, P., Tajik, H., Abubakar, S. and Zandi, K., 2014. A review on antibacterial, antiviral, and antifungal activity of curcumin. Biomed. Res. Int., 2014. https://doi.org/10.1155%2F2014%2F186864.

CHAPTER 9

Characterization Techniques for Herbal Products

Rakhee*, Jigni Mishra†, Raj K. Sharma*, Kshipra Misra‡

*Department of Chemistry, University of Delhi, Delhi, India, †Chemistry Division, Department of Biochemical Sciences (DBCS), Defence Institute of Physiology and Allied Sciences (DIPAS), Delhi, India, ‡Department of Biochemical Sciences (DBCS), Defence Institute of Physiology and Allied Sciences (DIPAS), Delhi, India

Abbreviations

^{13}C NMR	Carbon-13 nuclear magnetic resonance
^{1}H NMR	proton nuclear magnetic resonance
^{1}H NMR	proton nuclear magnetic resonance
2D NMR	two-dimensional nuclear magnetic resonance spectroscopy
ABTS$^+$	2,2′-azino-bis (3-ethylbenzothiazoline-6-sulphonic acid)
API	atmospheric pressure ionization
ASE	accelerated solvent extraction
BMOEA bst	bis (2 methoxyethyl) ammonium bis (tri-uoromethylsulfonyl) imide
CE-MS	capillary electrophoresis-mass spectrometry
d^6-DMSO	hexadeuterodimethyl sulfoxide
DEPT-20	distortionless enhancement by polarization transfer-20 type ^{13}C NMR
DMAE	dynamic microwave-assisted extraction
DMEA oct	*N,N*-dimethylethanolammonium octanoate
DNA	deoxyribonucleic acid
DOSY	diffusion ordered-NMR spectroscopy
DPPH	2,2-dipheyl-1-picryhydrazyl
EOF	electroosmotic flow
ESI-MS	electrospray ionization-mass spectrometry
ESI-MS/MS	electrospray ionization-mass spectrometry/mass spectrometry
FC	flash chromatography
FRAP	ferric-reducing antioxidant power
FT-IR	Fourier transform-infrared spectroscopy
GC	gas chromatography
GC-FTIR	gas chromatography-Fourier transform infrared spectroscopy
GC-MS	gas chromatography-mass spectrometry
H1-H1	proton-proton coupled nuclear magnetic resonance
HILIC	hydrophilic interaction liquid chromatography
HPLC	high pressure liquid chromatography
HPLC-^{1}H NMR	high-pressure liquid chromatography-proton-nuclear magnetic resonance

Management of High Altitude Pathophysiology. https://doi.org/10.1016/B978-0-12-813999-8.00009-4

HPLC-DAD/UV	high-performance liquid chromatography-diode array detector/ultraviolet spectroscopy
HPLC-DAD-ESI-MS/MS	high-performance liquid chromatography-diode array detector-electron spray ionization-mass spectroscopy/mass spectroscopy
HPLC-IR	high-pressure liquid chromatography-infrared spectroscopy
HPLC-NMR-MS	high-pressure liquid chromatography-nuclear magnetic resonance-ultraviolet spectroscopy
HPLC-UV	high-pressure liquid chromatography-ultraviolet spectroscopy
HPLC-UV-IR	high-pressure liquid chromatography-ultraviolet spectroscopy-infrared spectroscopy
HPTLC	high-performance thin layer chromatography
IR	infrared spectroscopy
JRES	J-resolved spectroscopy
KBr	potassium bromide
LAMP	loop-mediated isothermal amplification
LC-FTIR	liquid chromatography-Fourier transform infrared spectroscopy
LC-MS	liquid chromatography-mass spectroscopy
LC-MS-MS	liquid chromatography-mass spectroscopy-mass spectroscopy
LC-NMR-MS	liquid chromatography-nuclear magnetic resonance-mass spectroscopy
LC-PDA-MS	liquid chromatography-photo diode array-mass spectrometry
LC-PDA-NMR-MS	liquid chromatography-photo diode array-nuclear magnetic resonance-mass spectrometry
LC-TSP-MS	liquid chromatography-thermospray-mass spectrometry
LC-UV-MS	liquid chromatography-ultraviolet-mass spectrometry
LPLC	low-pressure liquid chromatography
LVI-GC-MS	large volume injection-gas chromatography-mass spectrometry
MAE	microwave-assisted extraction
MALDI-TOF/TOF	matrix-assisted laser desorption/ionization-time of flight mass spectrometer
MDA	malondialdehyde
MS	mass spectrometry
Multiplex PCR	multiplex polymerase chain reaction
NMR	nuclear magnetic resonance
NPMAE	nitrogen-protected microwave-assisted extraction
PCR	polymerase chain reaction
PPC	preparative planar chromatographic
ROD	silica rod
RP-HPLC	reversed-phase high-performance liquid chromatography
SARP	several ankyrin repeat protein
Semi-prep-HPLC	semi-preparative high-performance liquid chromatography
SFE	supercritical fluid extraction
SOD	superoxide dismutase
SPE-LC-MS	solid-phase extraction-liquid chromatography-tandem mass spectrometry
SSCP-PCR	single-strand conformation polymorphism-polymerase chain reaction
TLC	thin-layer chromatography
TMS	tetra methyl silane
TSP	thermospray soft ionization technique
UAE	ultrasound-assisted extraction
UMAE	ultrasonic microwave-assisted extraction
UV	ultraviolet spectroscopy

9.1 INTRODUCTION

Despite extensive development in the field of extraction and characterization of medicinally valued and pharmacologically active natural products from various sources, there is a need to consolidate all the available characterization data in one place. Several extraction methods using different solvents, ranging from low to high polarity or vice versa, have been used to isolate pure bioactive compounds (Adou, 2005). Separation and identification of the bioactive ingredients of the herbs have been carried out by both conventional techniques, including qualitative methods such as proximate and phytochemical analysis, and nonconventional techniques, including GC, HPLC, LC, NMR, UPLC, GC-MS, HPLC-MS, HPLC-SPE-NMR, HPLC-UV-DAD, HPTLC-MS, UP-HPLC, and UPLC-DAD-TOF-MS. (Bucar et al., 2013; Chauhan, 2014; Godevac et al., 2015; Kang, 2012; Mahrous and Farag, 2015; Yan et al., 2007). This chapter majorly focuses on the details underlying these techniques, with special reference to their applications in characterizing herbs growing at high altitudes. Such herbal sources, including *Cordyceps sinensis*, *Ganoderma lucidum*, *Hippophae* sp., *Rhodiola rosea*, and *Valeriana wallichii*, are rich in bioactive metabolites that impart endurance to the extreme climatic conditions (high pressure, low oxygen, and low temperature) prevailing at HA (Bhardwaj et al., 2016; Geetha et al., 2005; Rakhee et al., 2016, 2017; Sharma et al., 2012). The most prominent metabolites in these herbal sources are flavonoids, lipids, phenolics, polysaccharides, proteins, saponins, sugars, and tannins. (Azmir et al., 2013). The structures of some of the most commonly occurring phytoconstituents are depicted in Fig. 9.1.

9.2 CHARACTERISTICS OF HA HIMALAYAN MEDICINAL PLANTS

The HA Himalayan regions, until recently, were less explored because of difficulties with accessibility and harsh climatic conditions. In the last few years, however, HA medicinal plants and herbs have gained substantial interest among researchers because of their unique and rich content of numerous bioactive constituents (Theis and Lerdau, 2003). Considering the abundance of medicinally important pharmacological compounds present in HA herbs, around 70%–80% of the world's population depend on drug preparations from such herbal sources to treat several health problems.

Some of these pharmacological constituents along with their bioactivities are summarized in Table 9.1.

Some of the medicinal plants and herbs, such as *Aconitum heterophyllum*, *Acorus calamus*, *Asparagus racemose*, *Atropa belladonna Linn.*, *Bergenia ligulate*, *Cinchona*

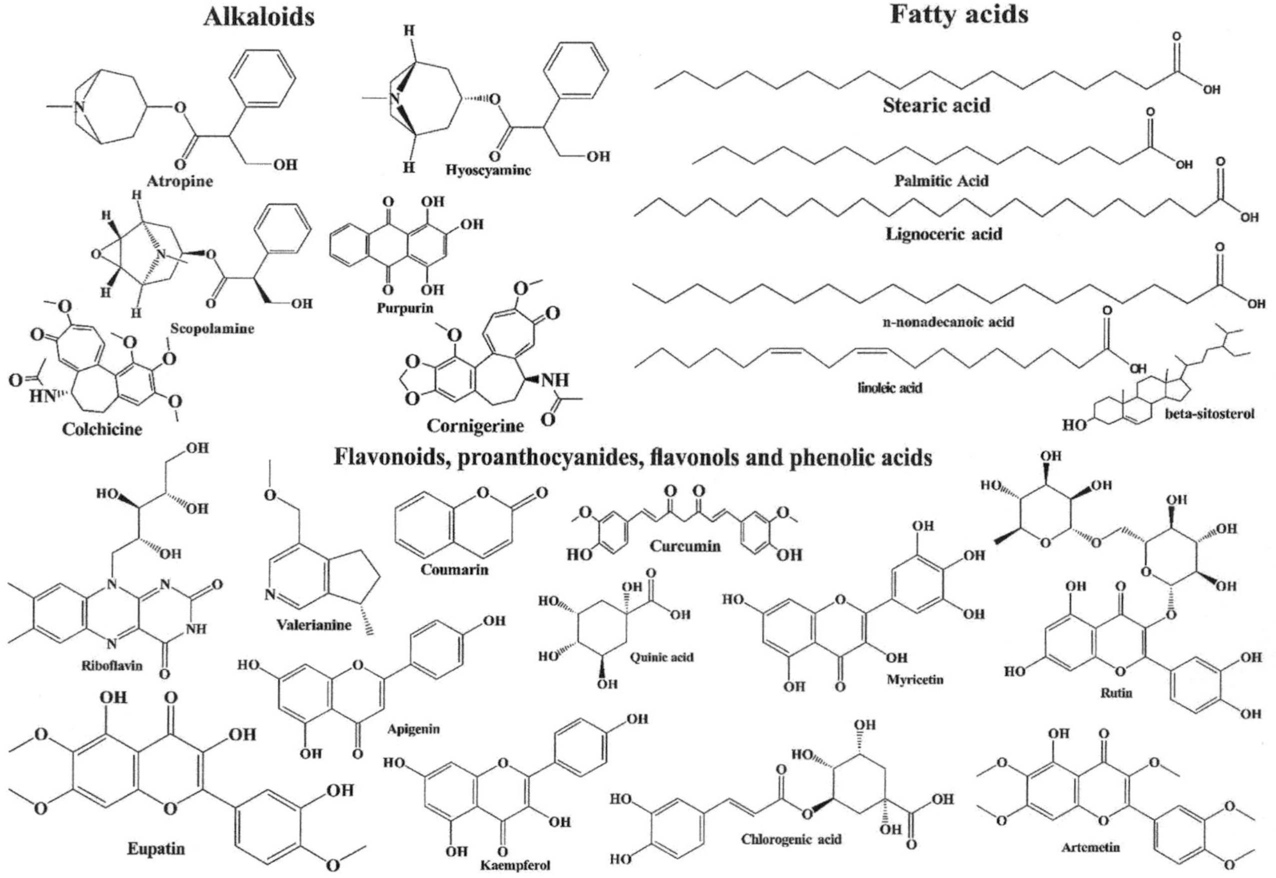

FIG. 9.1

Major classes of bioactive constituents found in medicinal plants.

(Continued)

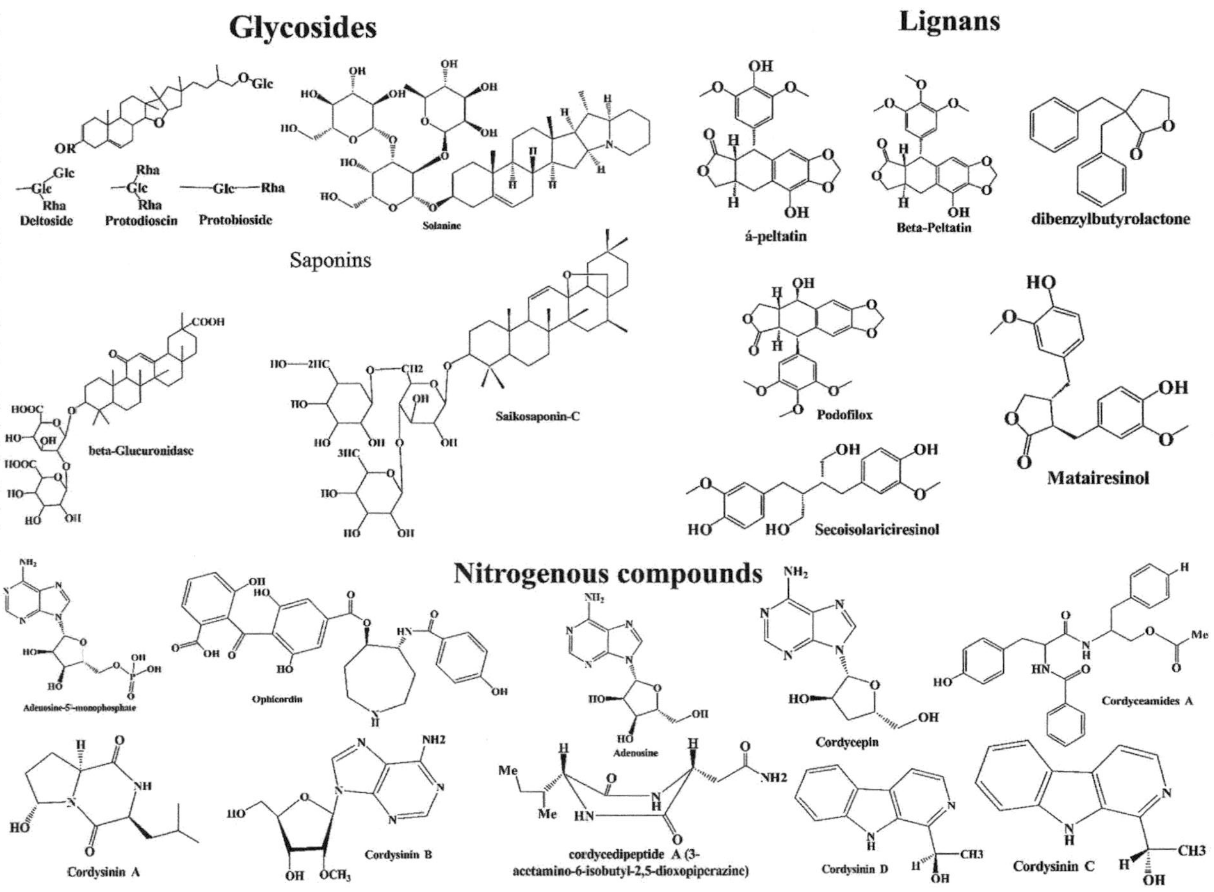

Lignans

dibenzylbutyrolactone

Matairesinol

Beta-Peltatin

á-peltatin

Podoflox

Secoisolariciresinol

Glycosides

Solanine

Protodioscin

Protobioside

Rha

Glc

Rha

Glc — Rha

O-Glc

Deltoside

Glc

Rha

OR

Saponins

Saikosaponin-C

CH2

beta-Glucuronidase

COOH

HOOC

HO

Nitrogenous compounds

Cordyceamides A

Cordysinin C

CH3

OH

Cordycepin

Cordysinin D

CH3

OH

Adenosine

NH2

cordycedipeptide A (3-acetamino-6-isobutyl-2,5-dioxopiperazine)

NH2

Me

Me

Ophicordin

OH

HO

OH

Cordysinin B

NH2

OCH3

OH

HO

Adenosine-5'-monophosphate

NH2

OH

Cordysinin A

NH

HO

H

O

(Continued)

FIG. 9.1—cont'd

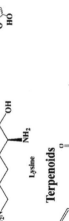
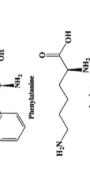
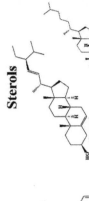

Resins

(+)-Pinoresinol

(+)-Pinoresinol 4-glucoside

Protein, peptide and amino acid

Isoleucine

Threonine

Phenylalanine

Lysine

Polysaccharide and sugar derivatives

Glucose

Cordycepic acid

Trehalose

Sterols

stigmasterol

Cholesteryl palmitate

Ergosterol

Ergosterol peroxide

Terpenoids

Myrtenyl acetate

Germacrene-D

P-cymene

1H-Cycloprop(e)azulen-7-ol

Caryophyllene

Valeranic acid

ursolic acid

Ganoderic acid A

Tannins

Acertannin

Hamamelitannin

penta-o-galloyl-beta-D-glucose

FIG. 9.1—cont'd

Table 9.1 High-Altitude Medicinal Plants With Their Active Components, Distribution, and Therapeutic Activities

Medicinal Plants	Images	Active Components	Distribution	Therapeutic Activities	References
Aconitum heterophyllum		Aconitine, benzoylmesaconine, mesaconitine, hypaconitine, hetidine, atidine	Uttaranchal, Himachal Pradesh, Sikkim (common in the subalpine and alpine zones of Himalayas at altitudes between 1800 and 4500)	Antiperiodic, anodyne, antidiabetic, antipyretic, narcotic, powerful sedative	Kaul (1997); http://senthuherbals.blogspot.in/2015/03/aconitum-heterophyllumadhividayam.html?view=snapshot
Acorus calamus		Myristic, palmitic, stearic, oleic, linoleic, arachidic, asarone, beta-asarone, (−)-4-terpineol, 2-allyl-5-ethoxy-4-methoxyphenol, epieudesmin, lysidine, (−)-spathulenol, borneol, furyl ethyl ketone, nonanoic acid, 2,2,5,5-tetramethyl-3-hexanol, bornyl acetate	Uttaranchal, Himachal Pradesh (found throughout the Indian Himalayas up to 6000 m and in Sri Lanka)	Intellect promoting, thermogenic, alexeteric	Balakumbahan et al. (2010)
Artemisia annua Linn.		Coumarin, apigenin, artemetin, cllorogenic acid, quinic acid, coumaric acid, myricetin, kaempferol, rutin, isorhanetin, eupatin, axillarin	Himachal Pradesh	Antimalarial, antibacterial, anticancer, antioxidant activity, antiasthma	Gupta et al. (2009), Chen et al. (2003), Kale et al. (2008); https://en.wikipedia.org/wiki/Artemisia_annua#cite_note-39

Continued

Table 9.1 High-Altitude Medicinal Plants With Their Active Components, Distribution, and Therapeutic Activities—cont'd

Medicinal Plants	Images	Active Components	Distribution	Therapeutic Activities	References
Asparagus racemosa		Steroidal saponins, shatavaroside A, shatavaroside B, filiasparoside C, shatavarins, immunoside, 8-methoxy-5,6,4′-trihydroxyisoflavone 7-O-β-ᴅ-glucopyranoside	Arunachal Pradesh, Meghalaya	Antiseptic, dysentry, diuretic	Hayes et al. (2008), Saxena and Chourasia (200⁻); https://en.wikipedia.org/wiki/Asparagus_racemosus
Atropa belladona Linn.		Atropine, l-hyoscyamine, l-hyoscine, 1-arginine, l-ornithine, scopolamine, hyoscyamine	Assam (District of Goalpara)	Antiinflammatory, anticholinergic/antispasmodic action, mild sedative	Mathews et al. (1990); Kaul (2010); Basumatary et al. (2004); https://en.wikipedia.org/wiki/Atropa_belladonna
Bergenia ligulata		Bergenin, tannic acid, gallic acid, stigmesterol, P-sitosterol, catechin, (+)-afzelechin, 1, 8-cineole, isovalaric acid, (+)-(6S)-parasorbic acid, arbutin, phytol, caryophyllene, damascenone, β-eudesmol, 3-methyl-2-buten-l-ol, (Z)-asarone, terpinen-4-ol, paashaanolactone	Uttaranchal, Arunachal Pradesh, Meghalaya, Himachal Pradesh (found throughout Himalayas, Kashmir and Khasi Hills of Assam)	Pulmonary affections, dysentery, antiinflammatory, antidiuretic, eye disease, cut and burns, fever, antidiabetic, astringent, cardiotonic, wound healer, antipyretic, hepatoprotective, anticancer, antiprotozoal, antiinflammatory	https://www.bimbima.com/ayurveda/pashanbhedabergenia-ligulata-a-great-herb-to-dissolve-kidney-stone-more/392/; Verma et al. (2014)

Cinchona ledgeriana		Quinine, anthraquinones, purpurin, anthragallol-1,2-dimethylether, anthragallol-1,3-dimethylether, rubiadin, 1-hydroxy-2-hydroxymethylanthr aquinone, 1-hydroxy-2-methylan thraquinone, morindone-5-methylether, anthraquinones	Meghalaya	Antispasmodic, astrigent, antibacterial, antimalarial	Maehara et al. (2013); https://books.google.co.in/ books? id=g2Pt45ET.JNnkC& lpg=PA297&ots= bv8FvfuJ3i&dq= anthraquinones%20%20in %20natural%20 Cinchona% 20Iedgeriana&pg= PA297#v=onepage&q= anthraquinones%20%20in %20natural%20Cinchona % 20Iedgeriana&f=false; Wijnsma et al. (1985)
Colchicum luteum Baker		Colchicine, cornigerine, purpurin, anthragallol-1,2-dimethylether, anthragallol-1,3-dimethylether, rubiadin, 1-hydroxy-2-hydroxymethylan thraquinone, 1-hydroxy-2-methylanthraquinone	Ladakh, Kargil, Kashmir to Chamba (Suru valley)	Antigout, stomachache	Singh and Chaurasia (1998); http://www. catalogueoflife.org/annual- checklist/2014/ details/ species/id/9766304

Continued

Table 9.1 High-Altitude Medicinal Plants With Their Active Components, Distribution, and Therapeutic Activities—cont'd

Medicinal Plants	Images	Active Components	Distribution	Therapeutic Activities	References
Cordyceps sp.		Cordycepin, adenosine, cytosine, thymine, adenine, guanine, uracil, hypoxanthine, uridine-5-mono-phosphate, adenosine-5-monophosphate, guanosine-5-monophosphate	Arunachal Pradesh, Himachal Pradisesh, Uttarakhnnd, Sikkim	Antioxidant, antitumor, anti-inflammatory, anticancer activity, immunomodulatory, antifibrotic, antidiabetic, antiarteriosclerosis, radioprotective activity antithrombotic activity	Rakhee et al. (2016); Mamta et al. (2015)
Curcuma sp.		Curcumin, curcuminoids, 4-hydroxybenzaldehyde	Tripura (Jampui Hill, Suryamaninagar, Baramura hill)	Anticancer activity, cytotoxicity, Bis-demethoxycurcumin	Tripathi et al. (2002), Saha et al. (2016), Kuttan et al. (1985), Nalli et al. (2017)
Ganoderma sp.		Ganoderic acids, riboflavin, ascorbic acid, mannitol, mannitol-alpha, alpha-trehalose, stearic acid, palmitic acid, lignoceric acid, n-nonadecanoic acid, behenic acid, tetracosanol, hentriacontane, choline, lycine	Thrissur District, Kerala, Eastern Himalayas, Kashmir valley, Garhwal, South East Maharashtra	Antimutagenic, anticarcinogenic, antiperoxidative activity, antiinflammatory activity	Bhardwaj et al. (2016), Lakshmi et al. (2003), Jones and Kinghorn (2012)

Species		Constituents	Distribution	Properties/Uses	References
Hippophae sp.		Threonine, valine, methionine, leucine, lysine, trytophan, isoleucine, phenylalanine, ethyl 2-methylbutanoate, ethyl 3-methyl butanoate, alpha-tocopherol	North-eastern Himalayas of India	Cytoprotective, antistress, immunomodulatory, hepatoprotective, radioprotective, antiatherogenic, antitumor, antimicrobial, antiulcerogenic, tissue regeneration	Gao et al. (2000), Geetha et al. (2003)
Nardostachys jatamansi		E-2-methyl, 3-(5,9 dimethylbicyclo[4.3.0] nonen-9-yl)-2-propenoic acid, 2′,2′-dimethyl-3′-methoxy-3′,4′-dihydropyranocoumarin	Uttaranchal, Himachal Pradesh Arunachal Pradesh, Meghalaya (Grows at heights up to 5000 m in Eastern Himalayas, in Nepal, Bhutan and Sikkim)	Palpitation of heart, intestinal tonic	https://indianbiodiversitytalk.blogspot.in/2012/08/jatamansi-under-threat-in-Himalaya-due-to-illegal-trade.html
Origanum vulgare		P-cymene, γ-terpinene, caryophyllene, spathulenol, germacrene-D, β-fenchyl alcohol, carvacrol, thymol, δ-terpineol	Uttaranchal, Himachal Pradesh (Native to Europe, naturalized in the Middle East)	Volatile oil used as aromatic stimulant in rheumatism	http://powo.science.kew.org/taxon/urn:lsid:ipni.org:names:453395-1; Mockute et al. (2001)
Polygonatum verticilatum		α-Bulnesene, linalyl acetate, eicosadienoic, pentacosane, piperitone, docasane, diosgenin, santonin and calarene	Himachal Pradesh, Uttaranchal, Meghalaya, Arunachal Pradesh (Found in Himalayas at height of 1800 to 3900 m)	Fever, glactagogue, vitiated condition of pitta and vata, aphrodisiac, emollient	Saboon et al. (2016)

Continued

Table 9.1 High-Altitude Medicinal Plants With Their Active Components, Distribution, and Therapeutic Activities—cont'd

Medicinal Plants	Images	Active Components	Distribution	Therapeutic Activities	References
Rhodiola sp.		L-methionine, L-phenylalanine, L-lysine, L-leucine, and L-histidine, palmitoleic acid, oleic acid, linoleic acid	Leh, Ladakh	Adaptogenic, antifatigue, antidepressant, antioxidant, anti-inflammatory, antinociception, anticancer activities, prevent cardiovascular, neuronal, liver, skin disorders	Dhar et al. (2013)
Valeriana wallichii		Boryl isovalerianate, chatinine, formate, glucoside, isovalerenic acid, 1-camphene, 1-pinene, resins, terpincol and valerianine	Uttaranchal, Himachal Pradesh (Distributed throughout drier parts of India)	Alexiteric, cures epileptic seizures, head troubles, disease of eye, blood disease, suppression of urine, astringent carminative, hypnotic, aphrodisiac, pain in joint, disease of liver, clear the voice	Devi and Rao (2014)

Valeriana sp.		Bromazepam, clonazepam, diazepam, valerenic acid, beta-sitosterol, ursolic acid, 4,4′,8,8′-tetrahydroxy-3,3′-dimethoxyl-dibenzyl-ditetrahydrofuran and caryophyllene acide, valerane, naphthalene, linoleic acid, ethyl ester, myrtenyl acetate, ursolic acid, 4,4′,8,8′-tetrahydroxy-3, 3′-dimethoxyl-dibenzyl-ditetrahydrofuran	Himachal Pradesh, it grows profusely in Bharmour division of Chamba, Kanda area of Karsog, and Chansil of Rohru forest division	Treatment of habitual constipation, insomnia, epilepsy, neurosis, anxiety, diuretic, hepatoprotective, analgesic, and cytotoxic, antispasmodic, anticonvulsant, antiinflammatory	Patočka and Jakl (2010), Saklani et al. (2012)
Withania somnifera		Withanolides, withaferin A	Uttaranchal, Himachal Pradesh.	Rheumatism, consumption, coagulating milk, antioxidant, anxiolytic, adaptogen, memory enhancing, antiparkinsonian, antivenom, antiinflammatory, antitumor, immunomodulation, hypolipidemic, antibacterial, cardiovascular protection, sexual behavior	Red Data Book; Wijnsma et al. (1985)

ledgeriana, Colchicum luteum Baker, Cordyceps sp., *Curcuma* sp., *Ganoderma* sp., *Hippophae* sp., *Nardostachys jatamansi, Origanum vulgare, Polygonatum verticilatum, Rhodiola* sp., *Valeriana* sp., and *Withania somnifera* are unique in respect to their biological activities (Balakumbahan et al., 2010; Geetha et al., 2003; Hayes et al., 2008; Kaul, 1997; Mathews et al., 1990; Mishra et al., 2017a, b; Mockute et al., 2001; Patočka and Jakl, 2010; Rakhee et al., 2016; Saboon et al., 2016; Saha et al., 2016; Srimal, 2001; Wijnsma et al., 1985).

Most of the bioactive compounds in the HA plants encompass secondary metabolites (as seen in Table 9.1). For instance, many HA herbal sources consist of substantial amounts of volatile organic compounds (VOCs), which are primarily responsible for imparting aroma and fragrance. Such compounds are used as chief ingredients in essential oils and oil-based healing formulations used to treat arthritis, muscle contractions, and sciatica. Several other secondary metabolites in these herbs are alkaloids, amino acids, fatty acids, flavonoids, glycosides, lignans, nucleotides, nucleoside, phenolic acids, phenols, proteins, resins, sterols, tannins, and terpenoids (Azmir et al., 2013; Ingle et al., 2017).

In light of the abundance of phytoconstituents in herbal sources and their varied bioactivities, it is essential to compile all the analytical techniques that enable isolation and characterization of these compounds. This compilation could further support their application as therapeutic agents.

9.3   EXTRACTION, ISOLATION, AND CHARACTERIZATION OF BIOACTIVE COMPOUNDS FROM HERBAL SOURCES

HA herbal sources have unique salubrious phytoconstituents that offer adaptability to extreme pressure and temperature (Brusotti et al., 2014; Rakhee et al., 2017). A number of advanced and sophisticated techniques that are commonly used to isolate and characterize these bioactive compounds are explained in the following sections (Brusotti et al., 2014; Rakhee et al., 2017).

The extraction method is the first step involved during sample preparation. Two types of extraction methods generally are employed: conventional (boiling under reflux, decoction, infusion, maceration) and nonconventional (microwave-assisted extraction (MAE), ultrasound-assisted extraction (UAE), supercritical fluid extraction (SFE), pressurized-assisted extraction (PAE), hydrotropic extraction, enzyme-assisted extraction, liquid-liquid extraction, liquid phase extraction, solid phase extraction, microporous absorption resin) (Ingle et al., 2017; Sasidharan et al., 2011). The second step focuses on isolation and concentration of the desired natural product using different chromatographic techniques and solidifying methods. The third step includes the analyses of isolated bioactive compounds for chemical characterization (antioxidant assays, enzymatic assays, immunoassays, in vivo and in vitro assays, phytochemical screening, and toxicity evaluation), and genetic characterization (conventional PCR, DNA probe, LAMP, multiplex PCR, SARP, and SSCP-PCR) (Wangchuk, 2014). The isolated herbal products then are subjected to various structure elucidation techniques, including DEPT-20, ESI-MS, FT-IR, ^{1}H NMR, ^{13}C NMR, ^{1}H—^{1}H coupled NMR, 2D-NMR, and MALDI-TOF/TOF (Mishra et al., 2017a, b; Seger et al., 2013; Siddiqui et al., 2017) and, if required, to hyphenated spectroscopic techniques such as HPLC-NMR-MS, HPLC-IR, HPLC-UV-IR, LC-PDA-NMR-MS, LC-UV-MS, LC-TSP-MS, LVI-GC-MS, and SPE-LC-MS (O'Connor and Gibbons, 2013; Potterat and Hamburger, 2013; Sasidharan et al., 2011; Seger et al., 2013; Siddiqui et al., 2017). These techniques have been reported extensively in literature for characterization of biomolecules, even on a nano scale. A schematic representation describing these extraction, isolation, and characterization techniques is given in Fig. 9.2.

9.3.1 Extraction Techniques

Extraction of bioactive compounds, especially secondary metabolites from HA medicinal plants and herbs, depends on several factors, such as polarity and stability of the sample, toxicity, volatility, viscosity, and purity of the extraction solvent (Sasidharan et al., 2011). Several secondary metabolites are present intracellularly so the raw material needs to be broken down or hydrolyzed to improve the extraction yield (Bennett and Wallsgrove, 1994).

9.3.1.1 Conventional Extraction Methods

Some classical and conventional methods of extraction mainly depends on the polarity of solvents and nature of the targeted class of compounds to be isolated (Pereira and Meireles, 2010; Sasidharan et al., 2011).

Distillation

Distillation is the most common method for separating mixtures of secondary metabolites found in medicinal herbs where compounds are separated in different phases (Jones and Kinghorn, 2012). Separation of various bioactive constituents in a mixture is carried out by heating the aqueous form of mixture, wherein different components are separated by their different boiling points, into the gaseous phase. The gas then is condensed back into liquid form and collected. Distillation is used for several purposes, such as production of aromatic components, and phenolic and terpenoid rich constituents. Four types of distillation methods—simple distillation, vacuum distillation, fractional distillation, and steam distillation—are used (Lea and Swoboda, 1962).

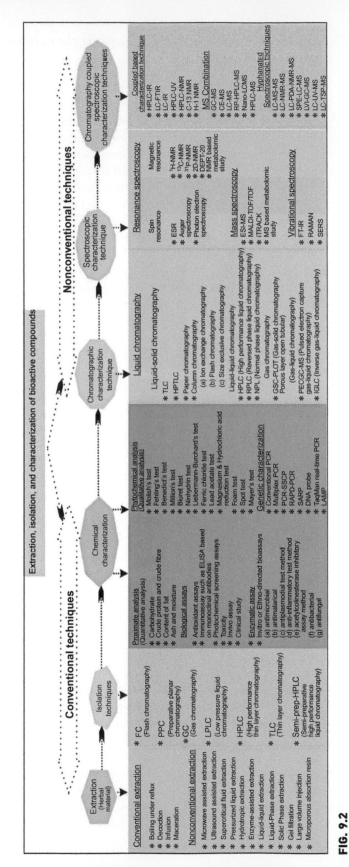

FIG. 9.2

Extraction and characterization techniques of bioactive herbal products.

Maceration

Maceration has been used in domestic preparation of medicinal drugs since ancient times. It gained popularity because of its convenient and cost-effective approach in obtaining essential oils and bioactive compounds. On small-scale extraction, maceration consists of several steps (Jones and Kinghorn, 2012). First, raw material is ground into smaller particles in order to increase the surface area for proper mixing with solvent. Second, an appropriate solvent called "menstruum" is added in a closed vessel to the sample. Third, the liquid is strained off and the "marc, or solid residue, is pressed to recover a large amount of occluded solutions. Infrequent shaking to improve the extraction yield facilitates extraction by increasing diffusion and by removing concentrated solution from the sample surface (Jones and Kinghorn, 2012).

Soxhlet Extraction

Soxhlet extraction has been used widely for extracting valuable bioactive compounds from various natural sources. In this extraction, a small amount of dry sample is placed in a thimble, which is placed in a distillation flask containing the solvent of particular interest. After reaching an overflow level, the solution of the thimble-holder is aspirated by a siphon, which unloads the solution back into the distillation flask. This solution carries the extracted solutes into the bulk liquid. The solute remains behind in the distillation flask, and the solvent passes back to the solid bed of samples. The process is repeated until complete extraction takes place (Saim et al., 1997).

9.3.1.2 Nonconventional Solvent Extraction Methods

Most of the isolation and extraction procedures meant for medicinal plants still use organic solvents of different polarities, water, or their mixtures. Some of the nonconventional extraction methods that are commonly performed are extraction on solid phase, ultrasound-assisted extraction (UAE) (Pingret et al., 2013), extraction with ionic liquids, microwave-assisted extraction (MAE) (Jones and Kinghorn, 2012), accelerated solvent extraction (ASE) (Rakhee et al., 2016), supercritical fluid extraction (SFE) (Paulaitis et al., 1983), and turbo-extraction (Hoberg et al., 1999).

Accelerated Solvent Extraction

ASE is applied to extract compounds of interest from solid and semisolid matrices, using either polar or nonpolar solvents, depending on the type of extract to be prepared. The extraction conditions generally are maintained at an elevated pressure and temperature to maintain the state of the solvent and to achieve enhanced extraction efficiency, respectively. In this extraction method, sequential extraction with solvents of different polarity and mixing of solvents is possible (Rakhee et al., 2016). This mode of extraction has been widely followed to prepare extracts rich in bioactive constituents, including flavonoids, nucleobases, and sterols from numerous herbal sources, such as medicinal mushrooms, sea buckthorn, and *V. wallichii* (Bhardwaj et al., 2015; Mamta et al., 2015; Sharma et al., 2012).

Ionic Liquid Extraction

In recent years, application of ionic liquids for UAE, MAE, or simple batch extraction of plant metabolites at room temperature or elevated temperature has gained increasing attention. Ionic liquids, such as *N,N*-dimethylethanolammonium octanoate (DMEA oct), and bis (2 methoxyethyl) ammonium bis (triuoromethylsulfonyl) imide (BMOEA bst) (Cutler et al., 2006; Wu et al., 2015) are used frequently for extraction of natural products from HA medicinal plants and herbs (Qi et al., 2006).

Microwave-Assisted Extraction

Extraction methods employing either diffused microwaves in closed systems or focused microwaves in open systems are being used now for their better efficiency (Jones and Kinghorn, 2012). Many types of microwave-assisted extraction methods are vacuum microwave-assisted extraction (VMAE), nitrogen-protected microwave assisted extraction (NPMAE), ultrasonic microwave-assisted extraction (UMAE), and dynamic microwave-assisted extraction (DMAE) (Ericsson and Colmsjö, 2000; Wang et al., 2008; Yu et al., 2009).

Solid-Phase Extraction

Solid-phase extraction procedures are based on adsorption of analytes or unwanted impurities on a solid phase to aid in purification of natural products. Solid-phase extraction employs a wide range of stationary phases with diverse chemical properties, for example, hydrophilic interaction liquid chromatography (HILIC) stationary phases, ion-exchange resins, mixed-mode material, reversed-phase material, and silica gel (Sasidharan et al., 2011).

Supercritical Fluid Extraction

SFE is a diffusion-based process of separating one component (extractant) from another (matrix) using supercritical fluids. Extraction conditions usually are maintained above the critical pressure (P_c) and critical temperature (T_c) of the extraction solvent in use. The advantages of such an extraction technique is that it offers faster separation, higher selectivity, and increased purity of the extract collected (Villanueva-Bermejo et al., 2017). Water and CO_2 are preferred as extraction solvents for preparing herbal extracts by supercritical fluid extraction because of their ease in handling and nontoxic nature. Additionally, the use of modifiers, such as ethanol in varying percentages, is at times incorporated to accentuate the extraction efficiency by alteration of polarities (Paulaitis et al., 1983). SFE recently has been proven to be highly beneficial in preparing bioactive extracts from plant sources. For example, extracts prepared from leaves of *Hippophae rhamnoides* using supercritical CO_2 have shown appreciable adjuvant activities against *diphtheria* and *tetanus* toxoids (Jayashankar et al., 2017).

Ultrasound-Assisted Extraction
In UAE, the plant material is placed in a glass container, covered with the appropriate extraction solvent, and put into an ultrasonic bath. This extraction technique involves considerable mechanical stress that induces cavitation and cellular breakdown, resulting in an increased extraction yield. The extraction time also is substantially reduced (Pingret et al., 2013).

9.3.1.3 Specialized Extraction Methods
Apart from the conventional and nonconventional extraction methods, a few specialized techniques also are adopted.

Preparative High-Performance Liquid Chromatography
Preparative high-performance liquid chromatography generally is used to obtain a high yield of analytes at a good resolution. This technique is appropriate for characterizing samples belonging to different classes of bioactive compounds, such as flavonoids, phenolic compounds, and vitamins. The different chemical components in the samples are fractionated according to their retention times. The fractions are subjected to further characterization studies such as antioxidant assays, biochemical analyses, and other miscellaneous biological assays (Aravind et al., 2008).

Preparative Planar Chromatography
Preparative planar chromatography is a type of liquid chromatography used to isolate bioactive compounds in a small sample size (range 1–100 mg). This technique is typically advantageous to purify small amounts of samples to be used in preparative and isolation purposes (Hostettmann et al., 1986).

9.3.2 Solvent Removal Techniques for the Preparation of Herbal Extracts

After carrying out any of the extraction methods, the sample extract is introduced to an appropriate drying process. The dried sample extract then is recovered by means of several concentration procedures while discarding residual solvent from the prepared extract (Adou, 2005). The concentrated solid extract is subjected to appropriate chromatographic techniques in order to identify specific and nonspecific types of phytoconstituents (Sasidharan et al., 2011).

9.3.2.1   Solvent Evaporation Techniques

Some of the frequently applied techniques to dry the extracted sample for solid-ification, purification, and concentration use hot air, microwave, and oven along with a freeze-drying method to purify and concentrate the sample (Saim et al., 1997).

Drying by Conventional Techniques

The extracted sample bearing high medicinal value is subjected to conventional methods of drying by use of hot air, microwave, or oven to increase its concentration and also to aid in removal of any residual impurities. The temperature in this process is controlled to ensure that no chemical decomposition takes place in the sample while the dried product is retrieved.

Lyophilization

Freeze drying or lyophilization of extracted samples from medicinal plants is used to remove solvents (usually water) through a process called sublimation. In sublimation, the solid frozen material is converted directly to a gaseous phase without passing through a liquid phase (Wang, 2000). For this, samples are frozen, and all the components are precooled to attain a frozen state. The ice then is sublimed in two steps. In the primary drying phase, about 98% of the water gets sublimed; the secondary drying phase aims to sublime the bound water molecules, resulting in a final purified extracted sample (Wang, 2000).

Rotary Vaporization

This type of drying is well-suited for the concentration of samples containing primary and secondary plant metabolites, which could be separated by evaporation (Mukherji et al., 2013). In rotary vaporization, the extracted sample is evaporated at a certain temperature under vacuum to remove the solvent from the sample, thereby leaving behind a dried, solvent-free concentrated sample. The dried extract is subsequently dissolved in deionized water and defatted with petroleum ether to remove fats. If required, successive partitioning with solvents of different polarities is carried out using a separating funnel to obtain a more purified sample (Mukherji et al., 2013).

Centrifugal Evaporation

This type of drying is achieved by speed-vacuum concentrator, where organic solvents such as alcohols, ethers, and chloroform, are dried under vacuum, i.e., low pressure and ambient temperatures. This facilitates shifting of the vapor-liquid equilibrium of the solvent toward the gas phase, while the sample being dried remains in solid phase. This is a gentler mode of drying than rotary vaporization, where drying occurs under elevated temperatures (Roth et al., 2010).

9.3.3   Characterization Techniques for Bioactive Constituents

The characterization methods to be adopted for any isolated herbal extract depends on the nature of the extract and on the type of extraction techniques that were followed to prepare the extract. Characterization largely falls under two categories: conventional and nonconventional techniques.

9.3.3.1   Conventional Characterization Techniques

Many chemical methods are used for qualitative evaluation of several phytochemicals, such as alkaloids, amino acids, fat and oils, glycosides, proteins, saponins, steroids and sterols, sugars, tannins, and terpenoids (Rakhee et al., 2016). Some of the specific qualitative tests that have been used for the primary level characterization of the pharmacologically active phytoconstituents are described below (Rakhee et al., 2017).

Proximate Analysis

HA herbs have gained widespread attention because of their richness in numerous types of phytoconstituents bearing pharmacological and nutraceutical properties. Proximate analysis is the most preferred analytical technique for evaluating the quantity of amino acids, ash, carbohydrates, fats, minerals, moisture, and protein contents present in extracts of HA medicinal plants (Rakhee et al., 2016).

Biochemical Characterization

Extracts prepared from HA medicinal herbs are subjected to estimation of biochemical activity in terms of standard curves wherein the standards comprise the desired class of bioactive compound under study. A number of biochemical assays such as DPPH (2,2-diphenyl-1-picryhydrazyl), $ABTS^{.+}$ (2,2′-azino-bis (3-ethylbenzothiazoline-6-sulphonic acid), FRAP (ferric-reducing antioxidant power), MDA (malondialdehyde), SOD (superoxide dismutase), and biological assays aid in the estimation of antibacterial activity, antiproliferative activity, and protein leakage (Geetha et al., 2003; Mishra et al., 2017a, b; Sharma et al., 2012). These assays have established the importance of HA plants in terms of antimicrobial, antioxidant, antitumor, immunomodulating, and other beneficial activities (Geetha et al., 2005; Mishra et al., 2017a, b).

9.3.3.2   Nonconventional Characterization Techniques

Nonconventional characterization techniques include relatively advanced and current techniques that are being implemented worldwide for the characterization of phytochemical constituents.

Chromatographic Characterization Techniques

Identification of secondary metabolites in herbal products usually is a tedious process because of their complex compositions. Nevertheless, a couple of chromatography-based techniques enable easy identification of metabolites, thereby facilitating characterization of any type of bioactive compound. Some chromatographic-based techniques that are generally adopted are described in this section.

Thin-Layer Chromatography. Thin-layer chromatography (TLC) is one of the most popular and simplistic chromatographic techniques used for separation of compounds. TLC is employed extensively in phytochemical evaluation of herbal drugs because of its advantages, such as rapid analysis of herbal extracts, minimum sample clean up requirement, and ease in acquiring qualitative and quantitative information of the resolved compounds (Siddiqui et al., 2017).

High-Performance Thin-Layer Chromatography. High-performance thin-layer chromatography (HPTLC) is employed widely in pharmaceutical industries for process development, identification, and detection of adulterants in herbal products; identification of pesticide content and mycotoxins; and in quality control of herbs and health food. HPTLC has been reported to offer several advantages over the more widely used high-performance liquid chromatography (HPLC) technique because of its high throughput (multiple samples can be analyzed simultaneously), economy in use of mobile phase, and cost effectiveness (Mishra et al., 2017a, b).

High-Performance Liquid Chromatography. Over the past decades, HPLC has become the most expansively applied analytical technique for the analysis of herbal medicines. Usually, reversed phase (RP) columns are used for this purpose. Both analytical and preparative modes of HPLC are used in pharmaceutical industries for isolation and purification of herbal compounds. The parameters that need to be considered while performing characterization by HPLC are rapid analysis time, resolution, sensitivity, and specificity (Yan et al., 2007).

Low-Pressure Liquid Chromatography. Column chromatographic methods that allow the use of mobile phase at atmospheric pressure, without applying additional forces, are used as a major tool in the fractionation protocols followed for isolation of natural products. This technique uses hydroxypropylated dextran gel Sephadex LH-20 that is based not only on principle of molecular sieves, but also on separation by adsorption (Scalbert et al., 1990).

Gas Chromatography. Gas chromatography (GC) is a well-established analytical technique that offers high resolution and sensitivity, particularly in the characterization, quantification, and identification of volatile organic compounds (VOCs). Thus, GC is used mainly in the industries dealing with cosmetics, environmental toxins, and pharmaceuticals (Chauhan, 2014). Essential oils, fatty acids, and other secretions present in herbal extracts can be analyzed easily using GC (Ashley et al., 1996).

Flash Chromatography. Flash chromatography (FC) is an advanced type of chromatography that is used for rapid separation of aromatic, organic, or phenolic compounds. This is an air pressure driven technique comprising medium and short column chromatography (Hostettmann et al., 1986).

Spectroscopic Techniques

Ultraviolet-Visible Spectroscopy. Ultraviolet-visible spectroscopy (UV-Vis spectroscopy) makes use of absorption spectroscopy in ultraviolet and visible wavelength ranges—180–380 nm and 380–750 nm, respectively—for characterizing molecules. This is one of the most basic techniques that need to be conducted while characterizing an analyte. All the major classes of biomolecules contain certain light absorbing functional groups known as chromophores. Upon absorbing UV/Vis light, these chromophores get excited from ground state to a higher energy level, thus giving out characteristic spectra, aiding in the identification of specific biomolecules (Perkampus and Grinter, 1992).

Fourier Transform-Infrared Spectroscopy. Fourier transform-infrared spectroscopy (FT-IR) is based on the fact that most molecules absorb light in the infrared region of the electromagnetic spectrum. The frequency ranges are measured as wave numbers typically over the range 4000–400 cm^{-1}. FT-IR is useful mainly to identify organic molecular groups and compounds because of the range of functional groups. Pellets of samples are prepared with KBr salts owing to their transparency, thus offering better resolution of the spectrum (Faix, 1991). FT-IR is the most commonly used technique for structure elucidation of bioactive herbal extracts. For example, the IR spectrum of the leaf extract of *Euphorbia thymifolia* from the Kumaon Himalayas exhibited absorption band at 3407 cm^{-1} for hydroxyl group, 1666 cm^{-1} for carbonyl group, and 1562 cm^{-1} for unsaturation (Prasad, 2008).

Electrospray Ionization-Mass Spectrometry. Electrospray ionization-mass spectrometry (ESI-MS) generates ions upon application of an electric potential to a flowing liquid, causing the liquid to get charged and then sprayed. This electrospray forms very small droplets of solvent containing the analytes. ESI permits the coupling of HPLC instruments directly with MS. ESI-MS technique is used to characterize phenolic-rich fractions of medicinal plants as well as for identification of flavonoids such as gentenstein 7-glucoside, iridin, orientin, dihexosyl quercetin, quercetin-3-O-rutinoside, rhamnosyl hexosyl methyl quercetin, C-(O-caffeoyl-hexosyl)-O-hexoside, and tricin 7-O-hexosyl-O-hexoside in *Tragia involucrate* (Sulaiman and Balachandran, 2016).

Nuclear Magnetic Resonance Spectroscopy

Proton NMR Spectroscopy. Proton nuclear magnetic resonance (^{1}H NMR) spectroscopy is used to identify different types of protons having a nuclear spin $(I) = 1/2\,$m, in herbal samples. This technique shows the presence of magnetically induced shielded and deshielded protons in terms of characteristic coupling constant (J) (Berger and Braun 2004; Pavia et al., 2008).

Carbon-13 NMR Spectroscopy. This spectroscopic technique is based on the magnetically active nuclei ^{13}C, which has very low natural abundance (approximately, 1.1%) in comparison to ^{12}C. The spectrum generated here is recorded as a proton decoupled form as singlet and gives an idea about the types of carbon atoms having different types of environment. The ^{13}C NMR delta values normally range between 0 and 220 ppm. Typically, tetramethylsilane is used as the internal reference compound and samples are prepared in d^6-DMSO (Pavia et al., 2008; Seger et al., 2013).

Two-Dimensional NMR Spectroscopy (2D-NMR). Complex overlapping signals in NMR spectra could be interpreted easily by two main strategies. The first strategy aims at simplifying the ^{1}H NMR spectrum through creation of a projection of the ^{1}H broad-band decoupled spectrum as in J-resolved experiment spectroscopy NMR (JRES-NMR), or through the selective suppression of signals from certain compounds using relaxation or diffusion filters as in diffusion ordered-NMR spectroscopy (DOSY) (Mahrous and Farag, 2015; Pavia et al., 2008). While both JRES and DOSY are 2D techniques, the most useful aspect in both the methods is the ability to extract simplified proton spectrum with less peak interference rather than collecting information displayed in the second dimension, such as coupling constant and translational diffusion coefficient, respectively (Pavia et al., 2008).

Matrix-Assisted Laser Desorption/Ionization-Time of Flight Mass Spectrometry (MALDI-TOF/TOF).

MALDI is a soft ionization technique for the characterization of different bioactive peptides and proteins. The sample to be analyzed by MALDI is uniformly mixed with a large quantity of matrix. The matrix absorbs nitrogen laser light of wavelength 337 nm and converts it into heat energy. A small part of the matrix gets heated rapidly and is vaporized, along with the sample (Rakhee et al., 2016; Rakhee et al., 2017).

Hyphenated Techniques

Recent advances in characterization techniques have witnessed the hyphenation of spectroscopic techniques with regular and/or complex types of chromatography such as GC, HPLC, HPTLC, and ion-exchange chromatography. Some are described below.

High-Pressure Liquid Chromatography With UV Detector.

High-pressure liquid chromatography with UV detector (HPLC-UV) technique is commonly used to isolate and characterize medicinally important compounds found in many HA herbal sources, such as rosarin and rosin. In one study, eight different chemotaxonomic markers, namely, 4-(+) catechin, gallic acid, 8-rhodionin, 5-rosarin, 6-rosavin, 7-rosin, 3-salidroside, and 2-tyrosol were characterized in different *Rhodiola* species using HPLC-DAD/UV technique (Liu et al., 2013).

High-Pressure Liquid Chromatography-Mass Spectrometry. HPLC with diode array detector coupled with ESI-MS (HPLC-DAD-ESI-MS/MS) has been well-adopted for the characterization of many natural products. This technique has been employed for simultaneous identification of flavonoids and other constituents in the **rhizomes** of three *Iris* species growing in Kashmir valleys of India (Bhat, 2014; Seger et al., 2013). About 27 compounds were identified in these *Iris* species based on UV spectra, MS/MS spectra and retention time. Among those 27 compounds, seven constituents from *Iris crocea* (major being hesperetin, iridal glycoside 5b, iridal glycoside 7, irilin A, irisolodone, irisolone and irisquin B) were reported (Bhat, 2014).

High-Pressure Liquid Chromatography-Nuclear Magnetic Resonance. NMR spectroscopy is by far the most powerful spectroscopic technique for obtaining structural information about organic compounds in solution. The direct coupling of LC with proton NMR has been reported in many papers. Early experiments of coupled HPLC-^{1}H NMR have been conducted in a stop-flow mode or with very minimal flow rates. This type of flow is necessary to accumulate a sufficient number of spectra per sample volume in order to improve the S/N ratio (Seger et al., 2013). One example of application of this technique is isolation of iridoid glycosides *in Picrorhiza kurroa* Royle, an important medicinal plant thriving in HA. Iridoid glycosides were isolated on a RP C-18 column using a mobile phase of methanol and water (40:60 v/v). Further characterization of these glycosides was performed by ^{13}C NMR, 2D-NMR, ^{1}H NMR, MS, and UV. These iridoid glycosides, namely, picrosides I, II and kutkoside, are used in various herbal drug formulations as strong hepatoprotective and immunomodulatory compounds (Sultan et al., 2016).

Liquid Chromatography-Mass Spectrometry. Liquid chromatography-mass spectrometry (LC-MS) has undergone tremendous technological improvements, thus permitting its application to analyze endogenous metabolites, such as carbohydrates, DNA, drugs, peptides, and proteins (Kang, 2012). This method has been used to characterize many bioactive components occurring in HA medicinal plants, such as in UAE-yielded extracts of *Actinocarya tibetica*, five bioactive flavonoids—gardenin, 5-methoxy-6,7-methylenedioxyflavone, mosloflavone, negletein, and 5,6,7-trimethoxyflavone (Balakumbahan et al., 2010; Singh et al., 2013).

Gas Chromatography-Mass Spectrometry. Gas chromatography-mass spectrometry (GC-MS) is a hyphenated analytical technique that combines the separation properties of GC with the detection feature of MS to identify different substances within a test sample (Bhardwaj et al., 2017). A large number of volatile organic compounds and essential oils of high medicinal value have been isolated by liquid chromatography or gas chromatography and further characterized using FTIR or MS techniques. In different varieties of the *Valeriana* genus, GC and LC have been applied effectively along with MS to confirm the presence of E-valerenyl isovalerate, furinol-1, furinol-2, hydroxyvalerenic acid, methyl valiant, pacifigorgiol, sesquiterpenoids, valerenic acid, valerenol, and Z-Valerenyl acetate (Bos, 1997; Chauhan, 2014).

Reversed Phase-High-Performance Liquid Chromatographic-Mass Spectrometry. Reversed phase-high-performance liquid chromatographic-mass spectrometry (RP-HPLC-MS) involves the separation of molecules based on hydrophobicity. Separation depends on the hydrophobic binding of the solute molecule from the mobile phase to the immobilized hydrophobic ligands attached to the stationary phase, the sorbent. In different HA medicinal plants such as *Hippophae rhamnoides*, *Prunus armeniaca*, and *Rhodiola imbricate*, RP-HPLC tagged with precolumn derivatization and CE have been used identify and quantify amino acids and vitamins (E and B-complex vitamins) (Xiong et al., 2013). RP-HPLC-MS has also been used to characterize essential and nonessential amino acids such as L-methionine, L-phenylalanine, L-lysine, L-leucine, and L-histidine in herbal extracts (Xiong et al., 2013).

Thin-Layer Chromatography-Mass Spectrometry. Advancement in TLC has witnessed its hyphenation with mass spectrometry for confirmation of biomolecules. In this technique, a particular band (corresponding to the analyte under study) is extracted from the TLC or HPTLC plate by an appropriate interface and is subjected to mass spectrometry to ensure better identification. This technique is particularly useful for resolving the kind of analytes that might not be distinguished because of closer retardation factor values but could be differentiated based on their varying m/z values. TLC-MS has been applied to identify phytoconstituents in herbal samples. In a recent study, TLC-MS aided in identification of nucleobases in *Hippophae* species extracts (Mishra et al., 2017a, b).

9.4   MULTIHYPHENATED CHARACTERIZATION TECHNIQUES

Hyphenation of spectroscopic and chromatographic techniques is not limited to two or three techniques. The coupling could involve more than one separation or detection techniques, such as LC-PDA-MS, LC-MS-MS, LC-NMR-MS, and LC-PDA-NMR-MS (Patel et al., 2010). The two key elements in natural product research are the isolation and purification of compounds present in crude extracts or fractions obtained from various natural sources, and the unambiguous identification of the isolated compounds (Patel et al., 2010). Thus, the online characterization of secondary metabolites in crude natural product extracts or fractions demands a high degree of sophistication and richness of structural information, sensitivity, and selectivity. The development of various hyphenated techniques has provided natural product researchers with extremely powerful tools that facilitate excellent separation efficiency as well as acquisition of online complimentary spectroscopic data on a GC or LC peak of interest within a complex mixture (Patel et al., 2010).

9.5 CONCLUSIONS

This chapter has described all the relevant extraction, solvent removal, and characterization techniques that are necessary to prepare therapeutic extracts from herbal sources. It could be well ascertained that both conventional as well as nonconventional techniques complement each other for facilitating complete characterization of any herbal extract.

Acknowledgments

The authors are thankful to the director, DIPAS, for providing facilities, support, and encouragement. Ms. Rakhee is grateful to dead, Department of Chemistry, University of Delhi, Delhi, India (09/045 (1463/2017-EMR-I)), for guidance and support. She also would like to acknowledge CSIR for her fellowship grant.

References

Adou, E., 2005. Isolation and Characterization of Bioactive Compounds from Suriname and Madagascar Flora. II. A Synthetic Approach to Lucilactaene. PhD Thesis, Blacksburg, Virginia. URI: http://hdl.handle.net/10919/29973.

Aravind, S., Arimboor, R., Rangan, M., Madhavan, S.N., Arumughan, C., 2008. Semipreparative HPLC preparation and HPTLC quantification of tetrahydroamentoflavone as marker in Semecarpus anacardium and its polyherbal formulations. J. Pharm. Biomed. Anal. 48 (3), 808–813.

Ashley, D.L., Bonin, M.A., Cardinal, F.L., Mccraw, J.M., Wooten, J.V., 1996. Measurement of volatile organic compounds in human blood. Environ. Health Perspect. 104 (Suppl 5), 871–877.

Azmir, J., Zaidul, I.S.M., Rahman, M.M., Sharif, K.M., Mohamed, A., Sahena, F., Jahurul, M.H.A., Ghafoor, K., Norulaini, N.A.N., Omar, A.K.M., 2013. Techniques for extraction of bioactive compounds from plant materials: A review. J. Food Eng. 117 (4), 426–436.

Balakumbahan, R., Rajamani, K., Kumanan, K., 2010. Acorus calamus: an overview. J. Med. Plants Res. 4 (25), 2740–2745.

Bennett, R., Wallsgrove, R., 1994. Secondary metabolites in plant defence mechanisms. New Phytol. 127 (4), 617–633.

Berger, S., Braun, S., 2004. 200 and More NMR Experiments. Wiley-VCH, Weinheim, pp. 305–307.

Bhardwaj, A., Gupta, P., Kumar, N., Mishra, J., Kumar, A., Rakhee, Misra, K., 2017. Lingzhi or Reishi medicinal mushroom, Ganoderma lucidum (Agaricomycetes), inhibits candida biofilms: a metabolomic approach. Int. J. Med. Mushrooms 19 (8), 685–696.

Bhardwaj, A., Pal, M., Srivastava, M., Tulsawani, R., Sugadev, R., Misra, K., 2015. HPTLC based che-mometrics of medicinal mushrooms. J. Liq. Chromatogr. Relat. Technol. 38 (14), 1392–1406.

Bhardwaj, A., Srivastava, M., Pal, M., Sharma, Y.K., Bhattacharya, S., Tulsawani, R., Sugadev, R., Misra, K., 2016. Screening of Indian lingzhi or reishi medicinal mushroom, Ganoderma luci-dum (agaricomycetes): a UPC2-MS approach. Int. J. Med. Mushrooms 18 (2), 177–189.

Bhat, G., 2014. HPLC-DAD-ESI-MS/MS identification and characterization of major constituents of Iris crocea, Iris germanica, and Iris spuria growing in Kashmir Himalayas, India. J. Anal. Bioanal. Tech. 5 (6), 1–10.

Bos, R., 1997. Analytical and Phytochemical Studies on Valerian and Valerian-Based Preparations (Dissertation). State University of Groningen, Department of Pharmaceutical Biology, Groningen, Netherlands, pp. 184–193.

Brusotti, G., Cesari, I., Dentamaro, A., Caccialanza, G. and Massolini, G., 2014. Isolation and char-acterization of bioactive compounds from plant resources: The role of analysis in the ethno-pharmacological approach. J. Pharm. Biomed. Anal., 87, pp.218–228. DOI: https://doi.org/10.1016/j.jpba.2013.03.007.

Bucar, F., Wube, A., Schmid, M., 2013. Natural product isolation—how to get from biological mate-rial to pure compounds. Nat. Prod. Rep. 30 (4), 525–545.

C T, S., Balachandran, I., 2016. LC/MS characterization of antioxidant flavonoids from Tragia invo-lucrata L. Beni-Suef Univ. J. Basic Appl. Sci. 5 (3), 231–235.

Chauhan, A., 2014. GC-MS technique and its analytical applications in science and technology. J. Anal. Bioanal. Tech. 5 (6), 1–5.

Chen, F., Tholl, D., D'Auria, J.C., Farooq, A., Pichersky, E., Gershenzon, J., 2003. Biosynthesis and emission of terpenoid volatiles from Arabidopsis flowers. Plant Cell 15 (2), 481–494.

Cutler, M., Lapkin, A., Plucinski, P.K., 2006. Comparative Assessment of Technologies for Extrac-tion of Artemisinin. Summary Report Commissioned Through the Malaria Medicines Venture. Geneva, August.

Devi, V.S., Rao, M.G., 2014. Valeriana wallichii-A rich aroma root plant. World J. Pharm. Pharm. Sci. 3, 1516–1525.

Dhar, P., Tayade, A.B., Kumar, J., Chaurasia, O.P., Srivastava, R.B., Singh, S.B., 2013. Nutritional profile of Phytococktail from trans-Himalayan plants. PLoS ONE. 8 (12). e83008.

Ericsson, M., Colmsjö, A., 2000. Dynamic microwave-assisted extraction. J. Chromatogr. A 877 (1), 141–151.

Faix, O., 1991. Classification of lignins from different botanical origins by FT-IR spectroscopy. Holz-forschung Int. J. Biol. Chem. Phys. Technol. Wood 45 (s1), 21–28.

Gao, X., Ohlander, M., Jeppsson, N., Björk, L., Trajkovski, V., 2000. Changes in antioxidant effects and their relationship to phytonutrients in fruits of sea buckthorn (Hippophae rhamnoides L.) during maturation. J. Agric. Food Chem. 48 (5), 1485–1490.

Geetha, S., Ram, M.S., Mongia, S., Singh, V., Ilavazhagan, G., Sawhney, R., 2003. Evaluation of anti-oxidant activity of leaf extract of Seabuckthorn (Hippophae rhamnoides L.) on chromium (VI) induced oxidative stress in albino rats. J. Ethnopharmacol. 87 (2), 247–251.

Geetha, S., Singh, V., Ram, M.S., Ilavazhagan, G., Banerjee, P., Sawhney, R., 2005. Immunomod-ulatory effects of sea buckthorn (Hippophae rhamnoides L.) against chromium (VI) induced immunosuppression. Mol. Cell. Biochem. 278 (1), 101–109.

Gođevac, D., Jadranin, M., Aljančić, I., Vajs, V., Tešević, V., Milosavljević, S., 2015. Application of Spectroscopic Methods and Hyphenated Techniques to the Analysis of Complex Plant Extracts. Vol. 1. Springer, Budapest, Hungary, pp. 1–459.

Gupta, P., Dutta, B., Pant, D., Joshi, P., Lohar, D., 2009. In vitro antibacterial activity of Artemisia annua Linn. growing in India. Int. J. Green Pharm. 3 (3), 255.

Hayes, P.Y., Jahidin, A.H., Lehmann, R., Penman, K., Kitching, W., De Voss, J.J., 2008. Steroidal saponins from the roots of Asparagus racemosus. Phytochemistry 69 (3), 796–804.

Hoberg, E., Orjala, J., Meier, B., Sticher, O., 1999. Diterpenoids from the fruits of Vitex agnus-castus. Phytochemistry 52 (8), 1555–1558.

Hostettmann, K., Marston, A., Hostettmann, M., 1986. Preparative Chromatography Techniques, second ed. Springer, Berlin, pp. 1–247.

Ingle, K.P., Deshmukh, A.G., Padole, D.A., Dudhare, M.S., Moharil, M.P., Khelurkar, V.C., 2017. Phytochemicals: extraction methods, identification, and detection of bioactive compounds from plant extracts. J. Pharmacogn. Phytochem. 6 (1), 32–36.

Jayashankar, B., Singh, D., Tanwar, H., Mishra, K.P., Murthy, S., Chanda, S., Mishra, J., Tulswani, R., Misra, K., Singh, S.B., Ganju, L., 2017. Augmentation of humoral and cellular immunity in response to Tetanus and Diphtheria toxoids by supercritical carbon dioxide extracts of *Hippophae rhamnoides* L. leaves. Int Immunopharmacol. 44, 123–136.

Jones W.P., Kinghorn A.D. (2012) Extraction of plant secondary metabolites. In: Sarker S., Nahar L. (eds Online ISBN: 978–1–61779-624-1), Natural Products Isolation. Methods in Molecular Biology (Methods and Protocols), Vol. 864. Publisher: Humana Press. DOI:https://doi.org/10.1007/978-1-61779-624-1_13.

Kale, A., Gawande, S., Kotwal, S., 2008. Cancer phytotherapeutics: role for flavonoids at the cellular level. Phytother. Res. 22 (5), 567–577.

Kang, J.S., 2012. Principles and applications of LC-MS/MS for the quantitative bioanalysis of analytes in various biological samples, Tandem Mass Spectrometry-Applications and Principles. Jeevan Prasain (Ed.), ISBN: 978–953–51-0141-3, Seoul, South Korea, Publisher: InTech, Available from: http://www.intechopen.com/books/tandem-massspectrometry-applications-and-principles/principles-and-applications-of-lc-ms-ms-for-the-quantitativebioanalysis-of-analytes-in-various-biol.

Kaul, M., 2010. High Altitude Botanicals in Integrative Medicine—Case Studies From Northwest Himalaya. ISSN: 0975–1068 (Online); 0972–5938 (Print), Vol. 09(1), Publisher: CSIR. pp.18–25. URL: http://hdl.handle.net/123456789/7149.

Kaul, M.K., 1997. Medicinal plants of Kashmir and Ladakh: Temperate and Cold Arid Himalaya. ISBN: 8173870616, 9788173870613,New Delhi, Publisher: Indus publishing, pp. 1–173.

Kuttan, R., Bhanumathy, P., Nirmala, K., George, M., 1985. Potential anticancer activity of turmeric (*Curcuma longa*). Cancer Lett. 29 (2), 197–202.

Lakshmi, B., Ajith, T., Sheena, N., Gunapalan, N., Janardhanan, K., 2003. Antiperoxidative, anti-inflammatory, and antimutagenic activities of ethanol extract of the mycelium of *Ganoderma lucidum* occurring in South India. Teratog. Carcinog. Mutagen. 23 (S1), 85–97.

Lea, C., Swoboda, P., 1962. Simple vacuum distillation procedure for determination of the volatile carbonyl content of autoxidising edible fats. J. Sci. Food Agric. 13 (3), 148–158.

Liu, Z., Liu, Y., Liu, C., Song, Z., Li, Q., Zha, Q., Lu, C., Wang, C., Ning, Z., Zhang, Y., 2013. The chemotaxonomic classification of Rhodiola plants and its correlation with morphological characteristics and genetic taxonomy. Chem. Central J. 7 (1), 1–118.

Maehara, S., Simanjuntak, P., Maetani, Y., Kitamura, C., Ohashi, K., Shibuya, H., 2013. Ability of endophytic filamentous fungi associated with *Cinchona ledgeriana* to produce Cinchona alkaloids. J. Nat. Med. 67 (2), 421–423.

Mahrous, E.A., Farag, M.A., 2015. Two-dimensional NMR spectroscopic approaches for exploring plant metabolome: a review. J. Adv. Res. 6 (1), 3–15.

Mamta, Mehrotra, S., Kirar, V., Vats, P., Nandi, S.P., Negi, P., Misra, K., 2015. Phytochemical and antimicrobial activities of Himalayan *Cordyceps sinensis* (Berk.) Sacc. Indian J Exp Biol. 53 (2015), 36–43.

Mathews, H., Bharathan, N., Litz, R., Narayanan, K., Rao, P., Bhatia, C., 1990. The promotion of Agrobacterium mediated transformation in Atropa belladona L. by acetosyringone. J. Plant Physiol. 136 (4), 404–409.

Mishra, J., Hande, P., Sharma, P., Bhardwaj, A., Rajput, R., Misra, K., 2017. Characterization of nucleobases in sea buckthorn leaves: an HPTLC approach. J. Liquid Chromatogr. Related Technol. 40 (1), 50–57.

Mishra, J., Rajput, R., Singh, K., Puri, S., Goyal, M., Bansal, A. and Misra, K., 2017. Antibacterial natural peptide fractions from Indian Ganoderma lucidum. Int. J. Pept. Res. Ther., pp. 1–12. https://doi.org/10.1007/s10989-017-9643-z

Mockute, D., Bernotiene, G., Judzentiene, A., 2001. The essential oil of Origanum vulgare L. ssp. vulgare growing wild in Vilnius district (Lithuania). Phytochemistry 57 (1), 65–69.

Mukherji, R., Joshi-Navare, K., Prabhune, A., 2013. Crystalline xylitol production by a novel yeast, Pichia caribbica (HQ222812), and its application for quorum sensing inhibition in gram-negative marker strain Chromobacterium violaceum CV026. Appl. Biochem. Biotechnol. 169 (2013), 1753–1763.

Nalli, M., Ortar, G., Moriello, A.S., Di Marzo, V., De Petrocellis, L., 2017. Effects of curcumin and curcumin analogues on TRP channels. Fitoterapia 122, 126–131.

O'Connor, S.E. and Gibbons, S., 2013. Editorial: Modern methods in plant natural products themed issue. Nat. Prod. Rep., 30(2013), pp.483–484. DOI: 10.1039/C3NP90008H.

Patel, K.N., Patel, J.K., Patel, M.P., Rajput, G.C., Patel, H.A., 2010. Introduction to hyphenated techniques and their applications in pharmacy. Pharm. Methods 1 (1), 2–13.

Patočka, J., Jakl, J., 2010. Biomedically relevant chemical constituents of Valeriana officinalis. J. Appl. Biomed. 8 (1), 11–18.

Paulaitis, M.E., Krukonis, V.J., Kurnik, R.T., Reid, R.C., 1983. Supercritical fluid extraction. Rev. Chem. Eng. 1 (2), 179–250.

Pavia, D.L., Lampman, G.M., Kriz, G.S. and Vyvyan, J.A., 2008. Introduction to Spectroscopy. ISBN: 9781111800628, Publisher: Cengage Learning, pp. 1–752.

Pereira, C.G., Meireles, M.A.A., 2010. Supercritical fluid extraction of bioactive compounds: fundamentals, applications, and economic perspectives. Food Bioprocess Technol. 3 (3), 340–372.

Perkampus, H.H., Grinter, H.C., 1992. UV-VIS Spectroscopy and its Applications ISBN: 978-3-642-77479-9(print), 978-3-642-77477-5(online). Publisher: SpringerBerlin-Heidelberg, pp.1–244. https://doi.org/10.1007/978-3-642-77477-5

Pingret, D., Fabiano-Tixier, A.S., Chemat, F., 2013. Natural product extraction: principles and applications. In: Mauricio, R.A., Juliana, P.M. (Eds.), Ultrasound-Assisted Extraction. RSC Publishing, pp. 89–112.

Potterat, O., Hamburger, M., 2013. Concepts and technologies for tracking bioactive compounds in natural product extracts: generation of libraries, and hyphenation of analytical processes with bioassays. Nat. Prod. Rep. 30 (4), 546–564.

Prasad, K., 2008. Phytochemical Investigation of Euphorbia, Pouzolzia, and Pavetta species from Kumaon Himalayas. PhD thesis, Department of Chemistry, Kumaun University, pp. 1–144.

Qi, S., Li, Y., Deng, Y., Cheng, Y., Chen, X., Hu, Z., 2006. Simultaneous determination of bioactive flavone derivatives in Chinese herb extraction by capillary electrophoresis used different electrolyte systems-borate and ionic liquids. J. Chromatogr. A 1109 (2), 300–306.

Rakhee, Sethy, N.K., Bhardwaj, A., Singh, V.K., Sharma, R.K., Deswal, R., Bhargava, K., Misra, K., 2017. Characterization of Ganoderma lucidum: phytochemical and proteomic approach. J. Proteins Proteom. 8 (1), 25–33.

Rakhee, Sethy, N.K., Singh, V.K., Sharma, S., Sharma, R.K., Deswal, R., Bhargava, K., Misra, K., 2016. Phytochemical and proteomic analysis of a high altitude medicinal mushroom Cordyceps sinensis. J. Proteins Proteom. 7 (3), 187–197.

Roth, S., Feichtinger, J., Hertel, C., 2010. Characterization of Bacillus subtilis spore inactivation in low-pressure, low-temperature gas plasma sterilization processes. J. Appl. Microbiol. 108 (2), 521–531.

Saboon, Bibi, Y., Arshad, M., Sabir, S., Amjad, M.S., Ahmed, E., Chaudhari, S.K., 2016. Pharmacology and biochemistry of Polygonatum verticillatum: a review. J. Coast. Life Med. 4 (5), 406–415.

Saha, L., Chakrabarti, A., Kumari, S., Bhatia, A., Banerjee, D., 2016. Antiapoptotic and neuroprotective role of curcumin in Pentylenetetrazole (PTZ) induced kindling model in rat. Indian J. Exp. Biol. 54 (2), 133–141.

Saim, N.a., Dean, J.R., Abdullah, M.P., Zakaria, Z., 1997. Extraction of polycyclic aromatic hydrocarbons from contaminated soil using Soxhlet extraction, pressurised and atmospheric microwave-assisted extraction, supercritical fluid extraction and accelerated solvent extraction. J. Chromatogr. A 791 (1), 361–366.

Saklani, N., Purohit, V.K., Andola, H.C., Rana, M., Chauhan, R., Nautiyal, A., 2012. Biological activities of *Valeriana wallichii* D. Med. Plants: Int. J. Phytomed. Relat. Ind. 4 (3), 115–120.

Sasidharan, S., Chen, Y., Saravanan, D., Sundram, K.M., Latha, L.Y., 2011. Extraction, isolation and characterization of bioactive compounds from plants' extracts. African J. Trad. Complem. Alt. Med. 8 (1), 1–10.

Saxena, V., Chourasia, S., 2001. A new isoflavone from the roots of *Asparagus racemosus*. Fitoterapia 72 (3), 307–309.

Scalbert, A., Duval, L., Peng, S., Monties, B., Du Penhoat, C., 1990. Polyphenols of Quercus robus L.: II. Preparative isolation by low-pressure and high-pressure liquid chromatography of heartwood ellagitannins. J. Chromatogr. A 502 (1990), 107–119.

Seger, C., Sturm, S., Stuppner, H., 2013. Mass spectrometry and NMR spectroscopy: modern high-end detectors for high resolution separation techniques: state of the art in natural product HPLC-MS, HPLC-NMR, and CE-MS hyphenations. Nat. Prod. Rep. 30 (7), 970–987.

Sharma, P., Kirar, V., Meena, D.K., Suryakumar, G., Misra, K., 2012. Adaptogenic activity of Valeriana wallichii using cold, hypoxia, and restraint multiple stress animal model. Biomed. Aging Pathol. 2 (4), 198–205.

Siddiqui, M.R., AlOthman, Z.A., Rahman, N., 2017. Analytical techniques in pharmaceutical analysis: a review. Arab. J. Chem. 10 (s1), 1409–S1421.

Singh, B., Chaurasia, O.P., 1998. Medicinal flora of Indian cold desert. In: Proc. XXV International Horticultural Congress, Part 13: New and Specialized Crops and Products, Botanic Gardens and Human-Horticulture Relationship. Vol. 523. pp. 65–74.

Singh, B., Sidiq, T., Joshi, P., Jain, S.K., Lawaniya, Y., Kichlu, S., Khajuria, A., Vishwakarma, R.A., Bharate, S.B., 2013. Anti-inflammatory and immunomodulatory flavones from Actinocarya tibetica Benth. Nat. Prod. Res. 27 (23), 2227–2230.

Srimal, R., 2001. Database of medicinal plants. Curr. Sci. 30 (3), 463.

Sultan, P., Jan, A., Pervaiz, Q., 2016. Phytochemical studies for quantitative estimation of iridoid glycosides in Picrorhiza kurroa Royle. Bot. Stud. 57 (1), 1–7.

Theis, N., Lerdau, M., 2003. The evolution of function in plant secondary metabolites. Int. J. Plant Sci. 164 (S3), 93–102.

Tripathi, A., Prajapati, V., Verma, N., Bahl, J., Bansal, R., Khanuja, S.P.S., Kumar, S., 2002. Bioactivities of the leaf essential oil of *Curcuma longa* (var. ch-66) on three species of stored-product beetles (Coleoptera). J. Econ. Entomol. 95 (1), 183–189.

Verma, P., Gauttam, V., Kalia, A.N., 2014. Comparative pharmacognosy of Pashanbhed. J. Ayurveda Integr. Med. 5 (2), 104.

Villanueva-Bermejo, D., Zahran, F., Garcia-Risco, M. R., Reglero, G. and Fornari, T., 2017. Supercritical fluid extraction of Bulgarian Achillea millefolium. J. Supercrit. Fluids, 119, pp.283–288. https://doi.org/10.1016/j.supflu.2016.10.005.

Wang, J.-X., Xiao, X.-H., Li, G.-K., 2008. Study of vacuum microwave-assisted extraction of polyphenolic compounds and pigment from Chinese herbs. J. Chromatogr. A 1198–1199 (2008), 45–53.

Wang, W., 2000. Lyophilization and development of solid protein pharmaceuticals. Int. J. Pharm. 203 (1), 1–60.

Wangchuk, P., 2014. Phytochemical Analysis, Bioassays, and the Identification of Drug Lead Compounds From Seven Bhutanese Medicinal Plants. PhD thesis, University of Melbourne.

Wijnsma, R., Go, J., Van Weerden, I., Harkes, P.V., Verpoorte, R. and Svendsen, A.B., 1985. Anthraquinones as phytoalexins in cell and tissue cultures of Cinchona species. Plant Cell Rep., 4(5), pp. 241–244.

Wu, L., Hu, M., Li, Z., Song, Y., Yu, C., Zhang, Y., Zhang, H., Yu, A., Ma, Q., Wang, Z., 2015. Determination of triazine herbicides in fresh vegetables by dynamic microwave-assisted extraction coupled with homogeneous ionic liquid microextraction high performance liquid chromatography. Anal. Bioanal. Chem. 407 (6), 1753–1762.

Xiong, X., Yin, Y., Huang, Y., Wang, Y., Wen, Q., Mu, Y., Shu, X., Zhan, Z., Zhou, Y. and Qiu, G., 2013. Methods of Amino Acid Analysis, Nutritional and Physiological Functions of Amino Acids in Pigs. Springer, Vienna, pp.217–229. https://doi.org/10.1007/978-3-7091-1328-8_15.

Yan, X., Wang, W., Zhang, L., Zhao, Y., Xing, D., Du, L., 2007. A high-performance liquid chromatography with UV-vis detection method for the determination of Brazilein in plant extract. J. Chromatogr. Sci. 45 (4), 212–215.

Yu, Y., Chen, B., Chen, Y., Xie, M., Duan, H., Li, Y., Duan, G., 2009. Nitrogen-protected microwave-assisted extraction of ascorbic acid from fruit and vegetables. J. Separation Sci. 32 (23–24), 4227–4233.

Further Reading

De Jong, G. (Ed.), 2016. Capillary Electrophoresis-Mass Spectrometry (CE–MS): Principles and Applications. John Wiley & Sons, pp. 1–368. ISBN: 978-3-527-33924-2.

Erickson, M.D., 1979. Gas chromatography/Fourier transform infrared spectroscopy applications. Appl. Spectrosc. Rev. 15 (2), 261–298.

Pal, M., Bhardwaj, A., Manickam, M., Tulsawani, R., Srivastava, M., Sugadev, R., Misra, K., 2015. Protective efficacy of the caterpillar mushroom, Ophiocordyceps sinensis (ascomycetes), from India in neuronal hippocampal cells against hypoxia. Int. J. Med. Mushrooms 17 (9), 829–840.

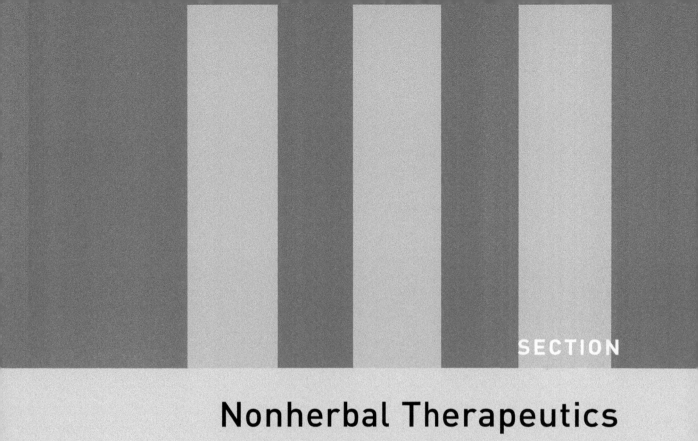

Nonherbal Therapeutics for High-Altitude Illness

CHAPTER 10

Allopathic Remedies

Anuja Bhardwaj*, Kshipra Misra†

*Chemistry Division, Department of Biochemical Sciences (DBCS), Defence Institute of Physiology and Allied Sciences (DIPAS), Delhi, India, †Department of Biochemical Sciences (DBCS), Defence Institute of Physiology and Allied Sciences (DIPAS), Delhi, India

Abbreviations

AMS	acute mountain sickness
CSF	cerebrospinal fluid
FDA	(US) Food and Drug Administration
FVC	forced vital capacity
HA	high altitudes
HACE	high-altitude cerebral edema
HAPE	high-altitude pulmonary edema
LLS	Lake Louise Score
NSAID	nonsteroidal antiinflammatory drug
PDE-5	phosphodiesterase-5
RCTs	randomized controlled trials
SpO_2	pulse oxygen saturation

10.1 INTRODUCTION

The term "allopathy" was coined by Samuel Hahnemann, who also is known as the founder of homeopathy. It was used to designate conventional medicine, which does not follow the principal of homeopathy, meaning "like cures like," but a system that subjugates disease with remedies opposite to patient's symptoms (Cuellar, 2006). Modern allopathic medicine owes its origin to Greco-Roman medicine and Northern European traditions (Go and Champaneria, 2002; Nandha and Singh, 2013). Allopathy is derived from Greek roots meaning "other than the disease" (Cuellar, 2006). Its foundation is laid on the sciences of anatomy, biochemistry, physiology, and the structure-function relationship between cells, tissues, and organs (Go and Champaneria, 2002; Nandha and Singh, 2013).

Management of High Altitude Pathophysiology. https://doi.org/10.1016/B978-0-12-813999-8.00010-0

Mainstream medicine, or allopathic medicine, relies greatly on pharmaceutical drugs (Clarke, 2008). Within this system of medical practice, pharmacotherapy aims to combat disease by the use of drugs or surgery that produce effects different from or incompatible with those produced by the disease itself. Pharmacotherapy is defined as the use of drugs in the treatment of disease (Clarke, 2008; Moscou and Snipe, 2014; Nandha and Singh, 2013). Consequently, allopathic pharmacology treats a disease because of the favorable effects of therapeutic dose of a drug on pharmacodynamics, making it a drug-based pharmacological approach applied on diseased physiology (Dobrescu, 2013). Broadly, allopathy focuses on diagnosis, treatment, and cure for acute illnesses via surgery, radiation, and other treatment modalities, including potent pharmaceutical drugs with major concern for toxicities (Borde et al., 2014; Go and Champaneria, 2002; Nandha and Singh, 2013).

Since its appearance in the nineteenth century, allopathy is the most acceptable medicine therapy among all the systems of medicine. This could be attributed to the fact that it works in conjunction with technology that enables the design of diagnostic procedures, drugs with specifications, vaccines, sophisticated surgical procedures, and transplants (Borde et al., 2014; Dobrescu, 2013; Nandha and Singh, 2013). Allopathy, therefore, is a conclusion of a proven scientific experiment (Nandha and Singh, 2013), which draws a rational and rapid progression (Dobrescu, 2013). Allopathy has contributed a great deal toward diagnosing infectious diseases and surgical interventions (Patwardhan, 2012), apart from its contribution in emergency care (Borde et al., 2014; Patwardhan, 2012). At the same time, allopathy also has several demerits. For example, its inefficacy in curing certain chronic diseases and unavoidable adverse effects, which need to be considered critically to develop an effective and safe healthcare system (Nandha and Singh, 2013).

In this chapter, we intend to provide detailed review on the allopathic remedies for the management of acute altitude illness and their practical recommendations for both prevention and treatment.

10.2   ALLOPATHIC REMEDIES FOR HIGH-ALTITUDE AILMENTS

The most critical environmental feature at high altitudes (HA) is diminished barometric pressure, which results in decreased partial pressure of oxygen at every point along the oxygen transport cascade from ambient air to cellular mitochondria. This hypobaric hypoxia triggers a series of physiological changes, collectively termed as acclimatization response, which allows adaption under a low-oxygen environment. Sometimes an individual, however, might fail to adapt or acclimatize and consequently develops acute mountain sickness (AMS) and more severe complications, such as high-altitude cerebral edema

(HACE) and high-altitude pulmonary edema (HAPE) (Luks et al., 2017). These HA-induced illnesses, their symptoms and pathophysiology, were detailed in Chapter 2, an understanding of which will help the reader better comprehend the pharmacological action and usage of each drug against HA-induced maladies.

Current prophylactic and treatment measures for altitude illness are based on patient risk factors, including a prior history of AMS, rate of ascent, and maximum altitudes for climbing and sleeping. Generally, low-risk patients should rely solely on nonpharmacologic strategies to prevent AMS, whereas moderate-to-severe risk patients might benefit from both behavioral approaches and pharmacologic prophylaxis (Callen et al., 2017). Several allopathic remedies are available to prevent and treat acute altitude illness, so we have focused mainly on pharmacological drugs; the recommended ones are listed in Table 10.1 and detailed in the following sections. Some of the pharmacological drugs available in the market to prevent and treat altitude sickness are depicted in the Fig. 10.1. It is important to note, however, that slow ascent allowing acclimatization is the best means to prevent altitude illness (Johnson and Luks, 2016).

10.2.1 Acetazolamide

Acetazolamide (*Diamox*, Cyanamid GmbH, Wolfratshausen, Germany) is the drug of choice for prevention of AMS, and is the only one approved by the US Food and Drug Administration (Netzer et al., 2013). It is a carbonic anhydrase (CA) inhibitor that is used to prevent symptoms associated with AMS.

Table 10.1 Pharmacological Drugs Recommended in the Prevention and Treatment of Altitude Sickness.

Recommended Drug	Mechanism of Action	Usage	Dosage
Acetazolamide	Hyperventillation via metabolic acidosis	AMS prevention	125 or 250 mg every 12 h
		AMS treatment	250 mg every 12 h
Dexamethasone	Unexplained	AMS prevention	2 mg every 6 h or 4 mg every 12 h
		AMS treatment	4 mg every 6 h
		HACE treatment	8 mg once then 4 mg every 6 h (22)
Nifedipine	Pulmonary vasodilation	HAPE prevention and treatment	30 mg (sustained release) every 12 h
Salmeterol	Alters alveolar epithelial sodium and fluid transport	HAPE prevention	125 µg every 12 h (inhaled)
Sildenfil	Pulmonary vasodilation	HAPE prevention and treatment	50 mg every 8 h
Tadalafil	Pulmonary vasodilation	HAPE prevention and treatment	10 mg every 12 h

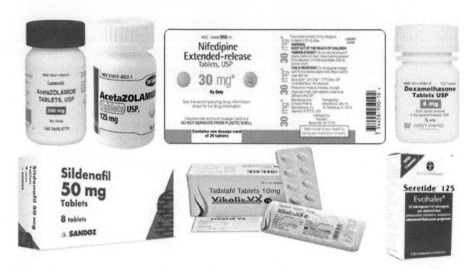

FIG. 10.1

Pharmacological drugs available in the market for the prevention and treatment of altitude sickness.

Treatment with acetazolamide accelerates acclimatization and ameliorates hypoxia (Luks et al., 2014; Schmidt, 2015; Zafren, 2014).

Acetazolamide decreases the risk of AMS by initiating a renal bicarbonate loss and causing a metabolic acidosis. This counteracts the dampening effect of hypocapnia on the full ventilatory response upon initial exposure to hypoxia. With the decrease in pH, peripheral chemoreceptor output increases, leading to a rise in minute ventilation (Luks, 2015; Zafren, 2014). Renal bicarbonate losses also lead to increased efflux of bicarbonate from the cerebrospinal fluid (CSF), leading to a decrease in CSF pH and further stimulation to ventilation via the central chemoreceptors. (Luks, 2015).

Multiple studies advocate the role of acetazolamide in prevention of AMS (Luks et al., 2014). An effective prophylactic dose of acetazolamide is 125 mg twice daily. The pediatric dose is 2.5 mg/kg twice daily up to 125 mg/dose (Luks et al., 2014; Zafren, 2014). Acetazolamide should be started the day before ascent and continued for 2 days after ascent. If ascent continues, it could be stopped after 2 days at the highest sleeping altitude or at the beginning of descent. The second dose should be taken at dinnertime rather than at bedtime because acetazolamide is a diuretic (Zafren, 2014).

In addition, acetazolamide (125 mg a night) (Netzer et al., 2013) maintains oxygenation during sleep and prevents periods of extreme hypoxemia (Callen et al., 2017; Luks et al., 2014; Schmidt, 2015; Zafren, 2014). Other sleep-improving drugs include temazepam (7.5–10 mg per night) and low-dose sustained-release theophylline (Netzer et al., 2013).

Although higher acetazolamide doses (up to 750 mg daily) are more effective at preventing AMS compared to a placebo, they are associated with more frequent or increased side effects and does not provide greater efficacy than lower doses, and therefore are not recommended for prevention (Luks et al., 2014). Acetazolamide, as a nonantibiotic sulfonamide, should not be used for individuals with a history of sulfa allergy, hepatic disease or marked renal impairment, adrenocortical insufficiency, hyponatremia, hypokalemia, or metabolic acidosis. The common adverse effects of acetazolamide include diuresis, dysgeusia (especially with carbonated beverages), glycemic and/or electrolyte changes, metabolic acidosis, and paresthesias (peritorbital region, hands, and feet). Elderly patients, those with diabetes mellitus, and those on concurrent high-dose aspirin might be more likely to experience these adverse effects (Callen et al., 2017; Netzer et al., 2013; Zafren, 2014). Acetazolamide taken for prevention affects patients with renal failure (metabolic acidosis), hepatic insufficiency (ammonium ion toxicity), chronic obstructive pulmonary disease (dyspnea), and pregnant hikers (dyspnea), and must be avoided (Netzer et al., 2013).

10.2.2 Dexamethasone

Dexamethasone is effective in preventing and treating both AMS and HACE, but it does not aid acclimatization like acetazolamide (Netzer et al., 2013; Schmidt, 2015; Zafren, 2014). If dexamethasone is discontinued at HA before acclimatization, a rebound effect is observed. Acetazolamide is preferred to prevent AMS while ascending, with dexamethasone reserved for treatment as an adjunct to descent (Schmidt, 2015; Zafren, 2014). When acclimatization is not possible, dexamethasone is a better choice for prophylaxis because it is relatively rapid acting, highly effective, and does not depend on acclimatization for its effect. Additionally, it might be used in patients who are allergic to acetazolamide. It is more effective than acetazolamide at rapidly relieving the symptoms of moderate to severe AMS (Schmidt, 2015).

Although its role in AMS prevention has been established in research studies and clinical experience, the mechanism by which it plays this role remains unexplained. In a study of HAPE-susceptible individuals ascending to 4559 m, dexamethasone has been shown to decrease pulmonary artery pressure and prevent HAPE. The mechanism underlying this finding is indistinct but might be associated to its effects on sympathetic activity, pulmonary capillary permeability, epithelial sodium transport, endothelial nitric oxide synthetase activity, and availability of nitric oxide (Luks, 2015). In AMS prevention, the recommended adult doses of dexamethasone are 2 mg every 6 h or 4 mg every 12 h (Luks et al., 2014; Zafren, 2014). Because altitude illness takes several hours to develop, dexamethasone should be started on the day of ascent and discontinued after 2–3 days at the highest sleeping altitude or when starting descent. Dexamethasone should not be used for more than 10 days in order to prevent steroid toxicity and adrenal suppression. Dexamethasone should not be used in children because of safety issues (Luks et al., 2014; Zafren, 2014).

Dexamethasone might have a role in HAPE treatment considering its potential role in HAPE prevention, and studies demonstrating effects on maximum exercise capacity, pulmonary inflammation, and ion-transporter function in hypoxia. Although reports document its clinical use in this regard, no study has established whether it is effective (Luks et al., 2014). When HAPE is associated with severe AMS, a combination therapy including nifedipine plus dexamethasone is strongly recommended (Netzer et al., 2013).

Dexamethasone is also an important rescue medication for HACE in extended excursions above 3000–4000 m when medical care is unavailable, provided it is used early (Schmidt, 2015). Extensive clinical experience supports the use of dexamethasone in patients with HACE. It is administered at a dose of 8 mg (intramuscular, intravenous, or orally) followed by 4 mg every 6 h until symptoms resolve. The pediatric dose is 0.15 mg/kg/dose every 6 h (Luks et al., 2014).

The usage of dexamethasone is contraindicated in patients with systemic fungal infections. It should be used cautiously in patients with cardiovascular disease, diabetes mellitus, gastrointestinal disease, psychiatric disorders, and hepatic or renal diseases. Common adverse effects of dexamethasone include fluid retention, dyspepsia, hyperglycemia, and mood changes (Callen et al., 2017).

10.2.3   Ibuprofen

Ibuprofen is a nonsteroidal antiinflammatory drug (NSAID) (Shah et al., 2015) that might prevent headaches on ascent to a moderate altitude, i.e., below 3500 m (Schmidt, 2015). Ibuprofen in a metaanalysis comprising only three randomized controlled trials (RCTs) has shown significant improvement in Lake Louise Score (LLS) over placebo. However, one might infer that ibuprofen would improve headache of any origin because it is a known headache medication. In another RCT, ibuprofen had effects equivalent to acetazolamide on the severity of AMS using LLS. Both studies used a dose of 600 mg three times daily (Luks et al., 2014; Netzer et al., 2013; Zafren, 2014), starting 6–24 h before ascent and continuing for a minimum of three doses (Callen et al., 2017). Considering these trials, the most recent international guidelines on treatment of acute altitude illness does not recommend the routine use of ibuprofen as prophylaxis against AMS (Shah et al., 2015). Therefore, it is unclear whether or not NSAID are useful as prophylaxis or treatment in high-risk situations or for more severe disease (Schmidt, 2015).

10.2.4 Nifedipine

Nifedipine is a calcium-channel blocker that acts by reducing pulmonary vascular resistance and pulmonary artery pressure, as well as systemic resistance and blood pressure (Luks, 2015; Schmidt, 2015). A single randomized, placebo-controlled study and extensive clinical experience have established a role for nifedipine in HAPE prevention in susceptible individuals (Zafren, 2014). It also has been used in the treatment of HAPE (Netzer et al., 2013). Nifedipine and other phosphodiesterase-5 (PDE-5) inhibitors are considered as the preferred first-line agents for HAPE prevention, because lowering pulmonary artery pressure appears to be a more effective prophylaxis strategy as compared to a beta-agonist (Luks, 2015).

The recommended dose is 30 mg of the extended-release preparation administered twice daily (Luks et al., 2014). It should be started the day of ascent and continued for 5 days at the highest sleeping altitude unless the individual descends sooner (Zafren, 2014). It should not be relied on as the sole therapy, however, unless descent is impossible and access to supplemental oxygen or portable hyperbaric therapy cannot be arranged (Luks et al., 2014). Although nifedipine appears to be an ideal drug for prevention or treatment of HAPE, travelers with underlying renal disease might develop complications while taking this drug. Those with significant underlying liver diseases might experience an increased risk of drug accumulation while taking nifedipine (Netzer et al., 2013).

10.2.5 Salmeterol

Salmeterol is a long-acting inhaled beta-agonist (Netzer et al., 2013; Schmidt, 2015; Zafren, 2014). It acts by increasing sodium and fluid transport out of the alveolar space through the stimulation of amiloride-sensitive sodium channels on the apical membrane and possibly by augmenting Na^+-K^+-ATPase activity on the basolateral membrane (Luks, 2015).

It is reported for its use in the prevention and treatment of HAPE (Luks et al., 2014; Zafren, 2014). The prophylactic inhalation of salmeterol (125 µg, twice daily) might be effective in reducing the incidence of HAPE in susceptible individuals (Schmidt, 2015; Zafren, 2014). In a single randomized, placebo-controlled study, salmeterol decreased the incidence of HAPE by 50% in susceptible individuals (Luks et al., 2014). Very high doses (125 µg twice daily), however, are more likely to cause significant side effects (Luks et al., 2014; Zafren, 2014). Salmeterol is not recommended as monotherapy and must be considered as a supplement to nifedipine in high-risk individuals with a clear history of recurrent HAPE (Luks et al., 2014; Schmidt, 2015; Zafren, 2014).

10.2.6   Sildenafil

Sildenafil is a selective PDE-5 inhibitor that, by augmenting the pulmonary vasodilatory effects of nitric oxide, decreases hypoxic pulmonary hypertension in HAPE (Luks et al., 2014; Schmidt, 2015; Shah et al., 2015) with less effect on systemic blood pressure (Schmidt, 2015). No systematic studies have evaluated the role of sildenafil in HAPE treatment (Luks et al., 2014), however, based upon its mechanism of action, sildenafil might be an effective adjunct treatments for established HAPE, when neither descent nor oxygen supplementation are feasible (Schmidt, 2015).

10.2.7   Tadalafil

Tadalafil, as a selective PDE-5 inhibitor, has an effect similar to sildenafil and thus it decreases pulmonary hypertension by reducing hypoxic pulmonary vasoconstriction in HAPE (Schmidt, 2015; Zafren, 2014). The suggested oral dose of tadalafil for prophylaxis and treatment of HAPE is 10 mg, every 12 h (Zafren, 2014). In a single randomized, placebo-controlled trial 10 mg twice daily of tadalafil was effective in preventing HAPE in susceptible individuals. However, the sample size in the study was insignificant, and two subjects experienced severe AMS on arrival at 4559 m, although they did not have HAPE prior to descent. Therefore, the clinical experience with tadalafil is lacking compared to nifedipine, and it cannot be recommended to prevent HAPE in the absence of further study (Luks et al., 2014; Zafren, 2014).

10.2.8   Others

Several other pharmacological drugs are emerging candidates, including benzolamide and budesonide, or have restricted regional usage, such as sorojchi pills.

10.2.8.1   Benzolamide

Benzolamide is a more potent CA inhibitor and is more hydrophilic than acetazolamide (Collier et al., 2016). These attributes lead to very limited membrane permeability, restricting its uptake into the central and peripheral nervous systems and other organs in addition to the kidneys (Swenson, 2014). Benzolamide has a strong renal action to engender a metabolic acidosis and ventilatory stimulus that improves oxygenation at HA and reduces AMS. Collier et al., 2016, in a comparative double-blind, placebo-controlled study, observed that benzolamide (100 mg twice daily) was equally effective as prophylaxis for AMS as acetazolamide (both at 125 mg twice daily, and 250 mg twice daily) on exposure to an altitude of 5340 m. The authors also demonstrated, via extensive and well-controlled psychometric examination, that benzolamide had lesser side effects on the central nervous system than acetazolamide at sea level

(Collier et al., 2016). This study, therefore, provides insight into the development of a new prophylactic solution for AMS with fewer side effects.

10.2.8.2 Sorojchi Pills

Soroche or *sorojchi* is a South American word for altitude illness. Sorojchi pills are composed of aspirin (160 mg), salophen (160 mg), and caffeine (15 mg) (Ballon-Landa, 2001; Zafren, 2014). Sold over the counter in Bolivia and Peru (Zafren, 2014), they are marketed for prevention and treatment of altitude illness. Salophen is an ester of acetylsalicylic acid and paracetamol, which decomposes within the gut into aspirin and paracetamol. Sorojchi pills have not been tested and have no known benefit in preventing AMS, but are suggested to be effective in reducing headaches (Zafren, 2014). Travelers have reported their recommendation over acetazolamide (Croughs et al., 2011).

10.2.8.3 Budesonide

Zheng et al., 2014, conducted a double-blind, randomized controlled trial to examine the prophylactic efficacy of inhaled budesonide and oral dexamethasone in comparison to placebo against AMS after acute HA exposure (3900 m). During the study, 138 healthy young male lowland residents were recruited and randomly assigned to receive inhaled budesonide (200 µg, twice a day), oral dexamethasone (4 mg, twice a day), or placebo (46 in each group). After HA exposure, significantly fewer participants in the budesonide (23.81%) and dexamethasone (30.77%) groups developed AMS compared with the placebo group (60.46%). Both the budesonide and dexamethasone groups had lower heart rate and higher pulse oxygen saturation (SpO_2) at altitude than the placebo group. Only the budesonide group demonstrated less deterioration in forced vital capacity (FVC) and sleep quality than the placebo group. Budesonide also caused fewer adverse reactions than dexamethasone. Those who develop AMS tend to have lower SpO_2, lower pulmonary diffusing capacity, and higher alveolar-arterial oxygen pressure difference (Zheng et al., 2014). This study highlights another candidate drug for combating the ill-effects of AMS with fewer side effects.

10.3 CONCLUSIONS

Altitude illness is a constellation of syndromes, such as AMS, HACE and HAPE. Allopathy, as the modern system of medicine, has progressively provided various pharmacological interventions for both prophylaxis and treatment of the health complications associated with HA. However, there still is a lot of room for further development of better-suited allopathic remedies with fewer adverse effects. This subsequently necessitates acquaintance with current research and development in the same field. With this perspective, we have tried to present most of the allopathic remedies (pharmacological drugs) being used.

Acknowledgment

The authors are thankful to the director, Defense Institute of Physiology and Allied Sciences, Delhi, India, for his constant support and encouragement.

References

Ballon-Landa, G., 2001. Practical lessons from a family sojourn to equatorial South America. Infect. Dis. Clin. Pract. 10 (6), 307–311. Zafren, K., 2014. Prevention of high-altitude illness. Travel Med. Infect. Dis. 12(1), pp.29–39.

Borde, M.K., Lalan, H.N., Ray, I.M., Goud, T.S.K., 2014. Health awareness and popularity of allopathic, ayurvedic, and homeopathic systems of medicine among Navi Mumbai population. World J. Pharm. Pharm. Sci. 3 (9), 783–788.

Callen, E.D., Brooks, K.G., Kessler, T.L., 2017. The pharmacist's role in the treatment and prevention of acute mountain sickness. US Pharmacist 42 (7), 22–26.

Clarke, S., 2008. Health, disease, and therapy. In: Clarke, S. (Ed.), Essential Chemistry for Aromatherapy, second ed. Elsevier Health Sciences, pp. 115–121.

Collier, D.J., Wolff, C.B., Hedges, A.M., Nathan, J., Flower, R.J., Milledge, J.S., Swenson, E.R., 2016. Benzolamide improves oxygenation and reduces acute mountain sickness during a high-altitude trek and has fewer side effects than acetazolamide at sea level. Pharmacol. Res. Perspect. 4(3).

Croughs, M., Van Gompel, A., Van den Ende, J., 2011. Acute mountain sickness in travelers who consulted a pretravel clinic. J. Travel Med. 18 (5), 337–343.

Cuellar, N.G., 2006. Conversations in Complementary and Alternative Medicine: Insights and Perspectives from Leading Practitioners. Jones & Bartlett Learning.

Dobrescu, D., 2013. The Biggest Mistake in the History of the Drug. First J. Homeopathic Pharmacol. I (2), 41–50. (Published under the auspices of the Romanian Academy of Medical Sciences).

Go, V.L., Champaneria, M.C., 2002. The new world of medicine: prospecting for health. Nihon Naika Gakkai zasshi. J. Japan. Soc. Internal Med. 91, 159–163.

Johnson, N.J., Luks, A.M., 2016. High-altitude medicine. Med. Clin. 100 (2), 357–369.

Luks, A.M., 2015. Physiology in medicine: a physiologic approach to prevention and treatment of acute high-altitude illnesses. J. Appl. Physiol. 118 (5), 509–519.

Luks, A.M., McIntosh, S.E., Grissom, C.K., Auerbach, P.S., Rodway, G.W., Schoene, R.B., Zafren, K., Hackett, P.H., 2014. Wilderness medical society practice guidelines for the prevention and treatment of acute altitude illness: 2014 update. Wilderness Environ. Med. 25 (4), S4–S14.

Luks, A.M., Swenson, E.R., Bärtsch, P., 2017. Acute high-altitude sickness. Eur. Respir. Rev. 26 (143).

Moscou, K., Snipe, K., 2014. Pharmacology for Pharmacy Technicians-E-Book. Elsevier Health Sciences.

Nandha, R., Singh, H., 2013. Amalgamation of ayurveda with allopathy: a synergistic approach for healthy society. Int. J. Green Pharm. 7 (3), 173–176.

Netzer, N., Strohl, K., Faulhaber, M., Gatterer, H., Burtscher, M., 2013. Hypoxia-related altitude illnesses. J. Travel Med. 20 (4), 247–255.

Patwardhan, B., 2012. Health for India: search for appropriate models. J. Ayurveda Integrat. Med. 3 (4), 173.

Schmidt, S., 2015. Altitude sickness prevention and treatment. SA Pharm. J. 82 (3), 24–27.

Shah, N.M., Hussain, S., Cooke, M., O'Hara, J., Mellor, A., 2015. Wilderness medicine at high altitude: recent developments in the field. Open Access J. Sports Med. 6, 319–328.

Swenson, E.R., 2014. Carbonic anhydrase inhibitors and high-altitude illnesses. Sub-Cell. Biochem. 75, 361–386.

Zafren, K., 2014. Prevention of high-altitude illness. Travel Med. Infect. Dis. 12 (1), 29–39.

Zheng, C.R., Chen, G.Z., Yu, J., Qin, J., Song, P., Bian, S.Z., Xu, B.D., Tang, X.G., Huang, Y.T., Liang, X., Yang, J., 2014. Inhaled budesonide and oral dexamethasone prevent acute mountain sickness. Am. J. Med. 127 (10), 1001–1009.

CHAPTER 11

Homeopathic Remedies

Anuja Bhardwaj*, Kshipra Misra†

*Chemistry Division, Department of Biochemical Sciences (DBCS), Defence Institute of Physiology and Allied Sciences (DIPAS), Delhi, India, †Department of Biochemical Sciences (DBCS), Defence Institute of Physiology and Allied Sciences (DIPAS), Delhi, India

Abbreviations

A. montana	*Arnica montana*
AMS	acute mountain sickness
AYUSH	Ayurveda, Yoga and Naturopathy, Unani, Siddha and Homeopathy
c	centesimal
C. gigantea	*Calotropis gigantea*
CAM	complementary and alternative medicine
Carbo veg	Carbo vegetabilis
cH	centesimal Hahnemann
HA	high-altitudes
HACE	high altitude pulmonary edema
HAPE	high altitude cerebral edema
ISM and H	Indian Systems of Medicine and Homeopathy
NO	nitric oxide
UV	ultraviolet
V/V	volume by volume
VEGF	vascular endothelial growth factor

11.1 INTRODUCTION

Complementary and alternative medicine (CAM) has been described as any diagnosis, treatment, or prevention that adjuncts conventional medicine. The application of CAM has increased tremendously during the past few years and has acquired medical, economic, and sociological importance. One component of CAM, homeopathy is (Molassiotis et al., 2005) a medical discipline intended to make the sick healthy and bring about an instant, mild, and permanent cure. At the outset, a homeopathic practitioner (homeopath) analyzes all the symptoms, common and uncommon, of an illness. The homeopath then prescribes an individualized remedy or a complex of two or more

217

remedies, whichever is most effective to alleviate that illness or disease (Morris, 2015).

Homeopathy, an important component of 19th century health care, recently has undergone a global revival (Jonas et al., 2003). This alternative therapeutic system is based on the Principle of Similars and the use of minimum doses (Morris, 2015; Jonas et al., 2003). Homeopathy uses preparations of substances whose effect, when administered in healthy individuals, correspond to the manifestations including the symptoms, clinical signs, and pathological states in the patient (Ernst, 2002; Jonas et al., 2003). These preparations of homeopathic medicines are prepared by serial dilutions and shaking and usually are given in potentized forms (Jonas et al., 2003; Ledermann, 1945). In the modern era, the clinical application of homeopathic research is in its infancy, including as pertains to high-altitude (HA) medicine. As more consideration is being focused on alternative methods of treatment, some clinical trials pertaining to homeopathy have been conducted (Shackelton et al., 2000).

This chapter summarizes scientific findings and suggests areas where further research is required for the establishment of homeopathic remedies against HA pathophysiological conditions, such as acute mountain sickness (AMS). It reviews key homeopathic remedies, including homeopathic coca, *Aloe vera*, and other potential candidates against HA maladies.

11.2   THE ORIGIN AND PRINCIPALS OF HOMEOPATHY

A German physician, Samuel Christian Hahnemann (1755–1843), is credited with the origin of homeopathy (Relton et al., 2017). In 1790, he came across an herbal text where he found that Cinchona bark (*China officinalis)* cured malaria because it was bitter, an explanation given by Scottish physician William Cullen (1710–90). However, Hahnemann was not convinced by this explanation and decided to perform few experiments. He started taking repeated doses of Cinchona bark powder to determine its effects, which appeared later. Those effects were noticeably similar to the symptoms of malaria. He concluded that Cinchona bark acted because it could produce similar symptoms to those of intermittent fever (malaria) in healthy individuals. He tested several other medicines in a similar manner, using himself, family members, and friends as research subjects. He repeatedly noticed that a drug given to a healthy person would cause symptoms similar to the ones it cured in sick patients. This procedure was called a "proving" or, in modern homeopathy, a "human pathogenic trial." After years of obsessively studying this phenomenon, Hahnemann concluded that he had discovered the first principle of homeopathy: *similia similibus curantur* or the "like cures like" principle. Another theory by Hahnemann stated that the effects of drugs are enhanced when given in minute doses, which are obtained by carrying dilution or trituration to the extreme. His third principle was the notion that most chronic diseases are only a manifestation of suppressed itch or psora (Ernst, 2016).

In the 1820s, Hahnemann noticed the paradoxical phenomenon that serial dilution combined with succussion (shaking of medication) rendered his remedies, not weaker, but stronger. He called this process of dilution and shaking as potentization or dynamization, and proposed that it was capable of awakening healing powers of even otherwise inert substances. Initially, Hahnemann followed these steps mainly because he needed to minimize the side effects of the frequently toxic substances he employed as medicines (Ernst, 2016), but later he claimed that this potentization process extracted the vital or "spirit-like" nature of these substances. The limit of molecular dilution (Avagadro's number) was not discovered until the later part of Hahnemann's life. By then, homeopaths all over the world were reporting that even very high potencies (dilutions lower than Avagadro's number) produced clinical effects. The implausibility of such claims has led many to dismiss any evidence of homeopathy's effectiveness as artifact or delusion (Jonas et al., 2003; Relton et al., 2017).

Therefore, in addition to selecting the best remedy for a disease, it is also equally important for a homoeopath to choose the appropriate potency. The potency of a remedy is defined by the number of dilutions and succussions that the remedy undergoes (Jonas et al., 2003; Morris, 2015). A cH potency is a centesimal potency, manufactured according to Hahnemann's methods, and one part of the crude substance or "mother tincture" to 99 parts of solvent. Usually, 6cH is a commonly used low potency. Low potencies act mostly on physical symptoms of a disease, directly within structures or function; higher potencies act on deeper pathology that affects the mental and emotional aspects of a disease, and its action lasts longer. Low potencies generally are given in complexes and in chronic illnesses that manifest at a physical level (Morris, 2015).

11.3 WORLDWIDE PREVALENCE AND FURTHER SCOPE OF HOMEOPATHY

Homeopathy could play a vital role in the treatment of acute and urgent aspects of illness resulting from mechanical, psychological injuries, or other causes (Roberts et al., 2016). Homeopathic medicines are formed from a wide variety of substances—plants, animals, minerals, or chemicals. To diminish toxicity, the medicinal substances are diluted successively and shaken vigorously between each dilution step (Relton et al., 2017).

Some researchers say homeopathic medicines are commonly considered to carry a low risk of causing serious side effects (Roberts et al., 2016).

Many researchers, however, claim that the principles upon which homeopathy is based are scientifically implausible. Despite this, treatment by homeopaths and the provision of homeopathic medicines remain popular, and it has been endorsed by a number of governments worldwide with inclusion of its provision in a number of publicly funded healthcare systems. India, for example, has an estimated 300,000 practitioners of homeopathy, and homeopathy is part of the Indian Ministry of Health. In France, 43.5% of the country's healthcare providers prescribe homeopathic medicines (mostly along with allopathic medicines). In the United Kingdom, homeopathy has been provided by the National Health Service since its inception in 1948 (Relton et al., 2017).

Homeopathy, although of German origin, is widely practiced in India, yet the government has recognized the disservice that homeopathy had done to traditional systems of medicine. To promote the use of traditional medicine, the government in 1995 created the Department of Indian Systems of Medicine and Homeopathy (ISM and H), which was renamed the Department of Ayurveda, Yoga and Naturopathy, Unani, Siddha and Homeopathy (AYUSH) in 2003. India thus has given official recognition to multiple systems of medicine, including homeopathy (Roy, 2015).

Some randomized, placebo-controlled trials and laboratory research report unexpected effects of homeopathic medicines. However, evidence on the effectiveness of homeopathy for specific clinical conditions is limited and poorer in quality than research done on allopathic medicine. More and better research, which must be unbiased by belief or disbelief in the system, is required. Until homeopathy is better understood, physicians must accept homeopathy's possibility and maintain communication with patients who use it (Jonas et al., 2003).

11.4 HOMEOPATHIC REMEDIES FOR HIGH-ALTITUDE AILMENTS

As described in earlier chapters, individuals ascending to altitudes higher than 2500 m are at risk of developing AMS, and the more serious high-altitude cerebral edema (HAPE) and high-altitude pulmonary edema (HACE). Individual susceptibility and the rate and extent of exposure to HA are the most important etiological factors for AMS, HAPE and HACE (Harrison et al., 2016). Several medicinal systems, including homeopathy, contribute toward the prophylaxis and treatment of the medical conditions related to HA exposure. Numerous homeopathic medicines, mostly plant-based, are being used to treat HA ailments. Some of these are discussed below.

11.4.1   *Aloe vera*

Aloe vera Linn. (Fig. 11.1) is a succulent herb that belongs to the *Liliaceae* family (Manickam et al., 2014). It exhibits numerous bioactivities, such as antiallergic, anticancer, antioxidant, antimicrobial, antiinflammatory, immunomodulatory, hepatoprotective, antiulcer, and antidiabetic (Sánchez-Machado et al., 2017). Some of these activities occur because of the presence of the polysaccharides, acemamman and glucomannan (Sánchez-Machado et al., 2017).

Frostbite is considered the severest from of cold injury and a major medical problem for people at high altitudes. Frostbite can involve loss of digits as a traumatic form of cold injury with tissue loss. In an investigation led by Sarkar et al. (1996), *Aloe vera* (mother tincture) was given both topically (Q) and orally (200 potency), followed by complementary and supplementary remedies in sequence for 7 days after cold exposure. The results demonstrated prevention of tissue damage by restoring the microcirculation and prevented necrosis of the affected areas. It also was observed that other adverse effects of cold exposure, such as reduced appetite and loss in body weight, also were improved (Sarkar et al., 1996; Selvamurthy and Basu, 1998). Later, an *Aloe vera*-based homeopathic topical formulation, DIP-1, was developed by Defense Institute of Physiology and Allied Sciences (DIPAS), India, as the sole homeopathic remedy to treat frostbite. Studies pertaining to this formulation suggested that the presence of prostaglandins, which are known inducers of angiogenesis, imparted it the ability to promote neovascularization at (ischemic) wound site via nitric oxide (NO) signaling, and subsequently, caused tissue regeneration. The topical application

FIG. 11.1
The succulent herb, *Aloe vera*.

of DIP-1 also has demonstrated its ameliorative effects against UV-induced erythema, edema, blistering, and hyperplasia observed during high-altitude exposure. This protective effect of DIP-1 is attributable to the bioactive compound, prostaglandin E_2. Prostaglandin E_2 is a potent inflammatory mediator involved in skin erythema and inflammation in animal/human skin in response to UV exposure. It controls vascular endothelial growth factor (VEGF), a potent vascular permeability element. Studies have indicated that DIP-1 mediated regeneration of dermal fibroblast and epidermal keratinocytes via upregulation of NO signaling in human endothelial cells. It is now well-established by a number of human trials that DIP-1 aids in generating thermogenesis in frostbite-affected areas and also protects against UV-induced skin damage as observed on exposure to high altitudes (Manickam et al., 2014).

11.4.2 *Arnica montana*

Arnica montana (*A. montana*) has been used for centuries in the homeopathic system of medicine for the treatment of 66 pathological conditions, frequently contusions, wounds, rheumatism, and inflammation. According to European *Pharmacopeia* (1809), *A. montana* tincture is produced from *A. montana* flowers (Fig. 11.2) with 0.04% sesquiterpene lactones expressed as dihydrohelenalin tiglate.

FIG. 11.2
The plant *Arnica montana* with yellow-colored flowers.

The tincture contains one part of the drug in 10 parts of ethanol [60% (V/V) to 70% (V/V)]. According to the European Union, herbal preparations containing *A. montana* are tincture (1:10) extracted with ethanol 70% (V/V), tincture (1:5) extracted with ethanol 60% (V/V) and liquid extract (1:20) extracted with ethanol 50% m/m, mainly of flowers. Tincture is dried by evaporation, and the extract is incorporated in numerous herbal drug products.

Several studies advocate the potential efficacy of homeopathic *A. montana* in ameliorating ill-effects associated with high-altitudes exposure. For instance, Paradise L. in US patent No. 5795573 A has established that homeopathic topical antiinflammatory preparations containing synergistic combination of extracts from *A. montana*, *Rhus toxicodendron*, and *Aesculus hippocastanum* and *belladonna* can be used to treat muscular cramps, soreness, and pain. Another study revealed that *Arnica*, in combination with *Ruta graveolens*, *Aconitum napellus*, *Bellis perennis*, *Hamamelis virginiana*, *Hypericum perforatum*, *Calendula officinalis*, *Ledum palustre*, and *Bryonia alba*, is effective for treating inflammation. All these medical conditions are seen with exposure to high altitudes. *Arnica* 6c has been investigated for its antiinflammatory potential on carrageenan and rat paw edema induced by nystatin. It also has considerably reduced inflammation of histamine-induced edema. The authors suggested that this effect came from the ability of *Arnica* 6c to inhibit the action of histamine and increase vascular permeability. Another research reported similar results when a solution of *A. montana* 6cH, dexamethasone or 5% hydroalcoholic solution was injected into male adult Wistar rats.

Homeopathic *A. montana* (*Arnica* 30cH) also exhibits antioxidant activity as reported by Camargo et al. in 2013. Their study revealed that homeopathic *A. montana* improved mitochondrial oxidative stress and lipid peroxidation in rat liver (Kriplani et al., 2017).

11.4.3 *Calotropis gigantea*

Calotropis gigantea is a weed that grows in the wastelands of Africa and Asia. It is commonly called "crown flower," "giant milkweed," and "shallow wort" (Kanchan and Atreya, 2016). In India, it is called "aak," "akauwa," or "arka" (Gautam and Bhadauria, 2009; Kanchan and Atreya, 2016). This plant could be identified through its thick oblong leaves and odorless flowers, which are purplish in color (Fig. 11.3). In general, *C. gigantea* is used to treat asthma, colds, coughs, diarrhea, fever, indigestion, leprosy, leukoderma, and rheumatism in Ayurveda, Chinese, and homeopathic medicines (Kanchan and Atreya, 2016). Further, as mentioned in homeopathic *Materia Medica*, it also is used to treat toothache, elephantiasis, purging, and vomiting (Gautam and Bhadauria, 2009). DIP-2, a product developed by DIPAS, India, from the alcoholic extract of *C. gigantea*, has been found to be a potential skin permeation enhancer. This formulation, in combination with DIP-1 as one of its active ingredients, has

FIG. 11.3
Calotropis gigantea plant with thick oblong leaves and purplish flowers.

been studied as a potential auto-debridement and tissue regenerative agent. DIP-2 has numerous effects, such as antiinflammatory, antioxidant, antibiotic, vasodilation, and wound healing (Manickam et al., 2014).

11.4.4   Carbo Vegetabilis

Carbo vegetabilis is a homeopathic preparation of vegetable (wood) charcoal prepared by burning under inadequate oxygen supply. During this form of, hydrocarbon molecules merge with oxygen to form carbon dioxide and water, a condition similar to that observed in energy metabolism under low oxygen availability at tissue level. In accordance to the first principle of homeopathy, it is expected that vegetable charcoal Carbo vegetabilis (Carbo veg) is homeopathic to the state of hypoxia induced at high altitudes where inadequate oxygen is available for metabolism in the tissues. Patients prescribed with Carbo veg in altitude illness generally would have an air hunger and a craving for air blowing on their face (either from wind or by fanning). They often feel faint and can have a sense of weight in the eyes, eyelids, head, or elsewhere. A gassy, distended abdomen also is common (Dooley, 2001).

11.4.5   Homeopathic Coca

The coca plant (*Erythroxylum coca*) grows wild across the regions of Central and South America, where it has been used traditionally for medicinal purposes. It is widely recognized for its action in enhancing work capacity, including the reduction of fatigue and the mitigation of thirst and hunger. Coca leaf tea is commonly used as an anecdote by the travelers and climbers to South American countries for its apparent symptomatic relief to acute mountain sickness (AMS) (Biondich and Joslin, 2015; Shackelton et al., 2000; Havryliuk et al., 2015).

Shackelton et al. (2000) performed a placebo-controlled, single-blind, nonrandomized study in which homeopathic coca preparations were provided to 11 members of the 1998 Everest Challenge Expedition, with a control group of 13 climbers. In this study, the homeopathic coca significantly reduced the effects of altitude on trekkers in the experimental group when compared with placebo, as evident from the significant increase in oxygen saturation and a decrease in AMS symptoms.

DIPAS, India, has developed a homeopathic formulation, named DIP-3, using the alcoholic leaf extract of the plant *Erythroxylum coca* as the mother tincture. This homeopathic remedy prevents the symptoms of AMS by imparting immediate acclimatization and thus improves the overall well-being of the body at any degree of altitude. It also can be used as a topical anesthetic, analgesic, diuretic, antispasmodic, antidepressant, and mood-elevating energizing tonic meant for normalizing bodily physiology (Manickam et al., 2014).

11.4.6 *Lycopodium clavatum*

Lycopodium clavatum (Fig. 11.4) is commonly known as club moss, clubfoot moss, foxtail, ground pine, sulfer, and wolf's claw. It is a pteridophyte abundantly found in tropical, subtropical and in many European countries. In homeopathy, it is used in the treatment of aneurisms, constipation, fevers, and chronic lung and bronchial disorders. It also reduces gastric inflammation, simplifies digestion, and helps in treatments of chronic kidney disorders. Several studies support the analgesic, antioxidant, anticancer, antimicrobial, antiinflammatory, neuroprotective, immunomodulatory, and hepato-protective activity of *Lycopodium clavatum*. It also can mitigate tiredness and chronic fatigue. The spores of *Lycopodium clavatum* are used routinely by native Americans in treating nose bleeds and in healing wound (Banerjee et al., 2014).

11.4.7 Shilajit (*Asphaltum punjabianum*)

Shilajit (also known as shilajatu or mineral pitch) is a highly recommended drug in the Ayurvedic medicine system of India (Jayawardene et al., 2010; Pande et al., 2017; Sharma and Chaudhary, 2015). It is also a homeopathic treatment used by inhabitants of India and Pakistan (Witcher et al., 2014). Shilajit is made by scraping the dried exudate of the plant *Asphaltum punjabinum* found on rock (Jayawardene et al., 2010). It is a pale-brown to the blackish-brown herbomineral drug (Sharma and Chaudhary, 2015) that contains humus, organic plant materials, and fulvic acid as the major components

FIG. 11.4
Lycopodium clavatum, a petridophyte growing in the wild.

FIG. 11.5
The herbomineral, Shilajit.

(Fig. 11.5). Fulvic acid is reported to facilitate the transportation of essential minerals into cells to maintain and restore their electrical potency, preventing their decay and death (Sharma and Chaudhary, 2015). Shilajit is reported to aid in metabolism by maintaining the equilibrium between catabolism and anabolism, promote energy production, and enhance the absorptive and detoxifying capacity of the body. It also stimulates the immune system and blood formation within the body (Sharma and Chaudhary, 2015). The health benefits of Shilajit also include promotion of longevity, rejuvenation, increase in physical strength, and antiaging activity (Pande et al., 2017).

11.4.8   Others

Agaricus muscarius 200 and Pulsatilla 200, are two other most-often indicated remedies, besides *Aloe vera*, for chilblains, covering about two-thirds of cases. *Agaricus* is worse for cold and has symptoms resembling frostbite. Pulsatilla is worse in the evening and from exposure to heat (Foubister, 1962). At high altitudes, anxiety and depression are commonly observed neuroendocrine symptoms among affected individuals (Strewe et al., 2017). Homeopathic L72 for depression (Pilkington et al., 2005) and Aconitum and *Argentum nitricum* (for adults) (Pilkington et al., 2006) are some of the homeopathy medicines that could be used for HA-associated anxiety and depression because they have a general and not individualized action.

11.5   CONCLUSIONS

Homeopathy is an alternative therapeutic medicinal system that provides numerous remedies for various medical conditions and/or diseases. The evidence of the effectiveness of homeopathy for specific clinical conditions such as HA-induced pathophysiologies is limited. Therefore, more research is required with better understanding of the concept of homeopathy and its potential role in ameliorating various health ailments, including high-altitude maladies.

Acknowledgment

The authors thank to the Director, Defense Institute of Physiology and Allied Sciences, Delhi, India, for his constant support and encouragement.

References

Banerjee, J., Biswas, S., Madhu, N.R., Karmakar, S.R., Biswas, S.J., 2014. A better understanding of pharmacological activities and uses of phytochemicals of *Lycopodium clavatum*: a review. J. Pharmacogn. Phytochem. 3 (1), 207–210.

Biondich, A.S., Joslin, J.D., 2015. Coca: high-altitude remedy of the ancient Incas. Wilderness Environ. Med. 26 (4), 567–571.

Dooley, T.R., 2001. Help for Altitude Illness. National Center for Homeopathy. Available at:http://www.homeopathycenter.org/homeopathy-today/help-altitude-illness. Accessed 17 December 2017.

Ernst, E., 2002. A systematic review of systematic reviews of homeopathy. Br. J. Clin. Pharmacol. 54 (6), 577–582.

Ernst, E., 2016. Definition and main principles of homeopathy. In: Homeopathy-The Undiluted Facts. Springer International Publishing, Switzerland, pp. 7–12.

Foubister, D.M., 1962. Therapeutic hints for students of homœopathy. British Homoeopath. J. 51 (4), 288–292.

Gautam, A.K., Bhadauria, R., 2009. Homeopathic flora of Bilaspur District of Himachal Pradesh, India: a preliminary survey. Ethnobot. Leafl. 2009 (1), 14.

Harrison, M.F., Anderson, P.J., Johnson, J.B., Richert, M., Miller, A.D., Johnson, B.D., 2016. Acute mountain sickness symptom severity at the south pole: the influence of self-selected prophylaxis with acetazolamide. PLoS One 11 (2), e0148206.

Havryliuk, T., Acharya, B., Caruso, E., Cushing, T., 2015. Understanding of altitude illness and use of pharmacotherapy among trekkers and porters in the Annapurna region of Nepal. High Alt. Med. Biol. 16 (3), 236–243.

Jayawardene, I., Saper, R., Lupoli, N., Sehgal, A., Wright, R.O., Amarasiriwardena, C., 2010. Determination of in vitro bioaccessibility of Pb, As, Cd and Hg in selected traditional Indian medicines. J. Anal. At. Spectrom 25 (8), 1275–1282.

Jonas, W.B., Kaptchuk, T.J., Linde, K., 2003. A critical overview of homeopathy. Ann. Intern. Med. 138 (5), 393–399.

Kanchan, T., Atreya, A., 2016. Calotropis gigantea. Wilderness Environ. Med. 27 (2), 350–351.

Kriplani, P., Guarve, K., Baghael, U.S., 2017. *Arnica montana* L.—a plant of healing. J. Pharm. Pharmacol. 69 (8), 925–945.

Ledermann, E.K., 1945. Homeopathy and natural therapeutics. British Homoeopath. J. 35 (1), 30–46.

Manickam, M., Tulsawani, R., Sarkar, B.B., Misra, K., 2014. Herbal medicines: A possible solution for high altitude maladies. In: Brar, S.K., Kaur, S., Dhillon, G.S. (Eds.), Nutraceuticals and Functional Foods: Natural Remedy. Nova Science Publishers, New York, pp. 457–476.

Molassiotis, A., Fernadez-Ortega, P., Pud, D., Ozden, G., Scott, J.A., Panteli, V., Margulies, A., Browall, M., Magri, M., Selvekerova, S., Madsen, E., 2005. Use of complementary and alternative medicine in cancer patients: A European survey. Ann. Oncol. 16 (4), 655–663.

Morris, M.P., 2015. The efficacy of the homeopathic complex (*Arnica montana* 6CH, *Bryonia alba* 6CH, *Kalmia latifolia* 6CH, *Rhus toxicodendron* 6CH *Calcarea fluorica* 6CH) and physiotherapy in the treatment of chronic low back due to osteoarthritis. (Doctoral dissertation). University of Johannesburg.

Pande, P.S., Patil, N.S., Mane, V.D., Malpani, M.O., 2017. Evaluation of antioxidant activity of *Chopchini ark* and *Shvitra tikia*. Cogniz. Online Int. J. Interdiscip. Res., 4–9.

Pilkington, K., Kirkwood, G., Rampes, H., Fisher, P., Richardson, J., 2005. Homeopathy for depression: a systematic review of the research evidence. Homeopathy 94 (3), 153–163.

Pilkington, K., Kirkwood, G., Rampes, H., Fisher, P., Richardson, J., 2006. Homeopathy for anxiety and anxiety disorders: a systematic review of the research. Homeopathy 95 (3), 151–162.

Relton, C., Cooper, K., Viksveen, P., Fibert, P., Thomas, K., 2017. Prevalence of homeopathy use by the general population worldwide: a systematic review. Homeopathy 106 (2), 69–78.

Roberts, E.R., Tournier, A.L., Chatfield, K., Viksveen, P., Mathie, R.T., 2016. How safe is homeopathy? An analysis of the Posadzki et al. 2012 safety paper and fresh review of the same literature. Homeopathy 105 (1), 20.

Roy, V., 2015. Time to sensitize medical graduates to the Indian systems of medicine and homeopathy. Indian J. Pharm. 47 (1), 1–3.

Sánchez-Machado, D.I., López-Cervantes, J., Sendón, R., Sanches-Silva, A., 2017. *Aloe vera*: ancient knowledge with new frontiers. Trends Food Sci. Technol. 61, 94–102.

Sarkar, B.B., Purkayastha, S.S., Selvamurthy, W., 1996. Homoeopathy: a possible curative measure for frostbite. Asian Homoeopath. J. 6 (1), 31–41.

Selvamurthy, W., Basu, C.K., 1998. High altitude maladies: Recent trends in medical management. Int. J. Biometeorol. 42 (2), 61–64.

Shackelton, M.F., Tondora, C.M., Whiting, S., Whitney, M., 2000. The effect of homeopathic coca on high-altitude mountain sickness: Mt. Everest Base Camp. Complement. Health Pract. Rev. 6 (1), 45–55.

Sharma, V., Chaudhary, A.K., 2015. Ayurvedic pharmacology and herbal medicine. Int. J. Green Pharm. 9 (4), 192–197. (Medknow Publications & Media Pvt. Ltd.).

Strewe, C., Zeller, R., Feuerecker, M., Hoerl, M., Kumprej, I., Crispin, A., Johannes, B., Debevec, T., Mekjavic, I., Schelling, G., Chouker, A., 2017. PlanHab study: assessment of psycho-neuroendocrine function in male subjects during 21 days of normobaric hypoxia and bed rest. Stress 20 (2), 131–139.

Witcher, P., Gregory, R.L., Windsor, L.J., 2014. In: Evaluating the effect of fulvic acid on oral bacteria and cancerous oral cells. Poster Session Presented at IUPUI Research Day 2014, Indianapolis, Indiana.

CHAPTER 12

Nanoformulations: A Novel Approach Against Hypoxia

Anuja Bhardwaj*, Rajesh Arora†

*Chemistry Division, Department of Biochemical Sciences (DBCS), Defence Institute of Physiology and Allied Sciences (DIPAS), Delhi, India, †Department of Biochemical Sciences (DBCS), Defence Institute of Physiology and Allied Sciences (DIPAS), Delhi, India

Abbreviations

AMPK-PKC-CBP	5′-adenine monophosphate-activated protein kinase-protein kinase C-cyclic adenosine monophosphate response element-binding protein
ATP	adenosine triphosphate
bcl	b-cell lymphoma
CaM kinase II	calmodulin-dependent protein kinase II
cGK-1	cGMP-dependent protein kinase type 1
cGMP	cyclic guanosine monophosphate
CHH	chronic hypobaric hypoxia
DOX	doxorubicin
EPO	erythropoietin
ET-1/2/3	endothelin-1/2/3
FIO_2	oxygen fraction in air
GATA-4	GATA binding protein 4
H_2O_2	hydrogen peroxide
H9c2	cardiomyoblast cell line
HAT	histone acetyl transferase
HH	hypobaric hypoxia
HIF-1α	hypoxia-inducible factor-1α
HR-NPs	hypoxia-responsive nanoparticles
HVCM	human ventricular cardiomyocytes
NAD	nicotinamide adenine dinucleotide
NADH	nicotinamide adenine dinucleotide hydrogen
nm	nanometer
nM	nanomole
NNI	National Nanotechnology Initiative
PEG-CNPs	polyethylene glycol-coated nanoceria
PLGA-EPO-NP	poly-DL-lactide-coglycolide nanoparticles
r-EPO	recombinant EPO
ROS	reactive oxygen species
RVH	right ventricular hypertrophy
µM	micromole

231

Management of High Altitude Pathophysiology. https://doi.org/10.1016/B978-0-12-813999-8.00012-4

12.1   INTRODUCTION

The advent of nanotechnology is regarded as one of the greatest engineering innovations since the industrial revolution (Sachan and Gupta, 2015). Nanotechnology is an advanced scientific technique in the 21st century (Pandey and Pandey, 2014) and a promising technology of the future (Safari and Zarnegar, 2014). Global emergence of nanotechnology was marked by the announcement of the National Nanotechnology Initiative (NNI) in January 2000 (Roco, 2003). It is the most potential and a far-reaching technology in the current era (Pandey and Pandey, 2014) which assures re-engineering of the manmade world by igniting a wave of novel revolutionary commercial products from machines to medicine (Sachan and Gupta, 2015).

Nanotechnology encompasses innovative biotechnology processes, the synthesis of new drugs and their targeted delivery, regenerative medicine, neuromorphic engineering, and developing a sustainable environment (Roco, 2003), with advancements in the fields of biology, chemistry, engineering, and robotics (Bowman and Hodge, 2007; Sahoo et al., 2007). Nevertheless, its novelty lies in the purposive and precise manipulation of atoms and molecules in order to use the unique properties of matter (or nanoblocks) that could materialize at the nanoscale (Bowman and Hodge, 2007; Ramsden, 2016).

Because of the potential technological applications of this nanoscaled technology, its applications already have begun, and numerous products incorporating nanotechnology are available commercially (Bowman and Hodge, 2007; Romig et al., 2007). Various nanotechnology-based products include simple and passive nanoscale particles, compounds or composites for use in automotive paints and coatings, cosmetics, digital cameras, foods, pesticides, stain-resistant clothing, sporting goods, and sunscreens, all products that are easily purchased through pharmacies, retail shops, and sports stores (Bowman and Hodge, 2007). Nanotechnology is developing rapidly with more than 300 products already in the market (Maynard et al., 2006).

Considering the current prevalence and advances in the field of nanotechnology, this chapter presents the basics of this nanoscaled technology and nanoformulations. It focuses on the use of nanoformulations concerning hypoxia-related diseases, including research carried in context of high-altitude (hypobaric) hypoxia maladies. The chapter also highlights prospective nanoformulations in managing the high-altitude induced pathophysiologies.

12.2   EMERGENCE OF NANOTECHNOLOGY

Nanotechnology appears to be a modern science, however, one might discover the use of nanomaterials for centuries. For example, Roman glass artifacts contained metal nanoparticles, which provided beautiful colors on reflection. In medieval times, nanoparticles were used to decorate cathedral windows, and the Chinese used gold nanoparticles as an inorganic dye to introduce red color into their ceramic porcelains more than 1000 years ago (Pokropivny et al., 2007).

In India, the application of nanotechnology can be found in the Ayurvedic *bhasma*, which are described as herbo-mineral-metallic compounds of nanodimensions (Chaudhary, 2011; Sharma and Prajapati, 2016). Nagarjuna, an 18th century Indian alchemist, was the first to introduce the use of metals and minerals such as abhrak (mica), makshika (pyrites), rajat (silver), rasa (mercury), swarna (gold), and tamra (copper) as medicinal agents (Kumar Pal, 2015). The Ayurvedic *Bhasma* generally are prepared by melting metals or minerals at high temperature and then cooling them in suitable media such as herbal juices or decoction for specified times. This process is repeated several times to obtain *bhasma* (incinerated metals), which have been transformed into biologically active nanoparticles, free from toxic effects of the metals (Kumar Pal, 2015; Sharma and Prajapati, 2016) with rapid and targeted actions (Sharma and Prajapati, 2016). Examples of such Ayurvedic nanoformulations are *swarna bhasma* (gold calx), *mukta shukti bhasma* (pearl-oyster calx), *mandura bhasma* (copper calx), *yashad bhasma* (zinc calx), etc. (Kumar Pal, 2015; Sharma and Prajapati, 2016). Ayurvedic *bhasma* have substantial potential for their use in the development of nanomedicine in health care and treatment (Sharma and Prajapati, 2016).

The term "nanotechnology" was coined in 1974 by Norio Taniguchi (Allhoff et al., 2010; Krukemeyer et al., 2015), a researcher at the University of Tokyo. In his paper "On the Basic Concept of Nanotechnology," he described nanotechnology as a means to produce extra high accuracy and ultrafine dimensions, specifically, the preciseness and fineness on the order of 1 nanometer (nm), 10^{-9} m in length (Allhoff et al., 2010). Conventionally, the concept of nanotechnology originated during a talk given by Richard Feynman at the California Institute of Technology in December 1959 titled "There's Plenty of Room at the Bottom" (Allhoff et al., 2010; Maynard et al., 2006). In his speech, this physicist and Nobel prize winner spoke about the principles of miniaturization and atomic-level precision and how these concepts do not violate any known law of physics. Later on, Taniguchi developed Feynman's concepts (Allhoff et al., 2010).

Various definitions, referring chiefly to the dimensions, have been proposed for nanotechnology. According to the US National Nanotechnology Insitute, "nanotechnology is the understanding and control of matter at dimensions between approximately 1 and 100 nm, where unique phenomena enable novel applications" (Adams and Barbante, 2013). The prefix "nano" is derived from the Greek word *nanos*, which means dwarf. One nanometer (nm) is equal to one-billionth of a meter, or about the width of six carbon atoms or 10 water molecules (Allhoff et al., 2010; Bowman and Hodge, 2007; Jena et al., 2017; Veerasekar et al., 2016). Mostly, only one or two dimensions are in the nano regime (quantum wells and nanowires), but sometimes all three dimensions are nanoscaled (quantum dots and nanocrystals) (Adams and Barbante, 2013). This branch of science and technology deals with materials with at least one spatial dimension in the size range of 1–100 nm (Fakruddin et al., 2012; Veerasekar et al., 2016). Therefore, the simplest definition of nanotechnology is "technology at nanoscale (1–100 nm)," which involves the designing, characterization, production, and application of materials, devices, and systems by controlling the shape and size within the nanoscale (Fakruddin et al., 2012; Ramsden, 2016).

The term "nanotechnology" encompasses an emerging family of heterogenous technologies, including nanosciences and nanotechnologies enabling the manipulation of matter at the atomic level (Bowman and Hodge, 2007; Sahoo et al., 2007; Whitesides, 2005). Conceptually, both nanoscience and nanotechnology are based on the manipulation of individual atoms and molecules to produce materials for applications well below the submicroscopic level. This often requires biological, chemical, and physical knowledge at scales ranging between 1 and 100 nm; sometimes the subject also might concern the integration of the resulting structures into larger systems (Adams and Barbante, 2013). In a broader view, nanotechnology is a complex multidisciplinary science that entails the application of fields of science as diverse as surface science, organic chemistry, molecular biology, semiconductor physics, and microfabrication (Fakruddin et al., 2012; Pokropivny et al., 2007). Fig. 12.1 illustrates interdisciplinary sciences included within nanotechnology.

Nanotechnology has its applications in most of the sciences probably because of the inclusion of various scientific fields. Nanotechnology research has demonstrated and promises huge possibilities in several sectors, ranging from health care to construction and electronics. In medicine, it has shown revolutionary promises in drug delivery, gene therapy, diagnostics, and many areas of research, development, and clinical applications (Jena et al., 2017; Roco, 2003; Safari and Zarnegar, 2014).

The applications of nanotechnology for treatment, diagnosis, monitoring, and control of biological systems have been called "nanomedicine" by the National Institutes of Health (Moghimi et al., 2005). Basically, nanomedicine is the appliance of nanotechnology to medicine. Wherein, nanomedicine focuses on the application of nanotechnology concepts to medical applications, while nanotechnology incorporates all the basic research at a nanoscale level of biological systems (Riehemann et al., 2009). The rationale to employ

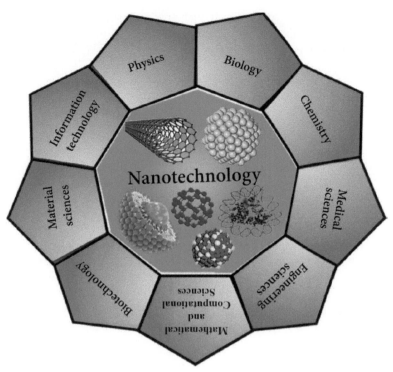

FIG. 12.1
Various interdisciplinary sciences included within nanotechnology.

nanotechnology in medicine stems from the fact that nanomaterials are similar in scale to biologic molecules and systems and could be designed to have various functions. Moreover, in comparison to atoms and macroscopic materials, nanomaterials have a high ratio of surface area to volume as well as tunable optical, electronic, magnetic, and biologic properties and can be engineered to have different sizes, shapes, chemical compositions, surface chemical characteristics, and hollow or solid structures. Such properties of nanomaterials are being incorporated into new generations of drug-delivery vehicles, contrast agents, and diagnostic devices, some of which are undergoing clinical investigation or have been approved by the US Food and Drug Administration for use in humans (Kim et al., 2010).

There is no doubt that nanotechnology has several advantages that have increased its applications, however, this field of science also has some disadvantages. The various advantages and disadvantages of nanotechnology are given in Fig. 12.2.

Advantages of nanotechnology	Disadvantages of nanotechnology
• Enhancement of solubility and bioavailability • Protection from toxicity • Enhancement of pharmacological activity • Enhancement of stability • Improve tissue macrophage distribution • Sustained delivery • Protection from chemical (e.g., pH), physical and biological (e.g., enzymes) degradation • Enhancement of permeability (e.g., through blood brain barrier) and retention effect • Decrease side-effects of conventional drugs • Improves therapeutic effect • Increase tolerability in the body • Carry larger payloads	• High-cost • Difficulty in scale-up processes • Easy inhalability of nanoparticles resulting into lung diseases • High immunogenicity • Chances of poor targeting • High ability to aggregate in biological systems

FIG. 12.2

Advantages and disadvantages of nanotechnology.

12.3   NANOMATERIALS

Nanomaterials are defined as materials with at least one external dimension in the size range from approximately 1–100 nm, with properties that often are unique because of their dimensions and are manufactured according to NNI. This definition implies mainly to engineered nanomaterials in contrast to naturally occurring and unintentionally produced nanomaterials, commonly called ultrafine particles. (Yokel and MacPhail, 2011). A nanomaterial has a number of structural characteristics contributed by its structure as a whole and by its individual elements, which are reflected in the functionality of a nanomaterial, a fundamental unit of a nanosystem (Pogrebnjak and Beresnev, 2012).

Nanomaterials can be classified in numerous ways, but the most important feature is its structure complexity, and therefore it can be classified into nanoparticles and nanostructures (Fig. 12.3). A nanoparticle is a nanosized complex of atoms and molecules, which are interrelated in a definite way; a nanostructured material is an ensemble of nanoparticles (Pogrebnjak and Beresnev, 2012). Nanoparticles generally are classified based on their dimensionality (one, two, and three dimensional), morphology (such as nanowires, nanoshells,

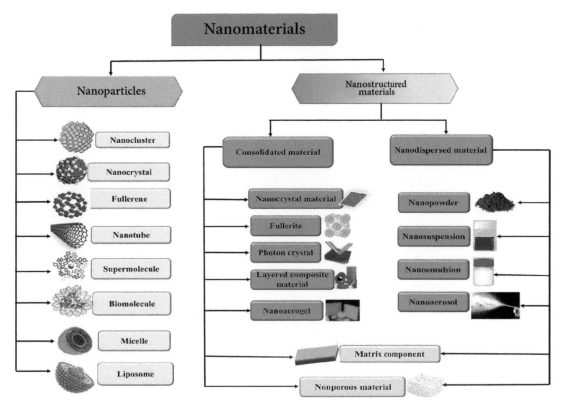

FIG. 12.3

Classification of nanomaterials.

and nanotubes), composition (hollow, spherical, and nanotubes), uniformity, and agglomeration. (Buzea et al., 2007).

There are principally two strategies to synthesize nanomaterials (Khan et al., 2017; Nikalje, 2015; Singh et al., 2011; Veerasekar et al., 2016): Top-down approach and bottom-up approach.

12.3.1 Top-Down Approach

The top-down approach includes a set of technologies that fabricate by removing certain parts from the bulk material substrate. The removing methods can be chemical, electrochemical, or mechanical, depending on the material of the base substrate and requirement of the feature size (Singh et al., 2011). Nanoscale structures are created by using larger, externally controlled devices to cut, carve, mill, and shape materials into the desired assembly (Nikalje, 2015; Veerasekar et al., 2016). Laser ablation, milling, nanolithography, hydrothermal technique, physical vapor deposition, and electrochemical method (electroplating) all use the top-down approach for nanoscale material manufacturing (Khan et al., 2017; Nikalje, 2015).

12.3.2   Bottom-Up Approach

The bottom-up approach also is called the building-up approach (Khan et al., 2017). In the bottom-up approach, molecular components arrange themselves into larger and more complex systems atom-by-atom, molecule-by-molecule, cluster-by-cluster from the bottom (Singh et al., 2011; Veerasekar et al., 2016). These technologies fabricate by stacking materials on the top of a base substrate (Singh et al., 2011). Laser pyrolysis, plasma or flame spraying synthesis, spinning, atomic or molecular, and chemical vapor deposition, as well as biological synthesis via algae, fungi, plants, and yeasts are some of the techniques for synthesis of nanoparticles that employ the bottom-up approach (Khan et al., 2017).

12.4   NANOFORMULATIONS

Nanotherapeutics, which is the application of nanotechnology in the medical and pharmaceutical sciences, has led to the development of nanoplatforms, such as nanobiomaterials and nanoformulations, of a particular shape, size, and surface properties that are crucial for biological interactions and their consequential therapeutic effects (Prasad et al., 2018).

Biomaterials described as natural or synthetic materials or constructs that interact with biological systems have been used widely in medical and pharmaceutical areas, such as dental and bone implants, drug delivery, tissue regeneration, and stem cell engineering. Drawbacks of classical biomaterials, however, have necessitated technological advancements in the development of novel biomaterials. This has been achieved by incorporating the benefits of nanotechnology, leading to the emergence of nanobiomaterials (Lee and Kim, 2014). Nanobiomaterials is a term for nanoscaled materials used for biomedical applications such as drug delivery, bio-imaging, tissue engineering, and biosensor, in order to develop unique functions required by these biomedical systems (Ali et al., 2013; Sitharaman, 2016; Shen, 2006; Yang et al., 2011). Accordingly, nanotechnology-based formulations are generally referred to as "nanoformulations" (Summerlin et al., 2015). Recently, nanoformulations of medicinal drugs have attracted the interest of several researchers, especially for drug delivery applications as they enhance the properties of conventional drugs and are specific to the targeted delivery site (Jeevanandam et al., 2016). Dendrimers, polymeric nanoparticles, liposomes, nanoemulsions, and micelles are some of the nanoformulations that are gaining importance in the pharmaceutical industry for enhanced drug formulation (Singh et al., 2016;

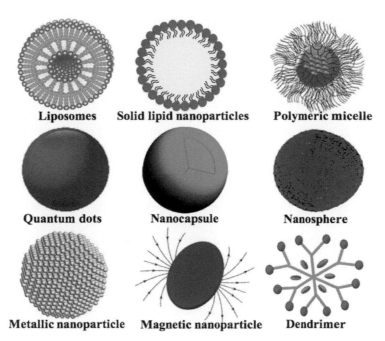

FIG. 12.4

Some of the nanoparticles used in the pharmaceutical sciences.

Jeevanandam et al., 2016). Some of these nanoparticles used to develop nanoformulations in pharmaceutical science are illustrated in Fig. 12.4.

12.5 NANOFORMULATIONS FOR HYPOXIA

Hypoxia implies reduced oxygen availability and constitutes significant effects on cellular and systemic physiology (Semenza, 2015). A variety of pathological conditions exist where the affected tissues are hypoxic (Airley et al., 2000), including obstructive sleep apnea, cerebral diseases, systemic hypertension, and congestive heart failure (Bhatia et al., 2017). Hypoxia is also an important factor where an inflammatory response exists, such as chronic inflammatory bowel disease, rheumatoid arthritis, and ischemia/reperfusion injury. It also is a vital factor for being considered in diseases accompanied with insufficient oxygen tensions because of poor respiratory functions, such as chronic bronchitis and cystic fibrosis, as well as in benign proliferative diseases such as diabetic vasculopathies, epilepsy, and psoriasis (Airley et al., 2000). Hypoxia also can develop in healthy individuals exposed to a low oxygen content environment, such as at high altitudes. Generally, an individual tends to acclimatize physiologically at the outset of HA-induced hypoxia. Inability to adapt at high altitudes often results in pulmonary edema, stroke, cardiovascular dysfunction, and even death. Hypoxia definitely is a life-threatening condition for both healthy individuals and those with cardiovascular, respiratory, and hemolytic diseases (Sun et al., 2016).

Targeting the activation and/or inhibition of hypoxia intermediates, therefore, could serve as a novel therapeutic strategy because there exists a nexus between hypoxia signaling intermediates and a set of responses originating from angiogenesis, glycolysis, erythropoiesis, and inflammation (Bhatia et al., 2017).

Numerous pharmaceutical drugs are available to curb most of these hypoxia-induced diseases and disorders. It is important to state, however, that current available pharmacological interventions have numerous shortcomings. For example, a major concern with conventional cytotoxic drugs is their nonselective delivery to normal vital organs and tissues, with inadequate delivery to tumor tissues (Nakamura et al., 2015). Another important problem is a low permeability and retention (EPR) effect associated with conventional drug formulations (Blanco et al., 2015). Herbal drugs comprise one option to overcome the drawbacks of the conventional therapeutics, but they also have few disadvantages for their delivery to be as efficient and effective as medicinal drugs. For example, most of the biologically active components of herbal extracts are highly hydrophilic, but demonstrate low absorption because they are unable to cross lipid membranes, possess high molecular sizes, and exhibit poor absorption, resulting in diminished bioavailability and efficacy (Bonifácio et al., 2014).

The innovation of therapeutic nanoparticles has revolutionized both the drug formulation and delivery (Arachchige et al., 2015). The various advantages of nanotechnology in drug formulation and delivery are depicted in Fig. 12.2. The application of nanotechnology to develop nanoformulations is one of the potential ways to overcome problems associated with conventional drugs and herbal formulations. Providing efficient and effective drugs for various pathophysiological conditions including respiratory disorders, cardiovascular diseases, cancer and high-altitude induced ailments. Some of the nanotechnology-based pharmaceutical drugs approved for their intended use are listed in Table 12.1.

12.6 NANOFORMULATIONS FOR HYPOBARIC HYPOXIA

As discussed in previous chapters, hypobaric hypoxia involves failure to acclimatize on HA exposure and is often associated with various pathophysiological conditions, such as pulmonary edema and cardiac hypertrophy (Wilkins et al., 2015). A few groups of researchers have been working in the field of nanotechnology with a scientific perspective to develop nanotherapeutics for

Table 12.1 Some of the Approved Nanopharmaceutical Drugs

S. No.	Trade Name	Active Ingredient	Application(s)	Company
Inorganic				
1	DexFerrum/ DexIron	Iron dextran	Iron deficiency in chronic kidney disease	American Reagent Inc.
2	Diafer	5% iron isomaltoside 1000 colloid	Iron deficiency anemia	Pharmacosmos
3	Feraheme/ Rienso	Ferumoxytol (iron polyglucose sorbitol carboxymethylether colloid)	Iron deficiency in chronic kidney disease	AMAG Pharmaceuticals/ Takeda
4	Ferrlecit	Sodium gluconate complex in sucrose injection (iron gluconate colloid)	Iron deficiency in chronic kidney disease	Sanofi-Aventis
5	INFeD/ CosmoFer	Iron dextran colloid	Iron deficiency in chronic kidney disease	Watson Pharmaceuticals Inc./ Pharmacosmos
6	Injectafer/ Ferinject	Iron carboxymaltose colloid	Iron deficiency in chronic kidney disease	American Reagent Inc./Vifor Pharma
7	Monofer	10% iron isomaltoside 1000 colloid	Iron deficiency anemia	Pharmacosmos
8	Venofer	Iron sucrose colloid	Iron deficiency in chronic kidney disease	American Reagent Inc.
Liposome				
9	Abelcet	Amphotericin B	Invasive severe fungal infections	Sigma-Tau Pharmaceuticals
10	Ambisome	Amphotericin B	Fungal/protozoal infections	Gilead Sciences
11	Amphotec	Amphotericin B	Severe fungal infections	Ben Venue Laboratories Inc.
12	Curosurf	Poractant alfa	Respiratory distress syndrome	Chiesi Farmaceutici S.p.A.
13	DaunoXome	Daunorubicin	HIV-related Kaposi's sarcoma	Gilead Sciences
14	Depocyt	Cytarabine	Malignant lymphomatous meningitis	Sigma-Tau Pharmaceuticals
15	DepoDur	Morphine sulfate	Postsurgical analgesia	Skyepharma Inc.
16	Doxil/Caelyx	Doxorubicin	Breast cancer, Kaposi's sarcoma, multiple myeloma, and ovarian cancer	Janssen
17	Epaxal	Inactivated hepatitis Λ virus (strain RGSB)	Hepatitis A	Crucell, Berna Biotech
18	Exparel	Bupivacaine	Pain management	Pacira Pharmaceuticals Inc.

Continued

Table 12.1 Some of the Approved Nanopharmaceutical Drugs—cont'd

S. No.	Trade Name	Active Ingredient	Application(s)	Company
19	Inflexal V	Inactivated hemagglutinin of influenza virus strains A and B	Influenza	Crucell, Berna Biotech
20	Marqibo	Vincristine	Acute lymphoblastic leukemia	Spectrum Pharmaceuticals
21	Mepact	Mifamurtide	High-grade, resectable, nonmetastatic osteosarcoma	Takeda Pharmaceutical Limited
22	Myocet	Doxorubicin	Combination therapy with cyclophosphamide in metastatic breast cancer	Teva Pharmaceuticals
23	Onivyde	Irinotecan	Pancreatic metastatic adenocarcinoma	Ipsen Biopharmaceuticals
24	Visudyne	Verteporphin	Choroidal neovascularization	Novartis
25	Vyxeos	Daunorubicin and cytarabine	Acute myeloid leukemia (AML); AML with myelodysplasia-related changes	Jazz Pharmaceuticals
Protein bound				
26	Abraxane	Paclitaxel	Metastatic breast cancer, nonsmall cell lung cancer, metastatic pancreatic cancer	Celgene
27	Ontak	Denileukin diftitox	Cutaneous T-cell lymphoma	Eisai Inc.
Micelle				
28	Estrasorb	Estradiol	Menopausal therapy	Novavax
29	Sandimmune	Cyclosporine	Immunosuppressive therapy	Novartis
Nanocrystalline				
30	Avinza	Morphine sulfate	Psychostimulant	Elan drug delivery Inc.
31	Emend	Aprepitant	Antiemetic	Merck
32	Focalin XR	Dexamethylphenidate hydrochloride	Attention deficit hyperactivity disorder (ADHD)	Novartis
33	Invega Sustenna	Paliperidone palmitate	Schizophrenia, schizoaffective disorder	Janssen Pharmaceuticals Inc.
34	Megace ES	Megestrol acetate	Antianorexic	Par Pharmaceuticals
35	Rapamune	Sirolimus	Immunosuppressant	Wyeth Pharmaceuticals
36	Ritalin LA	Methylphenidate hydrochloride	Psychostimulant	Novartis
37	Ryanodex	Dantrolene sodium	Malignant hypothermia	Eagle Pharmaceuticals Inc.

Table 12.1 Some of the Approved Nanopharmaceutical Drugs—cont'd

S. No.	Trade Name	Active Ingredient	Application(s)	Company
38	Tricor	Fenofibrate	Antihyperlipidemic	Abbott
39	Zanaflex	Tizanidine hydrochloride	Muscle relaxant	Acorda
Polymeric				
40	Adagen	PEG-adenosine deaminase	Severe combined immunodeficiency disease combined with ADA deficiency	Sigma-Tau Pharmaceuticals
41	Adynovate	Antihemophilic factor (recombinant), pegylated	Hemophilia	Shire Pharmaceuticals
42	Cimzia	Certolizumab pegol	Crohn's disease, rheumatoid arthritis, psoriatic arthritis, ankylosing spondylitis	UCB Inc.
43	Copaxone	Glatiramer acetate (L-glutamic acid, L-alanine, L-lysine, and L-tyrosine copolymer)	Multiple sclerosis	Teva Pharmaceuticals
44	Eligard	Leuprolide acetate	Prostate cancer	TOLMAR Pharmaceuticals
45	Genexol-PM	Methoxy-PEG-poly (D,L-lactide) taxol	Metastatic breast cancer	Samyang Genex
46	Gliadel	Carmustine	Brain cancer	Eisai Inc.
47	Krystexxa	Pegloticase	Chronic gout	Savient Pharmaceuticals
48	Macugen	PEG-anti-VEGF aptamer	Age-related macular degeneration	Eyetech Pharmaceuticals
49	Mircera	Methoxy polyethylene glycol-epoetin beta	Neovascular age-related macular degeneration	Hoffmann-La Roche
50	Neulasta	PEG-granulocyte colony stimulating factor	Neutropenia associated with cancer therapy	Amgen
51	Oncaspar	PEG-L-asparaginase	Acute lymphoblastic leukemia	Enzon Pharmaceuticals
52	Pegasys	PEG-alpha-interferon 2a	Hepatitis B/Hepatitis C	Nektar/Hoffmann-La Roche
53	PegIntron	Pegylated IFN alpha-2b	Hepatitis C	Merck
54	Pelgridy	Pegylated IFN beta-1a	Multiple sclerosis	Biogen Idec
55	Rebinyn	Coagulation factor IX (recombinant), glycopegylated	Hemophilia B	Novo Nordisk
56	Renagel	Poly (allylamine hydrochloride) (or Sevelamer hydrochloride)	End-stage renal disease	Genzyme
57	Somavert	PEG-Hepatocyte growth factor	Acromegaly	Nektar, Pfizer
58	Zilretta	Triamcinolone acetonide (extended release injectable suspension)	Osteoarthritis knee pain	Flexion therapeutics, Inc.
59	Zoladex	Goserelin acetate	Breast and prostate cancer	AstraZeneca

ameliorating the ill-effects of HA pathophysiologies. Such nanoformulations include nanocurcumin and nanoceria. The research is detailed as below.

12.6.1   Nanocurcumin for Hypobaric Hypoxia

The polyphenol curcumin has numerous pharmacological properties that were discussed in Chapter 8. Native curcumin, however, exhibits some constraints for it to be a pharmacologically effective drug, and different types of nanoformulations have been developed to overcome these shortcomings. The nanoformulations have demonstrated better permeability, prolonged blood circulation, good stability, and controlled release at the target site (Gera et al., 2017).

Two well-established facts are that, first, hypoxia induces cardiomyocyte hypertrophy and, second, that curcumin has antioxidant and antihypertrophic effects. It also is well-known that low bioavailability of curcumin limits the merits of its pharmacological ability, which has led to nanotization of curcumin. Nehra et al. (2015b) designed a study to determine ameliorative effect of nanocurcumin against hypoxia-induced (0.5% oxygen for 24 h) hypertrophy and apoptosis in H9c2 cardiomyoblasts and compared it to curcumin by examining hypoxia-induced oxidative damage, hypertrophy, and consequent apoptosis. They also studied the plausible underlying mechanism for such a protective role of nanocurcumin. Results of this study revealed that nanocurcumin significantly ameliorated hypoxia-induced hypertrophy and apoptosis in H9c2 cells via downregulation of atrial natriuretic factor expression, caspase-3/-7 activation, reducing oxidative stress, and stabilizing hypoxia-inducible factor-1α (HIF-1α) in comparison to curcumin. Overall, then, one might conclude that nanocurcumin has the potential to cure hypoxia-induced cardiac pathologies by restoring oxidative balance (Nehra et al., 2015b). During the same year, this group found that reduction of p-300 histone acetyl transferase (HAT) mediated histone acetylation and GATA-4-activated nanocurcumin prevented hypoxia-induced hypertrophy in primary human ventricular cardiomyocyte (HVCM) cells. Moreover, nanocurcumin prevented translocation of p53 to mitochondria by stabilizing mitochondrial membrane potential and de-stressed hypertrophied HVCM cells by normalizing lactate, acetyl-coenzyme A, pyruvate, and glucose content, as well as lactate dehydrogenase and 5′ adenosine monophosphate-activated protein kinase activities (Nehra et al., 2015a). Furthermore, Nehra et al. (2016b) highlighted the therapeutic role of nanocurcumin in improving cardiac damage because of chronic hypobaric hypoxia (CHH)-induced right ventricular hypertrophy (RVH) in comparison to curcumin. A sustained work load on the right heart on ascent to high altitudes promotes RVH. On CHH exposure (25,000 ft, effective oxygen fraction in air [FIO2] ~ 0.08, temperature $28 \pm 1°C$, relative humidity $55\% \pm 2\%$ for 3, 7,

14, and 21 days), cyclic guanosine monophosphate (cGMP)/cGK-1, calmodulin-dependent protein kinase II (CaM kinase II), and intracellular calcium levels were modulated. Free radical-induced damage and lipid peroxidation also were increased, and tissue damage because of apoptotic cell death as evident by cytochrome-c/caspase-3 activation was observed. These changes contributed together into RVH. Supplementation with nanocurcumin, however, reduced CHH-induced RVH and apoptosis while modulating cardiac cGMP/cGK-1 signaling and maintaining CaM kinase II, intracellular calcium levels, and redox status better than curcumin (Nehra et al., 2016b). Nanocurcumin, therefore, affirms its potential for preventing hypoxic cardiac damage induced at high altitudes and developed at lowlands.

The same group also has carried out a comparative study between curcumin and nanotized curcumin in order to evaluate its protective efficacy against hypobaric hypoxia (HH)-induced lung injury (Nehra et al., 2016a). HA exposure often leads to fluid accumulation in lungs, which can result in high altitude-induced pulmonary edema (Sagi et al., 2014). During the experiment, these researchers exposed rats to acute HH (6, 12, 24, 48 and 72 h). The HH-induced lung injury was examined using lung injury markers concerned with pulmonary vasoconstriction (ET-1/2/3 and endothelin receptors A and B) and transvascular fluid balance mediator (Na+/K+ ATPase), upregulated and downregulated, respectively. These results confirmed defective pulmonary fluid clearance, which promoted edema formation. Treatment with nanocurcumin, however, prevented lung edema formation and restored expression levels of ET-1/2/3 and its receptors. It also restored the blood analytes, circulatory cytokines, and pulmonary redox status altered by HH exposure better than native curcumin. Therefore, the study highlighted the protective efficacy of nanocurcumin in rat lungs under HH (Nehra et al., 2016a).

12.6.2 Nanoceria for Hypobaric Hypoxia

Cerium oxide nanoparticles, known as nanoceria, have reactive oxygen species (ROS) quenching effect in vitro and in vivo. Their efficacy in conferring lung protection during oxidative stress under hypobaric hypoxia was not known until 2013. Thus, with this perspective, spherical nanoceria (7–10 nm diameter) synthesized using microemulsion method were evaluated for their lung protective efficacy during hypobaric hypoxia. On repeated intraperitoneal injections of low micromole concentration (0.5 µg/kg body weight/week of nanoceria) for 5 weeks, nanoceria localized in rodent lungs without any inflammatory response. The lung-deposited nanoceria diminished ROS formation, lipid peroxidation, and glutathione oxidation. It also prevented oxidative protein modifications (nitration and carbonyl formation) during hypobaric hypoxia. Overall, the results revealed that by quenching ROS generated during hypobaric hypoxia, nanoceria deposited in lungs provided protection without inducing any inflammatory response (Das and Bhargava, 2013).

Arya et al. (2014) found that treatment with nanoceria at a concentration of 25 nM of primary neuronal culture challenged with 50 μM of hydrogen peroxide (H_2O_2) decreased ROS content and lowered the relative rise in cellular calcium flux. The treatment also reduced apoptosis that was ascribed to the maintenance of mitochondrial membrane potential and restoration of major redox equivalents, including NADH/NAD ratio and cellular ATP. This study further emphasized the defensive role of nanoceria against oxidative stress, a condition associated with several clinical complications (Arya et al., 2014). The neuroprotective and the cognition-enhancing activities of nanoceria during hypobaric hypoxia also have evaluated. Chronic hypoxic exposure encountered at high altitudes often affects the cortex and hippocampus regions of the brain, leading to memory impairment and cognitive dysfunction because of the generation of reactive nitrogen and oxygen species. Polyethylene glycol-coated nanoceria (PEG-CNPs) localized efficiently in rodent brain reduced oxidative stress and associated damage during hypoxia exposure. Morris water maze tests revealed that PEG-CNPs ameliorated hypoxia-induced memory impairment, promoted hippocampus neuronal survival, and augmented neurogenesis as evident from microscopic, flow cytometric, and histological studies. The studies revealed that the underlying molecular mechanism for enhanced neurogenesis by PEG-CNPs was through the 5′-adenine monophosphate-activated protein kinase-protein kinase C-cyclic adenosine monophosphate response element-binding protein binding (AMPK-PKC-CBP) protein pathway. This study, therefore, demonstrated that nanoceria could be a promising therapeutic agent for neurodegenerative diseases (Arya et al., 2016).

12.7 PROSPECTIVE NANOFORMULATIONS FOR HIGH-ALTITUDE PATHOPHYSIOLOGIES

Hypoxia also is seen commonly in patients with cardiovascular, respiratory, and hemolytic diseases, which frequently promote multiple end-organ damage and failure (Sun et al., 2016). Nanotherapeutics developed against these diseases might be valuable in the treatment of high-altitude induced pathophysiologies, as well as hypoxic signaling intermediates such as HIF-1α and erythropoietin (EPO). Such attempts already have been made concerning to HIF-1α, which is involved virtually in all aspects of the response to hypoxia (Arachchige et al., 2015).

Thambi et al. (2014) reported development of self-assembled hypoxia-responsive nanoparticles (HR-NPs), which effectively encapsulated doxorubicin (DOX), a model drug. The HR-NPs released DOX in a sustained manner under the normoxic condition, while the drug release rate noticeably increased under the hypoxic condition. In vitro cytotoxicity tests showed that the DOX-loaded HR-NPs had higher toxicity to hypoxic cells than to normoxic cells. HR-NPs could effectively deliver DOX into human squamous cell carcinoma cells under hypoxic conditions as revealed by the microscopic examination. In vivo biodistribution study demonstrated that HR-NPs were accumulated selectively at the hypoxic tumor sites. Overall, drug-loaded HR-NPs exhibited high anti-tumor activity in vivo. This study clearly emphasized the potential of HR-NPs as nanocarriers for drug delivery to treat hypoxia-associated diseases (Thambi et al., 2014). Another remarkable study was carried out by Chen et al. (2012), in which they compared the neuroprotective and beneficial effects of human recombinant erythropoietin poly-DL-lactide-coglycolide nanoparticles (PLGA-EPO-NP) with human recombinant EPO (r-EPO) against deficits after brain ischemia. Results demonstrated that PLGA-EPO-NP significantly reduced infarction volumes 72 h after injury compared with the same concentrations of r-EPO. Erythropoietin (EPO), the main endogenous cytokine in hypoxic physiological response, augments oxygen delivery and attenuates brain injury, and promotes cell survival through the *bcl* antiapoptotic gene subfamily. EPO, a large glycosylated moiety, is unable to across the blood-brain barrier and so requires specific carrier transport and/or endocytosis (Chen et al., 2012).

Other prospective candidates include herbal nanodrugs and UV protective nanoformulations. Some of the best examples of the many nanotechnology drugs available are found in Table 12.2 for consideration as valuable candidates in the management of high-altitude hypoxia pathophysiologies. Another possibility is that herbal plants and medicinal mushrooms already investigated for their protective efficacy against high-altitude induced pathophysiologies could be nanotized for better therapeutic effects, as initiated with curcumin.

12.8 CONCLUSIONS

Nanotechnology is an emerging technology that encompasses the knowledge and application of almost all the other fields of sciences for the betterment of human beings. HA-related pathophysiologies are threat to human physiology under low oxygen environment, high wind velocity, and UV exposure. A few pharmacological interventions are being used to manage pathophysiologies under these harsh environments. It would be more advantageous, however, if nanobased products are used in this context, because nanoformulations improve pharmacokinetics and thus the therapeutic value of both conventional allopathic and herbal drugs.

Table 12.2 Examples of Prospective Nanoformulations for Consideration as Valuable Candidates in the Management of High-Altitude Hypoxia Pathophysiologies

S. No.	Nanoformulation	Active Nanomolecule	Pharmacological Action(s)	Therapeutic Observations	References
1	Composite NP of glass embedded in chitosan-PEG polymeric matrix loaded with nitric oxide (NO)	NO	Vasodilator and antiinflammatory effect	Controlled and targeted antiinflammatory effect with sustained NO release from NPs Decreased blood pressure; increased vasodilation; prevented inflammatory response	Cabrales et al. (2010)
2	Superoxide dismutase (SOD) loaded in poly (D,L-lactide-co-glycolide) (PLGA) nanoparticles (NPs)	SOD	Antioxidant effect	Better neuronal uptake Neuroprotective effect of SOD-NPs was seen up to 6 h following H_2O_2-induced oxidative stress	Reddy et al. (2008)
3	Nuclear factor κB (NF-κB) decoy loaded poly-(ethylene glycol)-*block*-lactide/glycolide copolymer (PEG-PLGA)	NF-κB decoy	Blockade of NF-κB	Controlled local delivery of NF-κB decoy into lungs, targeting a battery of multiple important inflammatory cytokines through single intratracheal instillation Blockade of NF-κB by NP-mediated delivery of the NF-κB decoy attenuated inflammation and proliferation, thereby, attenuated the development of pulmonary arterial hypertension (PAH) and pulmonary arterial remodeling induced by monocrotaline	Kimura et al. (2009)
4	Chitosan encapsulated minocycline hydrochloride with tween 80 as surfactant	Minocycline hydrochloride	Antioxidant, antibiotic and antiinflammatory effect	Better neuronal uptake to have neuroprotective and neurorestorative effects ischemic injury/stroke Nootropic (cognition enhancer) effect and antioxidant effect	Nagpal et al. (2013)

#	Name (Company)	Drug/Material	Property/Effect	Outcome	Reference
5	Chitosan-gold nanocomposites	Gold nanoparticles	Antioxidant effect	Antioxidant and adaptogenic effects under hypoxic (8000 m) conditions	Koryagin et al. (2013)
6	Nanoselenium	Selenium	Protects against oxidative damage	Better effect than Selenium against pulmonary arterial hypertension at high altitudes. Prevented right articular hypertrophy; reduced levels of lipid peroxidation and improved gut functions	Moghaddam et al. (2017)
7	Respirocytes	Nanodeviced red blood cells (RBCs)	–	Greater efficiency than normal RBCs. Enhanced delivery of oxygen to tissues	Freitas Jr (1998)
8	Arikace (Company: Transvec Inc.; Phase III liposomal drug)	Amikacin	Antiobitic	Prolonged lung deposition with efficient, safe and tolerable effect against lung infections (*Pseudomonas aeruginosa*). Better lung deposition with efficient, safe and tolerable effect against lung infections	Bilton et al. (2013)
9	ABI-009 or nab-rapamycin (Company: AADi with Celgene; Phase II Albumin bound drug)	Rapamycin	Antiproliferative activity by inhibition of mammalian target of Rapamycin (mTOR)	Enhanced internalization of rapamycin into targeted (tumor) cells. Bladder cancer, PEComa, or pulmonary arterial hypertension	Anselmo and Mitragotri (2016)
10	Liprostin (Company: Endovasc Ltd.; Phase II/III liposomal drug)	Prostaglandin E-1 (PEG-1)	Vasodilator, platelet inhibitor and antiinflammatory effect	Liprostin improved the drug dynamics and improved the therapeutic index of various ailments including occlusive disease, limb salvage, claudication and arthritis	Bulbake et al. (2017)
11	Poly(glycerol-succinic acid) dendrimer encapsulated camptothecin	Camptothecin	Anticancer agent	Enhanced solubility, cellular uptake and cellular retention of Camptothecin. Various cancers	Morgan et al. (2006)

Continued

Table 12.2 Examples of Prospective Nanoformulations for Consideration as Valuable Candidates in the Management of High-Altitude Hypoxia Pathophysiologies—cont'd

S. No.	Nanoformulation	Active Nanomolecule	Pharmacological Action(s)	Therapeutic Observations	References
12	Polymeric nanoparticles encapsulating honokiol	Honokiol (isolated from *Magnolia officinalis*)	Antiinflammatory, antithrombotic, antirheumatic, antioxidant, anxiolytic, central nervous system depressant, antitumor	Enhanced vascular administration	Zheng et al. (2010)
13	*Ziziphus mauritiana* extract polymeric (chitosan) nanoformulation	Plant extract of *Ziziphus mauritiana*	Immunomodulatory activity	Immunomodulatory activity of the extract	Bhatia et al. (2011)
14	*Cuscutta chinensis* nanosuspension	Flavonoids (kaempferol, quercetin) and lignans	Anticancer, antiaging, immune-stimulatory	Enhanced solubility	Yen et al. (2008)
15	Curcumin encapsulated into methoxy poly (ethylene glycol)–palmitate nanocarrier	Curcumin (a natural polyphenol isolated from the root of *Curcuma longa*)	Antitumor, antioxidant, antiamyloidin, antiplatelet aggregation and antiinflammatory	Enhanced solubility and bioavailability of curcumin	Sahu et al. (2008)
16	Nanoformulation consisting of quercetin, EUDRAGIT and polyvinyl alcohol	Quercetin	Antiinflammatory, antitumor, antiviral, cardiovascular, antioxidant, hepatoprotective effects	Better antioxidant and free-radical scavenging activity than quercetin	Wu et al. (2008)
17	Liposome encapsulated breviscapine	Breviscapine (a flavonoid isolated from *Erigeron breviscapus* (vant.) Hand. Mazz)	Anticardiovascular	Cerebrovascular and cardiovascular diseases, pulmonary fibrosis	Chakraborty et al. (2016)
18	Nanoprecipitation of narigenin	Narigenin (a flavonoid of naringin, isolated from citrus fruits, tomatoes, chaerries, grapefruit and cocoa)	Antitumor, hepatoprotective effect	Antioxidant, antiinflammatory	Bilati et al. (2005)
19	Nanoemulsion and chitosan microsphere encapsulating Genistein	Genistein (an isoflavone isolated from soyabean, scoparius and other leguminous plants.)	Antioxidant; used in cardiovascular diseases; breast and uterine cancer; osteoporosis	Antioxidant	Si et al. (2010)

20	Nanoemulsion and ionic gelation formulations of berberine	Berberine (an alkaloid present in the roots of *Berberis vulgaris* L.)	Inhibits activator protein 1, cyclooxygenase-2 and DNA topoisomerase II; antitumor effect	Anticancer	Chakraborty et al. (2016)
21	Silymarin loaded solid lipid nanoparticles (Patented)	Silymarin (flavonoid isolated from *Silybum marianum* with silybinin, isosilybinin, silydianin and silychristin as active agents)	Antihepatotoxicity effect, blood lipid reduction, antidiabetic, antiplatelet and cardioprotective effect	Enhanced oral bioavailability with controllable release for 72 h	Xu et al. (2011)
22	Cryptotanshinone loaded solid-lipid nanoparticles	Cryptotanshinone (a quinoid diterpene isolated from the roots of *Salvia miltiorrhiza Bunge*)	Antiinflammatory, cytotoxic, antibacterial, antiparasitic, antiangiogenic effect	Enhanced bioavailability of cryptotanshinone	Hu et al. (2010)

Acknowledgment

The authors thank the Director of Defense Institute of Physiology and Allied Sciences, Delhi, India, for constant support and encouragement.

References

Adams, F.C., Barbante, C., 2013. Nanoscience, nanotechnology, and spectrometry. Spectrochim. Acta B At. Spectrosc. 86, 3–13.

Airley, R.E., Monaghan, J.E., Stratford, I.J., 2000. Hypoxia and disease: opportunities for novel diagnostic and therapeutic prodrug strategies. Pharm. J. 264 (7094), 666–673.

Ali, S.H., Almaatoq, M.M., Mohamed, A.S., 2013. Classifications, surface characterization, and standardization of nanobiomaterials. Int. J. Eng. Technol. 2 (3), 187–199.

Allhoff, F., Lin, P., Moore, D., 2010. What Is Nanotechnology and Why Does It Matter?: From Science to Ethics. Wiley-Blackwell, Chichester.

Anselmo, A.C., Mitragotri, S., 2016. Nanoparticles in the clinic. Bioeng. Transl. Med. 1 (1), 10–29.

Arachchige, M.C., Reshetnyak, Y.K., Andreev, O.A., 2015. Advanced targeted nanomedicine. J. Biotechnol. 202, 88–97.

Arya, A., Sethy, N.K., Das, M., Singh, S.K., Das, A., Ujjain, S.K., Sharma, R.K., Sharma, M., Bhargava, K., 2014. Cerium oxide nanoparticles prevent apoptosis in primary cortical culture by stabilizing mitochondrial membrane potential. Free Radic. Res. 48 (7), 784–793.

Arya, A., Gangwar, A., Singh, S.K., Roy, M., Das, M., Sethy, N.K., Bhargava, K., 2016. Cerium oxide nanoparticles promote neurogenesis and abrogate hypoxia-induced memory impairment through AMPK-PKC-CBP signaling cascade. Int. J. Nanomedicine 11, 1159–1173.

Bhatia, A., Shard, P., Chopra, D., Mishra, T., 2011. Chitosan nanoparticles as carriers of immunorestoratory plant extract: synthesis, characterization, and immunorestoratory efficacy. Int. J. Drug Deliv. 3 (2), 381–385.

Bhatia, D., Ardekani, M.S., Shi, Q., Movafagh, S., 2017. Hypoxia and its emerging therapeutics in neurodegenerative, inflammatory and renal diseases. In: Hypoxia and Human Diseases. InTech, Rijeka, Croatia.

Bilati, U., Allémann, E., Doelker, E., 2005. Nanoprecipitation versus emulsion-based techniques for the encapsulation of proteins into biodegradable nanoparticles and process-related stability issues. AAPS PharmSciTech 6 (4), E594–E604.

Bilton, D., Pressler, T., Fajac, I., Clancy, J.P., Sands, D., Minic, P., Cipolli, M., LaRosa, M., Galeva, I., Sole, A.A., Staab, D., 2013. Phase 3 efficacy and safety data from randomized, multicenter study of liposomal amikacin for inhalation (ARIKACE) compared with TOBI in cystic fibrosis patients with chronic infection due to pseudomonas aeruginosa. Pediatr. Pulmonol. 48 (S36), 207–453.

Blanco, E., Shen, H., Ferrari, M., 2015. Principles of nanoparticle design for overcoming biological barriers to drug delivery. Nat. Biotechnol. 33 (9), 941–951.

Bonifácio, B.V., da Silva, P.B., dos Santos Ramos, M.A., Negri, K.M.S., Bauab, T.M., Chorilli, M., 2014. Nanotechnology-based drug delivery systems and herbal medicines: a review. Int. J. Nanomedicine 9, 1–15.

Bowman, D.M., Hodge, G.A., 2007. A small matter of regulation: an international review of nanotechnology regulation. Columbia Sci. Technol. Law Rev. 8 (1), 1–36.

Bulbake, U., Doppalapudi, S., Kommineni, N., Khan, W., 2017. Liposomal formulations in clinical use: an updated review. Pharmaceutics 9 (2), 12.

Buzea, C., Pacheco, I.I., Robbie, K., 2007. Nanomaterials and nanoparticles: sources and toxicity. Biointerphases 2 (4), MR17–MR71.

Cabrales, P., Han, G., Roche, C., Nacharaju, P., Friedman, A.J., Friedman, J.M., 2010. Sustained release nitric oxide from long-lived circulating nanoparticles. Free Radic. Biol. Med. 49 (4), 530–538.

Chakraborty, K., Shivakumar, A., Ramachandran, S., 2016. Nanotechnology in herbal medicines: a review. Int. J. Herb. Med. 4 (3), 21–27.

Chaudhary, A., 2011. Ayurvedic Bhasma: nanomedicine of ancient India—its global contemporary perspective. J. Biomed. Nanotechnol. 7 (1), 68–69.

Chen, H., Spagnoli, F., Burris, M., Rolland, W.B., Fajilan, A., Dou, H., Tang, J., Zhang, J.H., 2012. Nanoerythropoietin is 10 times more effective than regular erythropoietin in neuroprotection in a neonatal rat model of hypoxia and ischemia. Stroke 43 (3), 884–887.

Das, M., Bhargava, K., 2013. Cerium oxide nanoparticles protect rodent lungs from hypobaric hypoxia-induced oxidative stress and inflammation. Int. J. Nanomedicine 8, 4507–4520.

Fakruddin, M., Hossain, Z., Afroz, H., 2012. Prospects and applications of nanobiotechnology: a medical perspective. J. Nanobiotechnol. 1 (10), 1–8.

Freitas Jr., R.A., 1998. Exploratory design in medical nanotechnology: a mechanical artificial red cell. Artif. Cells Blood Substit. Immobil. Biotechnol. 26, 411–430.

Gera, M., Sharma, N., Ghosh, M., Lee, S.J., Min, T., Kwon, T., Jeong, D.K., 2017. Nanoformulations of curcumin: an emerging paradigm for improved remedial application. Oncotarget 8 (39), 66680–66698.

Hu, L., Xing, Q., Meng, J., Shang, C., 2010. Preparation and enhanced oral bioavailability of cryptotanshinone-loaded solid lipid nanoparticles. AAPS PharmSciTech 11 (2), 582–587.

Jeevanandam, J., San Chan, Y., Danquah, M.K., 2016. Nanoformulations of drugs: recent developments, impact, and challenges. Biochimie 128 (129), 99–112.

Jena, M., Mishra, S., Jena, S., Mishra, S.S., 2017. Nanotechnology future prospect in recent medicine: a review. Int. J. Basic Clin. Pharmacol. 2 (4), 353–359.

Khan, I., Saeed, K., Khan, I., 2017. Nanoparticles: properties, applications and toxicities. Arab. J. Chem. https://doi.org/10.1016/j.arabjc.2017.05.011 (In Press).

Kim, B.Y., Rutka, J.T., Chan, W.C., 2010. Nanomedicine. N. Engl. J. Med. 363 (25), 2434–2443.

Kimura, S., Egashira, K., Chen, L., Nakano, K., Iwata, E., Miyagawa, M., Tsujimoto, H., Hara, K., Morishita, R., Sueishi, K., Tominaga, R., 2009. Nanoparticle-mediated delivery of nuclear factor κB decoy into lungs ameliorates monocrotaline-induced pulmonary arterial hypertension. Hypertension 53 (5), 877–883.

Koryagin, A.S., Mochalova, A.E., Salomatina, E.V., Eshkova, O.Y., Smirnova, L.A., 2013. Adaptogenic effects of chitosan-gold nanocomposites under simulated hypoxic conditions. Inorg. Mater. Appl. Res. 4 (2), 127–130.

Krukemeyer, M.G., Krenn, V., Huebner, F., Wagner, W., Resch, R., 2015. History and possible uses of nanomedicine based on nanoparticles and nanotechnological progress. J. Nanomed. Nanotechnol. 6 (6), 1–7.

Kumar Pal, S., 2015. The Ayurvedic Bhasma: The ancient science of nanomedicine. Rec. Pat. Nanomed. 5 (1), 12–18.

Lee, H., Kim, Y.H., 2014. Nanobiomaterials for pharmaceutical and medical applications. Arch. Pharm. Res. 37 (1), 1–3.

Maynard, A.D., Aitken, R.J., Butz, T., Colvin, V., Donaldson, K., Oberdörster, G., Philbert, M.A., Ryan, J., Seaton, A., Stone, V., Tinkle, S.S., 2006. Safe handling of nanotechnology. Nature 444 (7117), 267–269.

Moghaddam, A.Z., Hamzekolaei, M.M., Khajali, F., Hassanpour, H., 2017. Role of selenium from different sources in prevention of pulmonary arterial hypertension syndrome in broiler chickens. Biol. Trace Elem. Res. 180 (1), 164–170.

Moghimi, S.M., Hunter, A.C., Murray, J.C., 2005. Nanomedicine: current status and future prospects. FASEB J. 19 (3), 311–330.

Morgan, M.T., Nakanishi, Y., Kroll, D.J., Griset, A.P., Carnahan, M.A., Wathier, M., Oberlies, N.H., Manikumar, G., Wani, M.C., Grinstaff, M.W., 2006. Dendrimer-encapsulated camptothecins: increased solubility, cellular uptake, and cellular retention affords enhanced anticancer activity in vitro. Cancer Res. 66 (24), 11913–11921.

Nagpal, K., Singh, S.K., Mishra, D.N., 2013. Formulation, optimization, in vivo pharmacokinetic, behavioral, and biochemical estimations of minocycline loaded chitosan nanoparticles for enhanced brain uptake. Chem. Pharm. Bull. 61 (3), 258–272.

Nakamura, H., Jun, F., Maeda, H., 2015. Development of next-generation macromolecular drugs based on the EPR effect: Challenges and pitfalls. Expert Opin. Drug Deliv. 12 (1), 53–64.

Nehra, S., Bhardwaj, V., Ganju, L., Saraswat, D., 2015a. Nanocurcumin prevents hypoxia-induced stress in primary human ventricular cardiomyocytes by maintaining mitochondrial homeostasis. PLoS One 10 (9), e0139121.

Nehra, S., Bhardwaj, V., Kalra, N., Ganju, L., Bansal, A., Saxena, S., Saraswat, D., 2015b. Nanocurcumin protects cardiomyoblasts H9c2 from hypoxia-induced hypertrophy and apoptosis by improving oxidative balance. J. Physiol. Biochem. 71 (2), 239–251.

Nehra, S., Bhardwaj, V., Bansal, A., Saraswat, D., 2016a. Nanocurcumin accords protection against acute hypobaric hypoxia induced lung injury in rats. J. Physiol. Biochem. 72 (4), 763–779.

Nehra, S., Bhardwaj, V., Kar, S., Saraswat, D., 2016b. Chronic hypobaric hypoxia induces right ventricular hypertrophy and apoptosis in rats: Therapeutic potential of nanocurcumin in improving adaptation. High Alt. Med. Biol. 17 (4), 342–352.

Nikalje, A.P., 2015. Nanotechnology and its applications in medicine. J. Med. Chem. 5 (2), 081–089.

Pandey, A., Pandey, G., 2014. Nanotechnology for herbal drugs and plant research. Res. Rev. J. Pharm. Nanotechnol. 2 (1), 13–16.

Pogrebnjak, A.D., Beresnev, V.M., 2012. Nanocoatings Nanosystems Nanotechnologies. Bentham Science Publishers, Sharjah.

Pokropivny, V., Lohmus, R., Hussainova, I., Pokropivny, A., Vlassov, S., 2007. Introduction to Nanomaterials and Nanotechnology. Tartu University Press, Ukraine.

Prasad, M., Lambe, U.P., Brar, B., Shah, I., Manimegalai, J., Ranjan, K., Rao, R., Kumar, S., Mahant, S., Khurana, S.K., Iqbal, H.M., 2018. Nanotherapeutics: an insight into health care and multidimensional applications in medical sector of the modern world. Biomed. Pharmacother. 97 (2018), 1521–1537.

Ramsden, J., 2016. Nanotechnology: An Introduction. William Andrew.

Reddy, M.K., Wu, L., Kou, W., Ghorpade, A., Labhasetwar, V., 2008. Superoxide dismutase-loaded PLGA nanoparticles protect cultured human neurons under oxidative stress. Appl. Biochem. Biotechnol. 151 (2–3), 565–577.

Riehemann, K., Schneider, S.W., Luger, T.A., Godin, B., Ferrari, M., Fuchs, H., 2009. Nanomedicine: challenge and perspectives. Angew. Chem. Int. Ed. 48 (5), 872–897.

Roco, M.C., 2003. Nanotechnology: convergence with modern biology and medicine. Curr. Opin. Biotechnol. 14 (3), 337–346.

Romig, A.D., Baker, A.B., Johannes, J., Zipperian, T., Eijkel, K., Kirchhoff, B., Mani, H.S., Rao, C.N.R., Walsh, S., 2007. An introduction to nanotechnology policy: opportunities and constraints for emerging and established economies. Technol. Forecast. Soc. Chang. 74 (9), 1634–1642.

Sachan, A.K., Gupta, A., 2015. A review on nanotized herbal drugs. Int. J. Pharm. Sci. Res. 6 (3), 961–970.

Safari, J., Zarnegar, Z., 2014. Advanced drug delivery systems: nanotechnology of health design: a review. J. Saudi Chem. Soc. 18 (2), 85–99.

Sagi, S.S.K., Mathew, T., Patir, H., 2014. Prophylactic administration of curcumin abates the incidence of hypobaric hypoxia induced pulmonary edema in rats: a molecular approach. J. Pulm. Respir. Med. 4, 1000164.

Sahoo, S.K., Parveen, S., Panda, J.J., 2007. The present and future of nanotechnology in human health care. Nanomedicine 3 (1), 20–31.

Sahu, A., Bora, U., Kasoju, N., Goswami, P., 2008. Synthesis of novel biodegradable and self-assembling methoxy poly (ethylene glycol)–palmitate nanocarrier for curcumin delivery to cancer cells. Acta Biomater. 4 (6), 1752–1761.

Semenza, G.L., 2015. AJP-cell theme: Cellular responses to hypoxia. Am. J. Physiol. Cell Physiol. 309 (6), C349.

Sharma, R., Prajapati, P., 2016. Nanotechnology in medicine: leads from Ayurveda. J. Pharm. Bioallied Sci. 8 (1), 80–81.

Shen, J.C., 2006. Nanobiomaterials. *Zhongguo yi xue ke xue yuan xue bao*. Acta Academiae Medicinae Sinicae 28 (4), 472–474.

Si, H.Y., Li, D.P., Wang, T.M., Zhang, H.L., Ren, F.Y., Xu, Z.G., Zhao, Y.Y., 2010. Improving the anti-tumor effect of genistein with a biocompatible superparamagnetic drug delivery system. J. Nanosci. Nanotechnol. 10 (4), 2325–2331.

Singh, M., Manikandan, S., Kumaraguru, A.K., 2011. Nanoparticles: a new technology with wide applications. Res. J. Nanosci. Nanotechnol. 1 (1), 1–11.

Singh, K., Ahmad, Z., Shakya, P., Ansari, V.A., Kumar, A., Zishan, M., Arif, M., 2016. Nano formulation: A novel approach for nose to brain drug delivery. J. Chem. Pharm. Res. 8 (2), 208–215.

Sitharaman, B. (Ed.), 2016. Nanobiomaterials Handbook. CRC Press, Hoboken, NJ.

Summerlin, N., Soo, E., Thakur, S., Qu, Z., Jambhrunkar, S., Popat, A., 2015. Resveratrol nanoformulations: challenges and opportunities. Int. J. Pharm. 479 (2), 282–290.

Sun, K., Zhang, Y., D'Alessandro, A., Nemkov, T., Song, A., Wu, H., Liu, H., Adebiyi, M., Huang, A., Wen, Y.E., Bogdanov, M.V., 2016. Sphingosine-1-phosphate promotes erythrocyte glycolysis and oxygen release for adaptation to high-altitude hypoxia. Nature 7, 12086.

Thambi, T., Deepagan, V.G., Yoon, H.Y., Han, H.S., Kim, S.H., Son, S., Jo, D.G., Ahn, C.H., Suh, Y.D., Kim, K., Kwon, I.C., 2014. Hypoxia-responsive polymeric nanoparticles for tumor-targeted drug delivery. Biomaterials 35 (5), 1735–1743.

Veerasekar, B., Sathishkumar, G., Kumar, M.V., Krishnaveni, A., Ganesan, J., Selvaraj, D.E., 2016. Applications of nanotechnology in medicine. J. Chem. Pharm. Res. 8 (1S), 177–180.

Whitesides, G.M., 2005. Nanoscience, nanotechnology, and chemistry. Small 1 (2), 172–179.

Wilkins, M.R., Ghofrani, H.A., Weissmann, N., Aldashev, A., Zhao, L., 2015. Pathophysiology and treatment of high-altitude pulmonary vascular disease. Circulation 131 (6), 582–590.

Wu, T.H., Yen, F.L., Lin, L.T., Tsai, T.R., Lin, C.C., Cham, T.M., 2008. Preparation, physicochemical characterization, and antioxidant effects of quercetin nanoparticles. Int. J. Pharm. 346 (1), 160–168.

Xu, X., Yu, J., Tong, S., Zhu, Y. and Cao, X., 2011. Formulation of silymarin with high efficacy and prolonged action and the preparation method thereof. US Patent 20110201680. 18 August 2011, Jiangsu University.

Yang, L., Zhang, L., Webster, T.J., 2011. Nanobiomaterials: state of the art and future trends. Adv. Eng. Mater. 13 (6), B197–B217.

Yen, F.L., Wu, T.H., Lin, L.T., Cham, T.M., Lin, C.C., 2008. Nanoparticles formulation of Cuscuta chinensis prevents acetaminophen-induced hepatotoxicity in rats. Food Chem. Toxicol. 46 (5), 1771–1777.

Yokel, R.A., MacPhail, R.C., 2011. Engineered nanomaterials: exposures, hazards, and risk prevention. J. Occup. Med. Toxicol. 6 (1), 1–27.

Zheng, X., Kan, B., Gou, M., Fu, S., Zhang, J., Men, K., Chen, L., Luo, F., Zhao, Y., Zhao, X., Wei, Y., 2010. Preparation of MPEG–PLA nanoparticle for honokiol delivery in vitro. Int. J. Pharm. 386 (1), 262–267.

CHAPTER 13

Electrochemical Immunobiosensors for Point-of-Care Detection of Hypoxia Biomarkers

Chandran Karunakaran*, Paulraj Santharaman*, Murugesan Balamurugan*, Sushil Kumar Singh†, Jonathan C. Claussen‡

*Department of Chemistry, Biomedical Research Lab, VHNSN College (Autonomous), Virudhunagar, India, †Functional Materials Group, Solid State Physics Lab, Defence Research and Development Organization, Timarpur, India, ‡Mechanical Engineering, Iowa State University, Ames, IA, United States

Abbreviations

ADC	analog to digital converter
ALS	amyotrophic lateral sclerosis
CKD	chronic kidney diseases
CNPs	cerium oxide nanoparticles
CNT	carbon nanotubes
EDC	1-ethyl-3-(3-dimethylaminopropyl)carbodiimide
ELISA	enzyme linked immunosorbent assay
GPOx	glutathione peroxidase
GUI	graphical user interface
HPLC	high performance liquid chromatography
LabVIEW	Laboratory Virtual Instrumentation Engineering Workbench
MUA	11-mercaptoundecanoic acid
NHS	N-hydroxysuccinimide
NO	nitric oxide
NO_2^-	nitrite
$O_2^{-\cdot}$	superoxide anion radical
PCR	polymerase chain reaction
POC	point-of-care
PPy	polypyrrole
RONS	reactive oxygen or nitrogen species
SAM	self-assembled monolayer
SOD	superoxide dismutase
SOD1	copper, zinc superoxide dismutase
SPCE	screen-printed carbon electrodes
SPR	surface plasmon resonance
TFT	thin-film transistor
VI	virtual instrumentation
WE	working electrode

Management of High Altitude Pathophysiology. https://doi.org/10.1016/B978-0-12-813999-8.00013-6

13.1   INTRODUCTION

Hypoxia is a condition of oxygen pressure reduced below a critical threshold, which affects the function of organs, tissues, and cells. Especially at high altitudes, hypoxia induces a series of events in the cell, altering its metabolic potential and its growth-regulating mechanisms (Srinivasan et al., 2013). It constitutes a major clinical obstacle in terms of pathophysiological progression, which is often diagnosed by biopsies and serological profiling, and therapy. Although hypoxia is of importance physiologically and clinically, lacunae exist in our knowledge of the systemic and temporal changes in gene expression occurring in blood during the exposure and recovery from hypoxia. A group of transcription factors, designated hypoxia inducible transcription factors, are specifically induced by low tissue oxygen tension and likely have a role in the oxygen-sensing mechanism and reparative reaction (Hamed et al., 2012). The hypoxia-induced gene expression mechanism is present in almost all types of cells inducing metabolic and structural changes according to the tissue type, tissue requirements, and intensity of the hypoxic challenge. Although there is an extensive list of genes that respond to hypoxia, only the EPO receptor, CD36 (marker for early erythroid cells), and γ-globin have been suggested and used as biomarkers for hypoxia-induced erythropoiesis in blood cells (Mosqueira et al., 2012). While obtaining a whole blood sample is an easy and fast technique, the lack of additional biomarkers makes it difficult to differentiate between acute and chronic hypoxia exposure as well as from normoxic recovery stages.

Hypoxia also results in oxidative damage to biological molecules and impairment in signaling pathways in the physiological system. It has become widely viewed as an underlying condition in acute hypoxia diseases, such as ischemia-reperfusion injury, in a number of chronic diseases, such as aging, pulmonary disease, cancer, cardiovascular diseases, chronic kidney diseases (CKD), and in neurodegenerative diseases, including amyotrophic lateral sclerosis (ALS), Parkinson's disease, and Alzheimer's disease. Hypoxia, which can be caused, for example, by transplantation, bypass surgery, or exposure to high altitudes, induces significant metabolic changes because of the decrease/lack of oxygen supply. A significant number of xenobiotics increases reactive oxygen or nitrogen species (RONS) formation in major metabolizing organs, which also lead to induction of chronic hypoxia by inhibition and differential expression of mitochondrial enzymes, including NADPH oxidase, xanthine oxidase, cytochrome c, cytochrome c reductase, cytochrome c oxidase, superoxide dismutase (SOD), arginase, glutathione peroxidase (GPOx), aconitase, fumarase, catalase and paraoxonase. Novel markers/metabolic pathways are under investigation, often using modern analytical technologies for global metabolic profiling in the cell/organ of interest or a surrogate biofluid (Serkova et al., 2008; Santharaman et al., 2016). Therefore, measurements of these biologically important hypoxia biomarkers are imperative in human physiology because they provide valuable information regarding people at high altitude.

A variety of methods have been reported for the quantitative determination of hypoxia biomarkers, including spectrophotometry, high-performance liquid chromatography (HPLC) with linear ion-trap mass spectrometer, enzyme-linked immunosorbent assay (ELISA), Western blotting, surface plasmon resonance (SPR), differential scanning calorimetry analysis, and polymerase chain reaction (PCR). These methods are highly sensitive but have numerous disadvantages, including arduous sample preparation, test setup, complex operation that requires highly trained technicians, and slow test results from lengthy experimental analysis. They also involve expensive instruments that are difficult to automate and time-consuming shipping of samples not suitable for point-of-care (POC) applications (Santharaman et al., 2017). In this chapter, we discuss the design and fabrication of nanocomposite biofunctionalized onto screen printed electrode-based immuno-biosensing volume miniaturized platforms for the determination of some clinically important hypoxia biomarkers, including superoxide dismutase, cytochrome *c*, nitrite and its metabolites. We also describe the development of LabVIEW-based cost-effective virtual instrumentation and microcontroller-based portable electrochemical analyzer for point-of-care applications.

13.2 ELECTROCHEMICAL IMMUNOSENSOR FOR SOD1

Among the known antioxidative proteins, SOD is a first line of defense against hypoxia-induced oxidative stress (Batinić-Haberle et al., 2010; Sen and Chakraborty, 2011). Three types of SOD isozymes have been identified in human cells: cytosolic copper zinc SOD (Cu, ZnSOD; SOD1), mitochondrial manganese SOD (MnSOD; SOD2), and extracellular SOD (ECSOD; SOD3). SOD1, however, contributes to approximately 90% of cellular SOD activity and is one of the key antioxidant enzymes present in high concentration (μM range) in cells. It catalyzes the dismutation of superoxides ($O_2^{-\cdot}$) to water and molecular oxygen and reportedly elevates during oxidative stress as an adaptive cellular response (Weydert and Cullen, 2010). Therefore, the quantitative detection of trace amounts of SOD1 availability in biological samples is of great significance to preclinical diagnosis of hypoxia. Because of this biological significance, a highly sensitive and novel label-free immunosensor for the detection of SOD1 was developed, based on nitrite oxidase activity of SOD1 bound to the specific monoclonal antibody biofunctionalized to nanomatrix onto the screen-printed electrode surface. Especially, screen-printed electrodes-based sensors, which replace conventional electrochemical cell setups, reduce the required sample volume, simplify the apparatus, and make the point-of-care testing easy to handle and cost effective. The most significant advantage of these miniaturized SPE-based electrochemical biosensors is their compatibility with the microelectronics that allows the commercialization of laboratory prototypes into potential bioanalytical devices.

13.2.1   Fabrication and Biofunctionalization of the GNP-PPy-SPCE Electrode to Form the Anti-SOD1-SAM-GNP-PPy-SPCE Immunosensor

Fabrication of anti-SOD1-SAM-GNP-PPy-SPCE immunosensor platform is described briefly as follows (Santharaman et al., 2016). Prior to the surface modification, the bare SPCE was polished to remove the contaminants, unbounded ink constituents, and to increase the porosity and functionalities. The SPCE was pretreated in the presence of $1\,M\,H_2SO_4$ by cycling the potential between -0.5 to $1.0\,V$ at a scan rate of $100\,mV/s$ for 10 complete cycles. Then, the SPCE was washed with doubly distilled water and dried in room temperature. After pretreatment, pyrrole was electro-polymerized on the working surface of SPCE by the irreversible oxidation of $0.4\,M$ pyrrole in $0.1\,M$ KCl as supporting electrolyte by applying the potential from 0 to $0.9\,V$ vs Ag/AgCl at a scan rate of $50\,mV/s$ for 10 complete cycles. Then, the optimum of 5 cycles of GNP was deposited electrochemically on PPy-SPCE by a negative sweep of electrode potential between 0.9 and $0\,V$ at a scan rate of $100\,mV/s$ in $0.25\,mM$ $HAuCl_4 \cdot 3H_2O$ solution containing N_2 saturated $0.1\,M\,NaNO_3$ as an electrolyte. Further, the homogenous nanostructured GNP-PPy-SPCE was incubated with $1\,mM$ cysteine solution for $1\,h$ at room temperature to form the self-assembled monolayer (SAM) (Santharaman et al., 2016).

The resultant SAM-GNP-PPy-SPCE surface was rinsed with doubly distilled water to remove unbound/excess cysteine. Next, monoclonal antisuperoxide dismutase antibody (anti-SOD1) was covalently bound to the SAM-GNP-PPy-SPCE via an incubation step lasting $24\,h$. A quantity of $10\,\mu L$ of anti-SOD1 solution with $5\,\mu L$ of 2.5% glutaraldehyde cross-linking agent was incubated onto the SAM-GNP-PPy-SPCE. After the incubation of anti-SOD1 on the SAM-GNP-PPy-SPCE, the immunosensor surface was rinsed with PBS to remove any unbound, superficial SOD1 antibodies. A drop of SOD1 mixed in buffer solution or in cultured samples was applied onto the immunosensor surface and allowed to set for $20\,min$ at room temperature to form immunocomplex on the sensor for the electrochemical determination of SOD1 (Santharaman et al., 2016). The obtained anti-SOD1-SAM-GNP-PPy-SPCE immunosensor was stored at $4°C$ in a refrigerator when not in use. The stepwise fabrication of the immunosensor platforms are illustrated in Fig. 13.1.

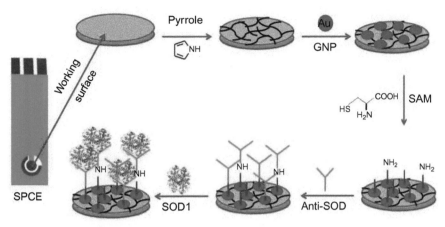

FIG. 13.1

Schematic representation of the label-free superoxide dismutase (SOD) 1 immunosensor. *From Santharaman, P., Das, M., Singh, S.K., Sethy, N.K., Bhargava, K., Claussen, J.C., Karunakaran, C., 2016. Label-free electrochemical immunosensor for the rapid and sensitive detection of the oxidative stress marker superoxide dismutase 1 at the point-of-care. Sens. Actuators B 236, 546–553.*

13.2.2 Electrochemical Immunoassay Performance

The detection modality of an electrochemical voltammetric sensor is based on the principle that the binding of antigen on the electrode surface will change the measured current with respect to the input voltage. In this case, the binding of anti-SOD1 and antigen can change the electrochemical characteristics of the electrode and result in a measurable change in both CV and EIS signals. The cyclic voltammetry curves were acquired before and after the formation of SOD1 immunocomplex with anti-SOD1. A peak obtained at the formal potential of 0.06 V because of the quasireversible nature of one electron reduction of Cu^{2+} to Cu^{1+}, confirming the binding of SOD1. The anti-SOD1-SAM-GNP-PPy-SPCE immunosensor next was studied with different SOD1 concentrations at the optimum pH 7.0 and antibody/antigen incubation time of 20 min. The CV curves overlap each other (not shown here) and show negligible current differences because of quasireversibility of the peak. Therefore, the inherent ability of nitrite oxidase-like activity of SOD1 as reported in our previous work was used to monitor the bound SOD1 concentration. The nitrite oxidase activity exhibited by SOD1 in this report is shown in Fig. 13.2. The electrochemical responses for the SOD1-bound immunosensor were obtained during the incremental increase of SOD1 concentrations from 0.5 nM to 5 μM in the presence of constant 100 μM nitrite solution as shown in Fig. 13.3. The results exhibited well-distinguished anodic peaks at 0.8 V because of the characteristic nitrite oxidase activity with significant increase in current response for increasing SOD1

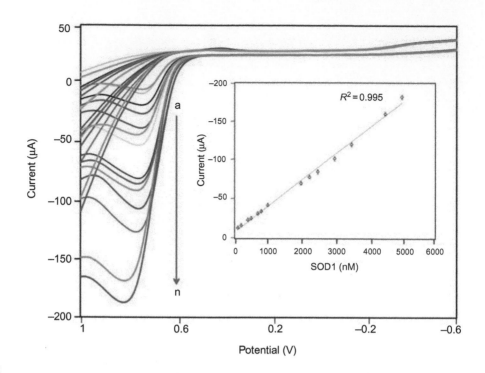

FIG. 13.2

Typical CV responses (a–n) of the anti-SOD-SAM-GNP-PPy-SPCE after the immunochemical incubation of 100, 200, 400, 500, 700, 800, 1000, 2000, 2250, 2500, 3000, 3500, 4500, 5000 nM of SOD1 in 0.1 M PBS (pH 7.0) containing constant (100 μM) nitrite solution and measured at a scan rate of 50 mV/s. Inset figure represents the linear graph of concentration of SOD1 vs current ($y = -0.033 \times -8.265$ and $r^2 = 0.995$) (PEG-CNPs—Polyethylene glycol coated 3 nm nanoceria). *From Santharaman, P., Das, M., Singh, S.K., Sethy, N.K., Bhargava, K., Claussen, J.C., Karunakaran, C., 2016. Label-free electrochemical immunosensor for the rapid and sensitive detection of the oxidative stress marker superoxide dismutase 1 at the point-of-care. Sens. Actuators B 236, 546–553. Reproduced with permission from Elsevier.*

concentrations. The calibration curve (inset Fig. 13.3) exhibits a linear range of response over the concentrations of 100 nM–5 μM ($r^2 = 0.995$ and $n = 3$) with a detection limit of 0.5 nM (the lowest quantity of a substance that can be distinguished from the absence of that substance within a stated confidence) and sensitivity of (46.6 ± 3.5 nA/nM) (the magnitude of the current response per unit concentration calculated from the slope of the calibration curve). The comparison of the electroanalytical parameters of the present SOD1 immunosensor with the earlier reported apta/immunosensors for various biomarkers reveals that the immunosensor showed comparable analytical characteristics with previously reported studies. These results demonstrate the potential capability of the developed immunosensor to sense SOD1 concentrations within the nanomolar and micromolar concentration ranges that are commonly found in the cytosol of neuronal cells, plasma, serum, and blood (Santharaman et al., 2016) (see Fig. 13.4).

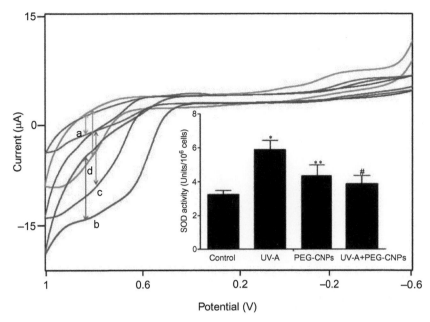

FIG. 13.3

Cyclic voltammograms of SOD1 normalized for control kerotinocyte cells, expressions in cells exposed to UV-A radiation, cells pretreated with CNPs and then exposed to similar UV-A dose immunocomplexed with anti-SOD-SAM-GNP-PPy-SPCE in 0.1 M PBS at a scan rate of 50 mV/s. Inset figure represents the estimation of superoxide dismutase activity for control kerotinocyte cells, expressions in cells exposed to UV-A radiation, cells pretreated with CNPs and then exposed to similar UV-A dose. *From Santharaman, P., Das, M., Singh, S.K., Sethy, N.K., Bhargava, K., Claussen, J.C., Karunakaran, C., 2016. Label-free electrochemical immunosensor for the rapid and sensitive detection of the oxidative stress marker superoxide dismutase 1 at the point-of-care. Sens. Actuators B 236, 546–553. Reproduced with permission from Elsevier.*

13.2.3 Biosensing of Superoxide Dismutase

The feasibility of the newly developed label-free SOD1 immunosensor for clinical analysis is demonstrated by measuring SOD1 in real samples and comparing the results with those from Western blot analysis and superoxide dismutase activity measurements. Densitometry immunoblots of superoxide dismutase normalized to beta actin showed 1.5-fold higher expressions in cells exposed to UV-A radiation. The same cells pretreated with cerium oxide nanoparticles (CNPs) as an antioxidant, protected the cells and tissues against the oxidative damage, after subsequent exposure to a UV-A dose and accordingly did not show elevated SOD1 levels. Also, CNPs alone had no effect on SOD1 expression ($*P<.05$, $**P<.01$, $^{\#}P < 0.02$ in comparison to control, $P<.05$ in comparison to UV treated). Further, SOD1 activity levels were measured in human epidermal keratinocyte cells and were found (inset Fig. 13.3) to be elevated twofold in cells on exposure to UV-A radiations for 30 min. Cells pretreated with CNPs and then exposed to such a UV dose, however,

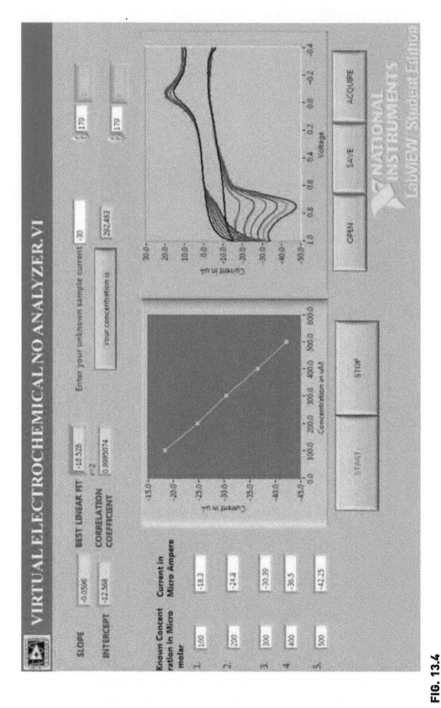

FIG. 13.4

Main front panel of the VI describing the electrochemical responses for various NO concentrations. *From Madasamy, T., Pandiaraj, M., Balamurugan, M., Karnewar, S., Benjamin, A.R., Venkatesh, K.A., Vairamani, K., Kotamraju, S., Karunakaran, C., 2012. Virtual electrochemical nitric oxide analyzer using copper, zinc superoxide dismutase immobilized on carbon nanotubes in polypyrrole matrix. Talanta 100, 168–174. Reproduced with permission from Elsevier.*

showed no appreciable change in SOD1 activity as compared to the control. Also, CNPs alone did not cause any elevated SOD1 activity of the cells ($*P < .05$, $**P < .01$, $^\#P < 0.02$ in comparison to control, $P < .05$ in comparison to UV-treated). These data indicate that CNP acts as a SOD1 mimetic. Also, the electrochemical determinations of SOD1 in the above-cultured human skin cells and the treated cells were carried out using the SOD1 immunosensor. In Fig. 13.3, the peak current ($15\,\mu A$) at the potential $0.8\,V$ for UV-A treated cells (curve b) was significantly higher than the control cells (curve a) showing the current response $5\,\mu A$. Further, the difference in amperometric response of cells on exposure to UV-A radiations was found to be nearly two-fold higher than in normal skin cells. These results agree with the SOD1 activity measurements and therefore confirm the suitability of the immunosensor for the measurement of cellular SOD1 levels in biological samples (Santharaman et al., 2016).

13.3 LABVIEW-BASED VIRTUAL INSTRUMENTATION FOR NITRIC OXIDE AND CYTOCHROME *C* BIOSENSING

The physical instruments for electrochemical techniques, including cyclic voltammetry, amperometry, impedimetric, and chrono-coulometry, are expensive, and the instrumentation systems are composed of predefined hardware components that are specific to their measurement function. Because of their hard-coded functions, these standalone instruments are limited in their versatility. Therefore, the development of inexpensive and easy-to-use analytical assays to measure clinically important biomarkers has been the focus of intensive research efforts. The graphical user interface (GUI) software LabVIEW (Laboratory Virtual Instrumentation Engineering Workbench) is ideal for creating low-cost and user-friendly virtual instruments.

Virtual instrumentation (VI) is an interdisciplinary field that merges sensing, hardware, and software technologies in order to develop flexible and sophisticated instruments for control and monitoring applications. Its ready-to-use libraries for integrating standalone instruments and data acquisition devices help to build a complete measurement and automation solution. The VI performing cyclic voltammetry was employed to measure NO using the copper, zinc superoxide dismutase (SOD1) immobilized onto carbon nanotubes (CNT) integrated polypyrrole (PPy) modified Pt electrode as a biosensor (Madasamy et al., 2012).

Before the addition of NO, no characteristic peak was obtained and no changes were observed in the current response. After the addition of NO, however, biosensors exhibited the significant increase in current anodically at the potential, $+0.8\,V$. It was attributed to the electrochemical oxidation of NO to NO_2^- via a cyclic redox reaction of SOD1 active site Cu(I/II) moiety. Then, the NO concentrations were varied at the same scan rate. The calibration curve thus plotted exhibited a linear range of response over the concentration of NO from 0.1 µM to 1 mM ($r^2 = 0.999$, $n = 3$) with a detection limit of 0.1 µM and the sensitivity of 1.1 µA/µM. Fig. 13.4 shows the measurement page (front panel) of the VI describing the electrochemical responses for various NO concentrations and a linear plot.

Cytochrome c (cyt c, a heme (Fe(III)) containing biologically important mitochondrial redox protein, plays an important physiological role in oxidative phosphorylation and as an electron carrier in the mitochondrial intermembrane space between cyt c reductase (complex III) and cyt c oxidase (complex IV). However, cyt c can be translocated out from mitochondria to cytosol under various pathological conditions, especially hypoxia triggering the activation of caspases and subsequent apoptotic cell death. Moreover, cyt c releases also have been identified in circumstances that can injure mitochondria, such as acute myocardial infarction, chemotherapy, a debilitating brain injury, and various neurological diseases. So, the quantitative detection of trace amount of cyt c release in biological samples is of great importance as preclinical diagnosis. We earlier developed cyt c immunosensor based on a specific cyt c monoclonal antibody immobilized on GNP-PPy nanocomposites functionalized screen-printed electrodes (SPE) (Pandiaraj et al., 2014). An earlier VI-based cost-effective measurement of cyt c was reported (Madasamy et al., 2017). The electrochemical determination of cyt c was performed using LabVIEW-based VI by using its direct electroactivity of heme of cyt c (Fe^{3+}/Fe^{2+}) specifically bound to anti-cyt c. Under controlled experimental conditions, the developed immunosensors were exposed to solutions of cyt c to form an immunocomplex. After the immunological binding of cyt c to anti-cyt c, a cyclic voltammetric curve was recorded. As expected, a pair of stable and well-defined reversible redox peaks (+0.298 and −0.085 V) were observed, with a formal potential (E^0) of 0.106 V, which could be ascribed to the electron transfer between the heme (FeIII)/(FeII) of cyt c and the underlying modified electrode. A series of cyt c solutions with different concentrations was prepared and applied on the working electrodes of the anti-cyt c-SAM-GNP-PPy-SPE immunosensors. The cyclic voltammetric curve responses are displayed in Fig. 13.5.

It can be seen from the figure that the redox peak current increased with increase in cyt c concentration because of the direct electron transfer of cyt c. A calibration plot based on the change in the cathodic peak current in virtual CV also was presented in the front panel of Fig. 13.5. The calibration plot shows a good linear relationship between the cathodic peak currents obtained by virtual CV and the concentrations of cyt c. Under optimal conditions, the electroanalytical parameters obtained for the cyt c determination using LabVIEW-based VI compared favorably with that of the standard electrochemical workstation.

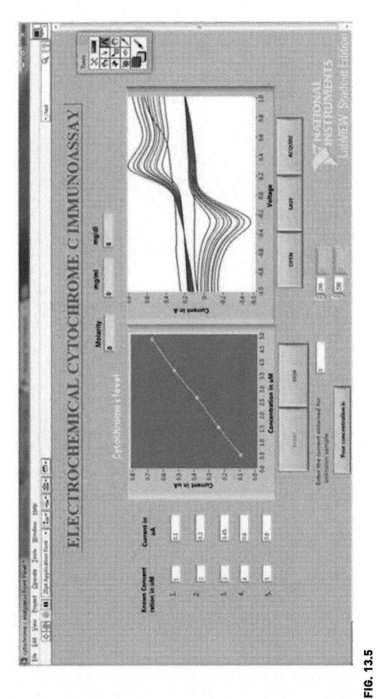

FIG. 13.5

Cytochrome *c* immunosensor measurement page using virtual instrument. *Courtesy of Dr. C. Karunakaran.*

13.4 ARM MICROCONTROLLER BASED PORTABLE ELECTROCHEMICAL ANALYZER FOR NITRITE BIOSENSING

Nitrite (NO_2^-) is a central homeostatic molecule in nitric oxide (NO) biology and serves as an important signaling molecule in its own right and also has a protective role against hypoxia, ischemia, or reperfusion injury. Nitrite supplementation also regulates HIF-1α stability during hypoxia and the subsequent hypoxia-responsive gene expression. It also extends the role of nitrite in modulating expression of genes, transcription factors, and signaling networks during hypoxia supporting the signaling potentials of nitrite. Therefore, sensitive and selective biosensors with point-of-care devices need to be explored to detect the physiological nitrite level because of its important role in human pathophysiology.

The recent development of software and hardware technology opens a new way to overcome the drawbacks of portability with the help of microcontroller-based portable devises. A microcontroller-based data acquisition unit integrated with potentiostat circuit is capable of performing electrochemical technique for the analysis. The acquired biosensor data are processed into digital form by the microcontroller and further transferred to analysis. The GUI-based system makes the analyzer easy to operate.

The design and construction of biosensor system based on the NO_2^- redox activities of cytochrome *c* reductase as novel biorecognition element, covalently coupled on MUA functionalized GNP-PPy nanocomposite modified screen-printed carbon electrode (SPCE) by using EDC and NHS has been reported (Fig. 13.6). In this approach, we have combined the distinct advantages of CcR biofunctionalized SPCE and ARM microcontroller-based portable sensing device for direct measurement of NO_2^- levels in cultured H9c2 cardiac cells under hypoxia conditions (Santharaman et al., 2017).

13.4.1 Circuit Design and Hardware Architecture

The overall circuit diagram of the microcontroller-based portable cyclic voltammetric analyzer for NO_2^- is shown in Fig. 13.7. A miniaturized three-electrode system was connected to an ARM Cortex M3 (LPC1768) microcontroller-based data acquisition unit containing a homemade potentiostat circuit, a level shifter, and an op-amp inverter (TL084). The necessary voltammetric waveform between the specified voltage ranges to be applied to the electrodes was produced by the built-in 12-bit digital to analog converter (DAC) of the LPC1768 microcontroller. The output current generated at the working electrode (WE) as a result of electrochemical reaction was converted to voltage using a current to voltage (I/V) converter and measured using an inbuilt analog to digital converter (ADC) of the microcontroller.

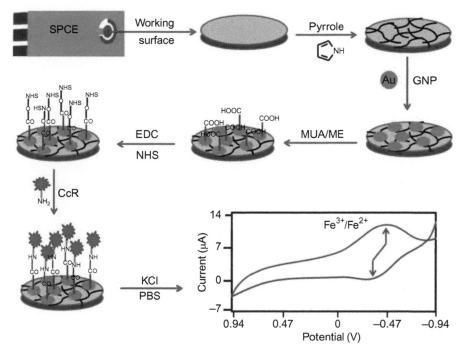

FIG. 13.6

Schematic representation of the step-wise fabrication of nitrite biosensor using CcR immobilized onto miniaturized disposable screen-printed carbon electrode. *From Santharaman, P., Venkatesh, K.A., Vairamani, K., Benjamin, A.R., Sethy, N.K., Bhargava, K., Karunakaran, C., 2017. ARM-microcontroller based portable nitrite electrochemical analyzer using cytochrome c reductase biofunctionalized onto screen-printed carbon electrode. Biosens. Bioelectron. 90, 410–417. Reproduced with permission from Elsevier.*

The ARM microcontroller has an built-in 12-bit ADC. Because the microcontroller can accept only positive voltages, the level of the output voltage is shifted by using an op-amp TL084 to accommodate negative voltages. Microcontroller sends the analog sweep voltage value to the potentiostat through the built-in DAC. Output voltage from the potentiostat circuit was fed to the ADC of the microcontroller. The microcontroller processed the data, which then was stored in an EEPROM (AT24LC04) and displayed in TFT (thin-film transistor; 320 × 480 high resolution) display. Switches were used to enter the range and select the calibration mode. The firmware program has been developed in C using IAR Embedded Workbench. The software program has used to perform and control all the electrochemical analyzer functions (Santharaman et al., 2017).

13.4.2 The Voltammetric Characterization of CcR Modified Biosensor

Fig. 13.8 shows the typical CVs of CcR-SAM-GNP-PPy-SPCE (curve *d*), SAM-GNP-PPy-SPCE (curve *c*), PPy-SPCE (curve *b*), and SPCE (curve *a*) electrodes in the presence of 0.1 M PBS pH 7.0 containing 0.1 M KCl as a supporting

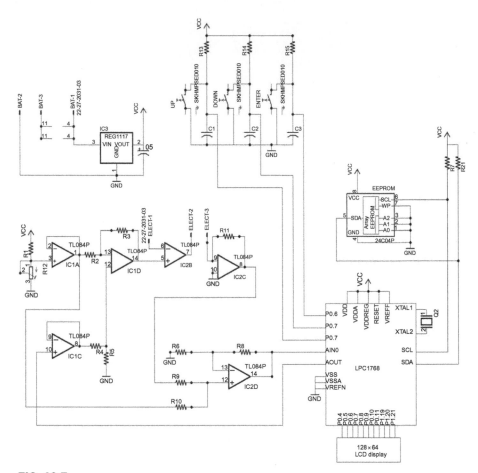

FIG. 13.7

Circuit diagram of the portable ARM Cortex M3 microcontroller-based electrochemical nitrite analyzer. *From Santharaman, P., Venkatesh, K.A., Vairamani, K., Benjamin, A.R., Sethy, N.K., Bhargava, K., Karunakaran, C., 2017. ARM-microcontroller based portable nitrite electrochemical analyzer using cytochrome c reductase biofunctionalized onto screen-printed carbon electrode. Biosens. Bioelectron. 90, 410–417. Reproduced with permission from Elsevier.*

electrolyte at a scan rate of 50 mV/s. No considerable redox peaks were obtained for bare SPCE (curve *8a*), PPy-SPCE (curve *8b*) and SAM-GNP-PPy-SPCE (curve *8c*) in the specified voltage range, indicating no Faradaic electron transfer process occurs. After the biofunctionalization of CcR on SAM-GNP-PPy-SPCE (curve *d*), a pair of characteristic reversible redox peaks at −0.34 and 0.45 V vs Ag/AgCl. This Faradaic redox peak clearly attributed to the electron transfer between the Fe(III)/Fe(II) redox couple of the CcR and the nanocomposite modified electrode surface. It clearly reveals that the CcR was coupled covalently on SAM-GNP-PPy nanocomposite modified electrode surface via EDC and NHS.

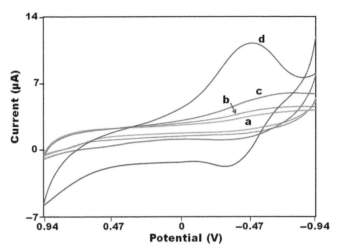

FIG. 13.8

Typical cyclic voltammetric responses of (a) bare SPCE, (b) PPy-SPCE, (c) SAM-GNP-PPy-SPCE, and (d) CcR-SAM-GNP-PPy-SPCE electrodes in 0.1 M PBS (pH 7.0) containing 100 μM DTPA in 0.1 M KCl using the scan rate of 50 mV/s vs Ag/AgCl. *From Santharaman, P., Venkatesh, K.A., Vairamani, K., Benjamin, A.R., Sethy, N.K., Bhargava, K., Karunakaran, C., 2017. ARM-microcontroller based portable nitrite electrochemical analyzer using cytochrome c reductase biofunctionalized onto screen-printed carbon electrode. Biosens. Bioelectron. 90, 410–417. Reproduced with permission from Elsevier.*

13.4.3   Electrochemical Response to NO_2^-

The electrochemical responses of the CcR-SAM-GNP-PPy-SPCE electrode in 0.1 M PBS (control) and various concentrations of NO_2^- using the scan rate of 50 mV/s are shown in Fig. 13.9. As the concentration increases, the anodic current responses also increase linearly at 0.8V. It is ascribed to the electrochemical oxidation of NO_2^- to NO_3^- via a cyclic redox reaction of an active site Fe(II/III) in the CcR moiety. The inset Fig. 13.9 represents the plot of observed anodic peak currents against NO_2^- concentrations. The calibration curve thus obtained exhibits a linear range of response over the concentration of NO_2^- from 100 nM to 1600 μM, but for clarity we have shown from 50 to 1600 μM ($r^2 = 0.998$ and $n = 3$) with a detection limit of 60 nM and the sensitivity of 172.57 nA/μM for CcR biosensor.

13.4.4   Analytical Performance of the Portable NO_2^- Analyzer

The electrocatalytic activity of the CcR-SAM-GNP-PPy-SPCE biosensor toward the determination of NO_2^- was assessed by the development of a portable, low-cost electrochemical analyzer. The accuracy of the instrument was evaluated by comparing its results with that of a commercial electrochemical

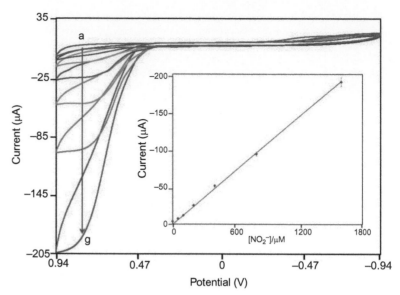

FIG. 13.9

The electrochemical responses of the CcR-SAM-GNP-PPy-SPCE electrode in (a) 0.1 M PBS, (b) 50 μM, (c) 100 μM, (d) 200 μM, (e) 400 μM, (f) 800 μM, and (g) 1600 μM of NO_2^- solution using the scan rate of 50 mV/s vs Ag/AgCl. Inset figure represents the linear calibration plot of anodic peak currents against NO_2^- concentrations ($y = -0.119 \times -4.2$, $r^2 = 0.998$). Each point represents the average of three measurements. *From Santharaman, P., Venkatesh, K.A., Vairamani, K., Benjamin, A.R., Sethy, N.K., Bhargava, K., Karunakaran, C., 2017. ARM-microcontroller based portable nitrite electrochemical analyzer using cytochrome c reductase biofunctionalized onto screen-printed carbon electrode. Biosens. Bioelectron. 90, 410–417. Reproduced with permission from Elsevier.*

analyzer (CHI1200B) under same experimental conditions. Before the addition of NO_2^-, characteristic redox peaks at -0.45 and -0.34 V vs Ag/AgCl because of the immobilized CcR were observed on the portable cyclic voltammetry analyzer as shown in Fig. 13.10. After the addition of 100, 200, 300, 400, 500 nM NO_2^- on the CcR biosensor surface, the current increased anodically at 0.8 V in the portable instrument, which was attributed to the redox reaction of NO_2^- by the CcR. These measurements confirmed that the portable electrochemical analyzer performs as good as a standard instrument for NO_2^- detection (Santharaman et al., 2017).

13.4.5   Measurement of NO_2^- in Hypoxia-Induced Cardiac Cell Lines

In order to further investigate the feasibility of the newly developed CcR-based biosensor being used for clinical analysis, NO_2^- levels were measured in H9c2 cells by using the biosensor and the reference kit (Measure iT, high-sensitivity

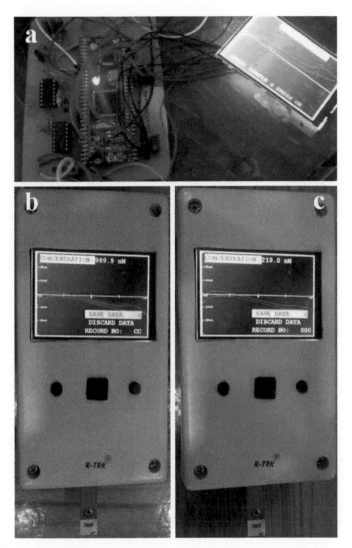

FIG. 13.10

Pictures of (A) actual physical appearance of portable ARM Cortex M3 microcontroller-based electrochemical analyzer using low power 320×480 pixels multicolor TFT LCD display showing the biosensor response of CcR peak in 0.1 M PBS containing 0.1 M KCl, (B) Experimental verification of electrochemical analyzer using the CcR biosensor at 1 µM NO_2^- concentration, (C) Measurement of NO_2^- in supernatant of hypoxia-induced H9c2 cell lysate (0.1% O_2 for 48h). *From Santharaman, P., Venkatesh, K.A., Vairamani, K., Benjamin, A.R., Sethy, N.K., Bhargava, K., Karunakaran, C., 2017. ARM-microcontroller based portable nitrite electrochemical analyzer using cytochrome c reductase biofunctionalized onto screen-printed carbon electrode. Biosens. Bioelectron. 90, 410–417. Reproduced with permission from Elsevier.*

Table 13.1 Measurement of NO_2^- in Hypoxia-Induced Cardiac Cell Lines

S. No.	Sample	Nitrite Conc. by Reference Assay Kit (µM)	Nitrite Conc. by Biosensor Using Standard Device (µM)	Nitrite Conc. by Biosensor Using ARM Based Analyzer (µM)
1	Normoxia	0.68±0.17	0.65±0.25	0.66±0.15
2	Hypoxia I	1.62±0.28	1.56±0.45	1.67±0.22
3	Hypoxia II	2.19±0.30	2.35±0.25	2.66±0.18

nitrite assay kit, Molecular Probes, Eugene, OR, United States) according to manufacturer's instructions. Exposure of H9c2 cells to hypoxia reportedly alters nitrite levels. Therefore, cells were cultured either at normoxia (CO_2 incubator maintained at 5% CO_2, 95% air, and 37°C), or hypoxia (CO_2 incubator maintained at 5% CO_2, 0.1% O_2, and 37°C) for 24 h (hypoxia I) and 48 h (hypoxia II). The cells were seeded at a density of 30% in 60 mm culture dish and allowed to reach a confluency of 60%–70% before treatment. Cells then were transferred to fresh growth media and cultured under the hypoxia conditions. Post-hypoxia exposure, both the normoxic and hypoxic cells were harvested using Trypsin-EDTA postexposure and washed with 1 × PBS (1500 rpm, 10 min, 4°C). The cell pellet was re-suspended in 150 µL 1 × PBS for the analysis using a biosensor and a nitrite estimation kit.

H9c2 cardiomyocytes cell lines were maintained under hypoxia to induce stress and vary the nitric oxide metabolites. The measurements of NO_2^- levels were performed on the normaxia, hypoxia-induced embryonic rat heart-derived H9c2 cardiomyoblasts using CA membrane coated CcR-SAM-GNP-PPy-SPCE biosensor and reference kit. The concentration of NO_2^- then was calculated by interpolating the obtained current response into the calibration plot prepared by the NO_2^- standard solutions; results were shown in Table 13.1. Each reading represents the average of three measurements. The measured values were validated by comparing the results obtained using high-sensitivity nitrite assay kit (Molecular Probes, Eugene, OR, United States). The NO_2^- levels present in H9c2 cells were recorded under normoxia (0.65 ± 0.25 µM), hypoxia I (1.56 ± 0.45 µM), and hypoxia II (2.35 ± 0.25 µM). Because these observations were comparable to the NO_2^- estimation done by reference kit, these results confirm the suitability of the electrode to measure cellular NO_2^- levels. In addition, the NO_2^- levels measured in hypoxia-induced H9C2 cells using the LPC 1768-based electrochemical analyzer shows a better analytical performance. The value obtained from the instrument is well-correlated with the standard CV instrument and reference assay (Santharaman et al., 2017).

13.5 CONCLUSION

This chapter focuses primarily on the design and development of volume miniaturized and cost-effective portable immuno-biosensors to measure hypoxia biomarkers, including NO, NO_2^-, cyt c, and SOD1. The nanocomposites used as host matrices enhanced the immobilization of enzymes without affecting its biological activity and direct electron transfer between the active site of the enzyme and the underlying electrode resulting in high sensitivity, wider linear range, lower limit of detection, and fast response. LabVIEW-based low-cost virtual instrumentation and microcontroller-based portable electrochemical analyzer for detecting hypoxia biomarkers with high sensitivity is described for point-of-care applications.

Acknowledgment

We are thankful for the support provided by the DIPAS-DRDO & DBT, New Delhi, and the managing board of Virudhunagar Hindu Nadar's Senthikumara Nadar College, Virudhunagar, Tamil Nadu, India.

References

Batinić-Haberle, I., Rebouças, J.S., Spasojević, I., 2010. Superoxide dismutase mimics: chemistry, pharmacology, and therapeutic potential. Antioxid. Redox Signal. 13 (6), 877–918.

Hamed, E.A., El-Abaseri, T.B., Mohamed, A.O., Ahmed, A.R., El-Metwally, T.H., 2012. Hypoxia and oxidative stress markers in pediatric patients undergoing hemodialysis: cross section study. BMC Nephrol. 13 (1), 136.

Madasamy, T., Pandiaraj, M., Balamurugan, M., Karnewar, S., Benjamin, A.R., Venkatesh, K.A., Vairamani, K., Kotamraju, S., Karunakaran, C., 2012. Virtual electrochemical nitric oxide analyzer using copper, zinc superoxide dismutase immobilized on carbon nanotubes in polypyrrole matrix. Talanta 100, 168–174.

Madasamy, T., Pandiaraj, M., Balamurugan, M., Santharaman, P., Venkatesh, A., Vairamani, K., Benjamin, A.R., Karunakaran, C., 2017. Virtual instrumentation for electrochemical biosensor applications. Sens. Lett. 15 (1), 1–10.

Mosqueira, M., Willmann, G., Zeiger, U., Khurana, T.S., 2012. Expression profiling reveals novel hypoxic biomarkers in peripheral blood of adult mice exposed to chronic hypoxia. PLoS One 7 (5), e37497.

Pandiaraj, M., Sethy, N.K., Bhargava, K., Kameswararao, V., Karunakaran, C., 2014. Designing label-free electrochemical immunosensors for cytochrome c using nanocomposites functionalized screen-printed electrodes. Biosens. Bioelectron. 54, 115–121.

Santharaman, P., Das, M., Singh, S.K., Sethy, N.K., Bhargava, K., Claussen, J.C., Karunakaran, C., 2016. Label-free electrochemical immunosensor for the rapid and sensitive detection of the oxidative stress marker superoxide dismutase 1 at the point-of-care. Sens. Actuators B 236, 546–553.

Santharaman, P., Venkatesh, K.A., Vairamani, K., Benjamin, A.R., Sethy, N.K., Bhargava, K., Karunakaran, C., 2017. ARM-microcontroller based portable nitrite electrochemical analyzer using cytochrome c reductase biofunctionalized onto screen-printed carbon electrode. Biosens. Bioelectron. 90, 410–417.

Sen, S., Chakraborty, R., 2011. The role of antioxidants in human health. In: Oxidative Stress: Diagnostics, Prevention, and Therapy. American Chemical Society, pp. 1–37.

Serkova, N.J., Reisdorph, N.A., Tissot van Patot, M.C., 2008. Metabolic markers of hypoxia: systems biology application in biomedicine. Toxicol. Mech. Methods 18 (1), 81–95.

Srinivasan, S., Spear, J., Chandran, K., Joseph, J., Kalyanaraman, B., Avadhani, N.G., 2013. Oxidative stress-induced mitochondrial protein kinase a mediates cytochrome c oxidase dysfunction. PLoS One 8 (10), e77129.

Weydert, C.J., Cullen, J.J., 2010. Measurement of superoxide dismutase, catalase, and glutathione peroxidase in cultured cells and tissue. Nat. Protoc. 5 (1), 51.

SECTION

IV

Nonmedical Therapies for High Altitude Ailments

CHAPTER 14

Performance Enhancement Through Physical Activity at High Altitudes

Deepti Majumdar

Ergonomics Division, Defence Institute of Physiology and Allied Sciences (DIPAS), Delhi, India

Abbreviations

AMS	acute mountain sickness
AVA	arterio-venous anastomoses
CBF	cerebral blood flow
CHD	congenital heart disease
CIVD	cold-induced vasodilatation
CMS	chronic mountain sickness
DIPAS	Defence Institute of Physiology & Allied Sciences
ECG	electrocardiography
HA	high altitudes
HACE	high altitude cerebral edema
HAPE	high altitude pulmonary edema
HR_{max}	maximum heart rate
IAAF	International Association of Athletics Federations, USA
LCe	load carriage ensembles
LV	left ventricular
MRI	magnetic resonance imaging
P50	PO_2 for 50% oxygen saturation
PAO_2	alveolar oxygen tension
PaO_2	oxygen pressure in the arterial blood
pCO_2	partial pressure of carbon dioxide
pO_2	partial pressure of oxygen
RBC	red blood cell
RV	right ventricular
VO_2	oxygen uptake
VO_2max	maximum aerobic capacity

14.1 INTRODUCTION

Sea level is the standardized geodetic reference point or an average level of surface of one or more of Earth's oceans from which elevations might be

279

Management of High Altitude Pathophysiology. https://doi.org/10.1016/B978-0-12-813999-8.00014-8

measured. The term "altitude" commonly is used to define the height of a location above or below sea level. In this context, we are concerned with human performance and habitability at high altitudes (HA), and, therefore, we will discuss only elevation or the height above sea level. Different zones in altitudes or mountainous regions describe the natural layering of ecosystems that occur at distinct altitudes because of varying environmental conditions. Temperature, humidity, soil composition, and solar radiation are important factors in determining altitudinal zones, which consequently support different vegetation and animal species. In addition to physical forces, biological forces also might produce zonation. The physical characteristics and relative location of the mountain are important in predicting altitudinal zonation patterns, apart from other factors, such as frequency of disturbances (e.g., fire, monsoons), wind velocity, type of rock, topography, nearness to streams or rivers, history of tectonic activity, and latitude. The altitudinal zonation of mountains always affects human habitability, directly or indirectly.

This chapter discusses how the application of different acclimatization strategies found in native highlanders could be used effectively to acclimatize lowlanders to the environmental adversaries of HA and enhance performance through physical activities. It would be of interest to apply different ergonomic interventions in identifying physical activities best suited for performance enhancement at HA, activities that might not result in any physical impairment from overexertion or be successful in minimizing occupational risk potentials of HA extreme climatic conditions. This chapter also discusses the various studies that have attempted to design methods to reduce HA injuries and maladies in sea-level dwellers and how to apply these strategies in improving occupational health, safety, and performance.

14.2   DEFINING ALTITUDE

Human beings rarely venture to altitudes above 6000 m, and no survival advantage is gained in being able to do so. No food or other commodity is worth gathering at these altitudes. Climbers ascend to great heights because of the adventure; climbing Mount Everest always has been seen as one of the ultimate human challenges (Grocott et al., 2009; Hornbein et al., 1989; Bock and Hultgren, 1986). In literature, altitude ranges according to human habitability are HA: 8000–12,000 ft (2438–3658 m); very HA: 12,000–18,000 ft (3658–5487 m); and extremely HA: more than 18,000 ft (more than 5500 m).

Until 8000 ft (2438 m), most people exhibit minimal HA effects. Above this height, however, no specific factors such as age, sex, or physical condition are known to correlate with the susceptibility to altitude sickness (Parmeggiani, 1983). People who are at significantly higher risk of getting HA maladies are native highlanders who re-enter HA after staying at low altitude, mountaineers and trekkers, miners, pilgrims, porters, and soldiers.

Soldiers who are deployed to HA conditions need to carry loads, which increase external stresses over and above his own body weight being lifted against gravity ascending a steep gradient. The effect from steepness of terrain has been found to be much greater than the effect of the load itself (Purakayastha and Selvamurthy, 2000; Ramaswamy et al., 1963; Tharion et al., 2005). The physiological optimum speed for level ground was reported by Ramaswamy and Majumdar (1955). Douglas et al. (1913) had experimentally showed that, for the same speed of walking, more energy is spent at HA than in lower altitudes (Douglas et al., 1913). The HA problems and maladies faced by soldiers have been discussed in detail by previous researchers (Tharion et al., 2005). Huang et al. (1984) studied the contribution of changes in metabolic rate to the increase in minute ventilation observed during exposure to HA (4300 m) and concluded that a substantial portion of the rise in minute ventilation could be attributed to increased metabolic rate at rest but not during exercise. Young et al. (1984) studied the influence of short-term HA (4300 m) residence on intramuscular pH and skeletal muscle enzyme activity of sea-level residents and found that limitations in exercise performance are mainly because of hypoxia that stimulates ventilation, not because of biochemical changes at HA.

The environmental risk factors leading to HA maladies include low oxygen partial pressure, temperature, high humidity, wind chill, and higher elevation. The summit of Mount Everest is in the death zone, which, in mountaineering, refers to altitudes above a certain point where the amount of oxygen is insufficient to sustain human life (Young and Reeves, 2002). This point generally is tagged as 8000 m (26,000 ft; less than 356 millibars of atmospheric pressure). All of the world's 14 summits in the death zone above 8000 m, called eight-thousanders, are located in the Himalaya and Karakoram mountain ranges in Asia (Wyss-Dunant, 1953). Fig. 14.1 gives a few examples of high-altitude peaks with respective oxygen partial pressure. Fig. 14.2 gives a schematic diagram indicating how different nonfreezing HA maladies could be handled to minimize further deterioration in condition of the patient.

Some residents of newly settled HA communities in the United States might be at increased risk for problems in adapting to HA living. They have not been genetically selected for HA living, unlike natives of older communities in the Andes mountains in South America or in Tibet, populations that have had millennia to adapt evolutionarily to this type of environmental stress (West, 2016). In comparison with Andean or Tibetan populations, differences in the adaptation to altitude are demonstrable. More remote and longer inhabiting populations perform better than the newly arriving lowlanders (Basu et al., 2007; Ge et al., 2009).

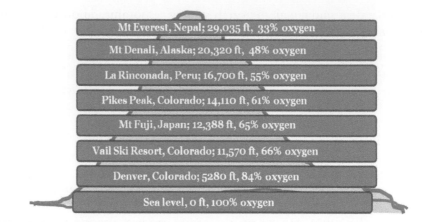

FIG. 14.1

Some examples of HA peaks with respective oxygen partial pressure.

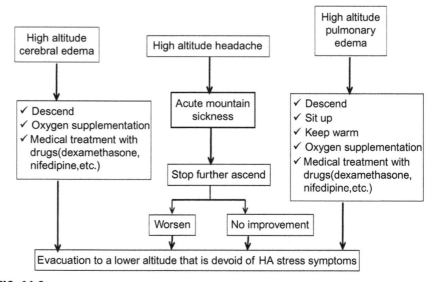

FIG. 14.2

Schematic diagram indicating how different nonfreezing high altitude maladies can be handled to minimize further deterioration in condition of the patient.

Each winter, millions of people ski at altitudes of 2500–3500 m (8200–11,500 ft) in Colorado. Each summer, more than 250,000 people visit the summit of Pikes Peak (4300 m or 14,100 ft). Upon arriving at HA, most individuals note a sensation of breathlessness because of hypoxia-induced hyperventilation and palpitations from an increased heart rate. These are normal physiological responses (Douglas et al., 1913). However, between 6 and 96 h of arrival, many individuals report headache, insomnia, anorexia, nausea, vomiting, dizziness, dyspnea, and loss of coordination (Norris et al., 2012). Data from hikers in Nepal indicated that the overall incidence of acute mountain sickness was 43%–53% (Bock and Hultgren, 1986; Furian et al., 2015; Lesham et al., 2008; Milner, 1998; Taylor, 2011).

14.3 ENVIRONMENTAL FACTORS AFFECTING PHYSICAL ACTIVITY AT HIGH ALTITUDES

At sea level, the partial pressure of oxygen (pO_2) available in the atmosphere and the oxygen demands of mitochondria are large. At each stage of the oxygen transport system, pO_2 decreases; this phenomenon is figuratively called the oxygen cascade. With decreasing pO_2, the S-shaped nature of oxyhemoglobin dissociation curve causes only a small change in the percentage saturation of hemoglobin until one reaches an altitude of 3048 m. At 1981 m (6500 ft) the alveolar pO_2 is reduced to 78 mmHg as compared to 100 mmHg at sea level, but oxygen saturation of hemoglobin remains at about 90%. This small change likely will not affect a person at rest or even carrying out mild exercise. It has been observed that short-term anaerobic power activities, such as sprint running, jumping, and shot-put and discus throws, are not adversely affected at altitude. Sometimes performance enhancement occurs because of reduced air density, resulting in reduced air friction as compared to sea level (Furian et al., 2015). Mountain climbers with an increased hypoxic ventilatory response are better able to climb to great heights, presumably because of the increased availability of alveolar oxygen. Hyperventilation is brought about by stimulation of the carotid bodies. With acute exposure even at heights below 3000 m (9840 ft), ventilation does not increase significantly. This situation corresponds to a oxygen pressure in arterial blood (PaO_2) of 60 mmHg. However, after 4 days of exposure to even modest increases in altitude, ventilation is consistently greater than normal ventilation at sea level. After a person acclimatizes, hyperventilation might occur at a PaO_2 as high as 90 mmHg. In contrary, low hypoxic ventilatory response occurs in acute mountain sickness, excessive polycythemia, and low birth weight.

The hypoxic ventilatory response persists for a sea-level resident who continues to remain at HA. At extreme altitudes, marked respiratory alkalosis develops to maintain a PaO_2 of more than 35 mmHg. In a decompression chamber with conditions equal to those at Mount Everest, $PaCO_2$ is 8 mmHg. In contrast, the native HA resident has a ventilatory response that is desensitized to hypoxia. Improved oxygen usage in peripheral tissues with decreased ventilatory effort has been postulated as an explanation for this phenomenon.

Vigorous aerobic activities in any form, however, are sensitive to altitude elevations. At HA, such as the Andes and Himalayas, the effects of reduced loading of hemoglobin with oxygen is very much evident, and it is hard to sustain physical activity. Acute exposure to 4300 m is found to reduce aerobic capacity by 32% as compared to that of sea level.

Studies have shown that the 140 million (approximately) people residing at elevations above 2500 m (8200 ft) have adapted to the lower oxygen levels. These adaptations are especially pronounced in people living in the Andes and the Himalayas. Compared with acclimatized newcomers, native Andean and Himalayan populations have better oxygenation at birth, enlarged lung volumes throughout life, and a higher capacity for exercise. Tibetans demonstrate a sustained increase in cerebral blood flow, lower hemoglobin concentration, and less susceptibility to chronic mountain sickness (CMS). These adaptations might reflect the longer history of HA habitation in such regions (Bastien et al., 2005).

Human beings spend their whole life by strict adherence to a range of internal body temperatures, that is, 0°C (ice crystal formation) to about 45°C (thermal coagulation of intracellular proteins). In order to maintain the internal temperature within this limit, humans have evolved very effective, and under some conditions very specialized, physiological responses to acute thermal stress, whether brought about by extreme heat or extreme cold. The hypothalamus responds to various temperature receptors located throughout the body and makes physiological adjustments to maintain a constant core temperature. For example, on a hot day, temperature receptors located in the skin send signals to the hypothalamus to cool the body by increasing the sweat rate.

Although thermoregulation is effective in regulating body temperature under normal conditions, exercise or physical activity in extreme cold or heat exerts heavy stress on the mechanisms that regulate the body temperature and the body's inherent ability of thermoregulation is not enough (Lewis, 1930; Little and Hanna, 1977). The human body loses 75% of metabolic energy as heat, and the human body is only 25% efficient, as far as use of energy is concerned. During exercise, heat is produced mainly from working muscle contractions, and a person's core temperature could reach above 40°C (104°F).

Among the many high mountain zones scattered throughout the Earth, only the Himalayas and Andes support large human populations at elevations greater than 3000 m, where hypoxia associated with reduced atmospheric oxygen pressure is acutely felt and where the interactions among hypoxia, cold ambient temperatures, and other altitude-related stresses are most intense. Permanent human habitation ranges up to 5000 m (and sometimes above) in both the Himalayas and Andes. The oxygen pressure of atmospheric air at 3000 m is about 70% of that at sea level, and at 5000 m is slightly greater than 50% of sea level values. This altitude range (3000–5000 m) is a natural zone of maximal hypoxic stress and, at the same time, it is a zone of great cold stress. Although microclimatic variations exist according to valley, slope, and peak sites in all HA zones, ambient temperatures always decline with increases in elevation. Depending upon elevation, season, and climatic zone, temperature declines ranging from 1°C drop/250 m elevation to a 1°C drop/100 m elevation (Little and Hanna, 1977).

The hands and feet are powerful agents for controlling the body's thermoregulation, serving as heat radiators and evaporators in hot environments and as thermal insulators in the cold. It has been shown in literature that each hand and foot can dissipate 150–220 W cm^{-2} through radiation and convection at rest in an ambient temperature of 27°C, with even greater heat dissipation through sweating. During hypothermia, the phenomenon of vasoconstriction is sufficiently strong to reduce heat flow to less than 0.1 W. Besides being powerful thermo-effectors, the nonhairy, glabrous skin of the extremities also could serve as powerful thermo-sensors that affect thermoregulatory behavior in a feed-forward and feed-back fashion. Blood flow to the extremities of the hands and feet respond rapidly upon exposure to cold, with a sympathetically mediated vasoconstriction reducing blood flow to the peripheries in favor of a central pooling of blood in the torso and deep body core. Because of this vasoconstriction and the high surface area-to-volume ratio, the skin temperature of the fingers and toes tends to rapidly and exponentially decrease to a level approaching to that of the ambient environment. Decrements in tactile sensitivity, manual dexterity, and gross motor function of the hands can lead to decreased overall performance in occupational settings, such as in mast and pole workers and divers. The rapid and sustained impairment of manual performance from local cold exposure also could degrade an individual's ability to operate emergency equipment (e.g., escape hatches, opening flares) or move into a safer situation (e.g., hauling oneself from the water into a life raft), turning a survivable situation into a critical one. Beyond immediate impairment, continued cold exposure and vasoconstriction also could lead to nonfreezing cold injuries such as immersion foot, and those from reduced nutritional blood flow leading to necrosis or other cold injuries, such as frostbite, from cell temperature dropping below the point of freezing and crystallization. The increased accessibility of outdoor recreational opportunities during winter, along with HA expeditions, extend the potential for injury beyond traditional occupational settings.

Despite an overall drive for vasoconstriction in the cold, a common observation in the toes and fingertips is that, after a brief period of lowered skin blood flow and temperature, a seemingly paradoxical and temporary increase in blood flow and rewarming occurs. During these episodes, skin temperature could rise by as much as 10°C, and such a rise and fall could occur repeatedly in a cyclic fashion. This pattern of periodic warming was first reported by Lewis in 1930 as the "hunting response," so called because of its apparent oscillatory pattern. This response also has been termed the "cold-induced vasodilatation" or CIVD phenomenon. In addition to the fingers and toes, CIVD has been observed in the face and forearms. Whole body thermal status is known to be an important determinant for CIVD prevalence and intensity, with an inverse relationship between CIVD responses and body temperature, along with the lack of any observed CIVD below a threshold body temperature. The mechanisms driving CIVD remain unclear, however, their anatomical endpoint of action likely revolves around arterio-venous anastomoses (AVA) within cutaneous microcirculation. AVA are shunts within the skin that permit blood to bypass the capillaries and instead flow directly from the arterioles to venules. The prevalence of AVA at the fingertips and toes seems to be the cause of an increased local blood flow resulting in CIVD (Taylor, 2011).

With the increasing operational deployment of troops in extreme HA areas where the effects of cold climate become more aggravated because of hypoxia, the incidence of cold injuries are on the rise. Cold injuries commonly faced by soldiers deployed at very high and extremely HAs include nonfreezing cold injuries (chilblains, trench foot) and freezing cold injuries (frostnip and frostbite). Among these, frostbite causes gross disability and maximum morbidity because it causes loss of extremities. The air at HA is dry and damages mucous membranes, which are kept moist by the air we breathe. More fluids are lost because one tends to sweat a lot while climbing. At HA, however, one tends to feel less thirsty, and the level of fluid intake reduces. All these factors simultaneously and synergistically result in a loss of fluids in the body and lead to increased blood viscosity. Because of the increase in viscosity, there is lack of oxygen supply to the capillaries and exposed body parts get cold, resulting in frostbite (Hashmi et al., 1998). How soon one suffers from frostbite depends on the climatic conditions and the duration of exposure. Frostbite occurs because the blood vessels can get frozen at extreme temperatures. Frostbite can be healed with proper medical care. The study further reported that lifestyle of some people make them more susceptible to frostbite than others. Tobacco smoking and peripheral vascular disease are definite factors affecting long-term prognosis. About 80% of the patients smoked 20–30 cigarettes per day while 10% habitually smoked *charas* or hashish (marijuana).

A significantly lower mortality because of cardiovascular disease, decreasing obesity prevalence, and higher rate of suicide in the United States at higher elevations are a few peculiarities of HA responses that have not been explained so far. The correlation between elevation and suicide risk was present even when the researchers control known suicide risk factors, including age, gender, race, and income. Research also has indicated that oxygen levels are unlikely to be a factor, considering that there is no indication of increased mood disturbances at HA in those with sleep apnea or in heavy smokers at HA (Taylor, 2011; West, 2016).

14.4 HUMAN HABITABILITY AND PHYSICAL ACTIVITY AT HIGH ALTITUDES

Most lowland people begin to develop hypoxia symptoms at 1–2 miles altitude. However, some permanent settlements in the Andes in South America and the Himalayas in Asia are 3 miles high. Mountain climbers have reached peaks that are more than 5 miles high, but only rarely without using oxygen tanks to assist in breathing. There is considerable variability between individuals and between populations in their ability to adjust to the environmental stresses of high mountain regions. The populations that are adapted most successfully usually are those whose ancestors have lived at HA for thousands of years. This is the case with some of the indigenous peoples living in the Andes of Peru and Bolivia as well as the Tibetans and Nepalese in the Himalayas. The ancestors of many people in each of these populations have lived above 13,000 ft (4000 m) for at least 2700 years. The implication is that natural selection over thousands of years has resulted in some people being genetically more suited to the stresses at HA (Mishra et al., 2015; Moore et al., 1998; Moore, 2001).

There are two major kinds of environmental stresses at HA for humans. First, the daily extremes of climate often alternate between hot, sun-burning days to freezing nights. Winds are often strong and humidity is low, resulting in rapid dehydration. Second, the air pressure is lower, usually the most significant limiting factor in high mountain regions. When we breathe air at sea level, the atmospheric pressure of about $14.7 \, \text{pounds in}^{-2}$ ($1.04 \, \text{kg cm}^{-2}$) causes oxygen to easily pass through selectively permeable lung membranes into the blood. At HA, the lower air pressure makes it more difficult for oxygen to enter our vascular systems, resulting in hypoxia, or oxygen deprivation. Hypoxia usually begins with the inability to do normal physical activities, such as climbing a short flight of stairs without fatigue. Other early symptoms of HA sickness include a lack of appetite, vomiting, headache, distorted vision, fatigue, difficulty with memorizing, and thinking clearly. In serious cases, pneumonia-like symptoms (pulmonary edema) because of hemorrhaging in the lungs and an abnormal accumulation of fluid around the brain (cerebral edema) develop. Pulmonary and cerebral edema usually result in death within a few days if the patient is not returned to normal air pressure levels at lower elevations. There is also an increased risk of heart failure because of the added stress placed on the lungs, heart, and arteries at HA (Moore et al., 1998; Ponsot et al., 2006).

When we travel to high mountain areas, our bodies initially develop inefficient physiological responses. The breathing rate and heart rate almost doubles, even while resting. Pulse rate and blood pressure rise sharply as our hearts pump harder to deliver more oxygen to the cells. Later, a more efficient response usually develops as acclimatization takes place. Additional red blood cells and capillaries are produced to carry more oxygen. The lungs increase in size to facilitate the osmosis of oxygen and carbon dioxide and the vascular network of muscles increases, enhancing the transfer of gases. Successful acclimatization, however, rarely results in the same level of physical and mental fitness that was typical of altitudes close to sea level. Strenuous exercise and memorization tasks still remain more difficult.

After returning to sea level after successful acclimatization to HA, the body usually has more red blood cells (RBCs) and greater lung expansion capability than needed. Because this situation provides athletes in endurance sports with a competitive advantage, the United States maintains an Olympic training center in the mountains of Colorado. Several other nations also train their athletes at HA for this reason. The physiological changes that result in increased fitness, however, are short term at low altitude. In a matter of weeks, the body usually returns to a normal fitness level.

The human body could adapt to HA through both immediate and long-term acclimatization (Fig. 14.3). At HA, in the short term, the lack of oxygen is sensed

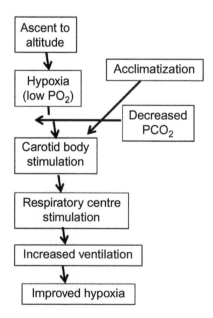

FIG. 14.3

Schematic diagram showing flowchart of sequential events for acclimatization to high altitudes and reduced oxygen pressure. PCO_2: partial pressure of carbon dioxide; PO_2: partial pressure of oxygen.

by the carotid bodies, which cause an increase in the breathing depth and rate (hyperpnoea). Hyperpnoea, however, also causes an adverse effect of respiratory alkalosis, inhibiting the respiratory center from enhancing the respiratory rate as much as would be required. Inability to increase the breathing rate could be caused by inadequate carotid body response or pulmonary or renal disease. In addition, at HA, the heart beats faster, the stroke volume is slightly decreased, and nonessential bodily functions are suppressed, resulting in a decline in food digestion efficiency (as the body suppresses the digestive system in favor of increasing its cardiopulmonary reserves).

Full acclimatization, however, requires days or even weeks. The body gradually compensates for the respiratory alkalosis by renal excretion of bicarbonate, allowing adequate respiration to provide oxygen without risking alkalosis. It takes about 4 days at any given altitude and the body undergoes physiological changes such as lower lactate production (because reduced glucose breakdown decreases the amount of lactate formed), decreased plasma volume, increased hematocrit (polycythemia), and increased RBC mass; a higher concentration of capillaries in skeletal muscle tissue; increased myoglobin, mitochondria number, aerobic enzyme concentration, 2,3-bisphosphoglyceric acid (2,3-BPG); hypoxic pulmonary vasoconstriction, and right ventricular hypertrophy. Pulmonary artery pressure increases in an effort to oxygenate more blood. Complete hematological adaptation to HA might be achieved when the increase in production of RBCs reaches a plateau and stops. The number of days required for full hematological adaptation can be approximated by multiplying the altitude in kilometers by 11.4 days. For example, adapting to an altitude of 4000 m (13,000 ft) requires 45.6 days. The upper altitude limit of this linear relationship has not been fully established. Appetite suppression can be severe during the early stages of HA stay, resulting in an average reduction in energy intake of approximately 40% and an accompanying loss of body mass. Diets low in salt and high in carbohydrates are well tolerated during early stages of HA acclimatization. In general, a high carbohydrate diet is beneficial for three reasons: It enhances HA tolerance, reduces severity of mountain sickness, and lessens the performance decrements during early stages of HA exposure.

The effect of altitude on human performance is complex. The HA imposes significant restrictions on work capacity and physiologic functions. Even at lower altitude, the body's adjustments do not fully compensate for the reduced oxygen pressure, and performance is compromised. Numerous variables are known to change from sea level. Even after several months of acclimatization, maximum aerobic capacity (VO_2max) remains at a value lower than at sea level because benefits of acclimatization are offset by a reduction in circulatory efficiency in both submaximal and maximal exercises. Though immediate exposure to altitude causes an increase in cardiac output during submaximal exercises, it reduces with days of exposure and does not improve with longer exposure. This reduction in cardiac output occurs because of a decrease in heart's stroke volume as the altitude stay lengthens.

A reduction in maximum cardiac output occurs after about 1 week of stay above 3048 m and persists throughout the stay. This reduction in blood flow during maximal exercise results from the increase in heart rate and stroke volume as a consequence of a decrease in plasma volume and an increase in total peripheral vascular resistance. The reduction in maximum heart rate might be influenced by enhanced parasympathetic tone induced by prolonged altitude exposure. Maximum aerobic power is depressed as ascent occurs, impairing the ability to work maximally, and although changes in hematological variables theoretically counterbalance the loss in aerobic power, practically they have not been shown to do so. Even moderate exercise becomes intolerable for people suffering with the effects of mountain sickness. The symptoms might subside with acclimatization, and some might even disappear. A person's ability to exercise improves and then he could do considerably more work. Mountain sickness usually can be prevented by acclimatizing slowly to moderate altitudes (below 3048 m), followed by slow progression to higher elevation. Physical activity also should be minimized during beginning of HA exposure. The length of acclimatization period depends on the altitude. Acclimatization at one altitude ensures only partial adjustment to higher elevation. Broadly, 2 weeks are required to adapt to altitude up to 2300 m, followed by an additional week of acclimatization required for every 610 m elevation to adapt to altitudes about 4600 m.

The environmental stress of cold might have positive effects on aerobic capacity at altitude, but this has been little investigated. Pulmonary ventilation increases with altitude and the measure of hypoxic ventilatory response holds some promise in predicting humans who might benefit from altitude conditioning. Cardiac function is well maintained, but lung function is not. The preferred fuel for exercise at altitude seems to be fat, while carbohydrate metabolism changes dramatically. Much is not known about HA anorexia and loss of muscle mass. Conditioning at altitude is known to benefit performance at altitude. It is not clear what effects of previous altitude exposure and training on aerobic capacity have on endurance performance immediately after return to sea level. Altitude adaptations in local circulation, cellular metabolism, and the compensatory increase in blood's oxygen-carrying capacity should facilitate subsequent sea level performance. The pulmonary adaptations and responses to prolonged hypoxic exposure are not lost immediately on descent from altitude. If tissue hypoxia is considered an important training stimulus, then altitude and training should act synergistically, so that total effect is greater than that experienced at sea level. So far, however, exercise-altitude research has not been designed adequately to evaluate this possibility. Often the activity level of the subjects is not controlled properly, making it difficult to determine whether an improved VO_2max or performance score after return from altitude represented a training effect, an altitude effect, or a synergism between altitude and training.

14.5 ENHANCING PERFORMANCE AT HA: AN ERGONOMICS OVERVIEW

Ergonomics is the scientific study of people at work, aiming to reduce stress, eliminate injuries, and work-related maladies, resulting in enhanced performance and safety. This might be achieved by designing tasks, workspaces, and equipment that are compatible with the physical capabilities of operators. These adjustments ensure the mental and physical wellness of individuals over longer duration of occupational exposure and increase their sustainability under adverse environmental conditions. Around the globe, variability within a population and across different populations are important phenomena that influence human-machine compatibility issues for different workforces operating under different working environments. Extensive variability and limitations in terms of body dimensions, strength, physical reachability, and cognitive abilities pose bottlenecks in the design of any equipment, workstation, or facility for the target population, apart from the extremes of environmental vagaries, such as those experienced at HA and extreme HA.

Studies need to be conducted along these lines to improve our understanding of such extreme situations that might be encountered normally at HA and during wartime and allow us to formulate evidence-based countermeasures to better equip personnel during such extreme activities. The daily work schedule of armed forces posted in HA areas involve a wide variety of workstations and operations. The increasing complexity of equipment-handling procedures and more stringent safety standards needs have led to a proliferation of instruments and information systems for these workstations and operational procedures to enable safe and effective operations. Designing appropriate uniforms, functional clothing, and personal protective equipment, along with design evaluation of different military workstations most typical of HA conditions, constitute important parts of current military low-intensity conflict scenarios. For activities at extreme altitudes, prior incorporation of ergonomic design principles are vital so that such clothing and equipment systems should not adversely affect human mobility and comfort while performing physical tasks.

The ability to work effectively at altitude is determined by three factors: altitude (high, very high, or extreme), duration of exposure (short, long, generations), and individual susceptibility. Individual susceptibility plays the biggest role in working capacity at altitude and the ability to adapt based on the duration of exposure. As a person goes to a higher altitude than to his initial exposure, the longer it takes for the body to acclimatize to the newer altitude. The time it takes to acclimatize must be taken into consideration when talking about work productivity. No one can expect to work at full capacity at the beginning. In fact, because the body will be continuously working under conditions with decreased oxygen, 100% work efficiency can never be reached. A short-term exposure to altitude reduces working capacity in proportion to the altitude height. Even after a few weeks of acclimatization, the environmental conditions still can reduce working capacity, and, especially, physical endurance. Descendants of natives living at HA have shown to have maximum adaptation. These people are able to perform very heavy work at HA with the same ease as other people performing similar tasks at sea level. Some people might never acclimatize to certain altitudes based on individual susceptibility. For instance, individuals with anemia, respiratory disease, and heart disease should not perform heavy work at moderate altitude or any work at HA (Parmeggiani, 1983). The combination of decreased oxygen and exercise increases blood pressure, making work at altitude dangerous for those with high blood pressure. Decreased oxygen level also might result in changes in senses, mood, and personality, increasing the probability for accidents and injuries. Some effects occur early and are temporary; others might persist after acclimatization or even for a period of time after descent. Vision is generally the sense most affected by altitude exposure, especially the ability to adapt to darkness. Mental effects are most noticeable at very high and extreme altitudes and include decreased perception, memory, judgment, and attention. Alterations in mood and personality traits are also common at altitude. Therefore, it is important to be aware of any signs and symptoms of altitude sickness and seek appropriate treatment. Ergonomics intervention in designing work-rest cycles might aid in sustaining HA stressors and preventing further deterioration of performance.

Several occupations are being undertaken at HA, such as mining operations, recreational facilities, modes of transportation, agricultural pursuits, and military campaigns, and all of these require physical and mental activities. All such activities involve increased requirements for oxygen. A major concern is that as one ascends higher and higher above sea level, both the total air pressure (the barometric pressure, PB) and the amount of oxygen in the ambient air (that portion of total pressure from oxygen, PO_2) progressively fall. As a result, the amount of work a person could have accomplished otherwise decreases progressively. These principles affect the workplace. For example, a tunnel in Colorado was found to require 25% more time to complete at an altitude of 11,000 ft than comparable work at sea level, and altitude effects were implicated in the delay. Not only is there increased muscular fatigue, but also deterioration of mental function. Memory, computation, decision-making, and judgment all become impaired. Scientists doing calculations at the Mona Loa Observatory at an altitude above 4000 m on the island of Hawaii noted that they required more time to perform their calculations and made more mistakes than at sea level. Because of the increasing scope, magnitude, variety, and distribution of human activities on this planet, more people are working at HA, and the effects of altitude have several occupational threats to pose on human performance. Thus, it is important to understand various factors affecting human performance at HA (Huang et al., 1984; MacDougall et al., 1991).

Deployment of large numbers of troops at HA and extreme HA conditions in current global scenarios makes it necessary to understand the physical activities involved in military operations. The activities and operations undertaken by soldiers at these heights are referred to as extreme military activities, which require personnel to possess extraordinary physiological and psychological capabilities resulting in significant human adaptation for survival and performance. Ergonomics evaluation in terms of direct assessment of physiology and cognitive demands during various operations might result in availability of quantified data that could help scientists to formulate interventions for improving human performance. In today's state-of-the-art multitasking military workstation layouts with multiscreen navigation, the operators need to process complex information coming through a number of displays and panels within a small time frame, notwithstanding the known cognitive debility at HA. During operational tasks, such flows of information received in the form of visual signals affect the visual scanning behavior of the operators, resulting in changes in the cognitive workload (Cymerman et al., 1981; Levine and Stray-Gunderson, 1997, 2005; Norris et al., 2012; Stray-Gunderson et al., 2001; Reynolds, 1996; Rose et al., 1988).

Studies on HA residents showed that for desensitization to occur, exposure to HA must occur in early childhood and last for several years. The decrease in hypoxic ventilatory response is first noted after 8 years of age, while vital capacity increases correspondingly. At sea level, offspring of lowlanders born and raised at HA exhibit the same phenomena as of native highlanders. The native highlander hyperventilates compared with the lowlander, and the HA resident hypoventilates compared with the newcomer to altitude. Therefore, native HA residents could perform a given physical activity with a relatively small ventilator requirement, so they have less dyspnea than others. This advantage increases their capacity to perform work at HA.

Exposures to HA have important implications for the cardiovascular system. On initial ascent, sympathetic activity markedly increases, resulting in an initial increase in heart rate and cardiac output. After prolonged exposure, however, maximal oxygen uptake decreases, stroke volume is lowered, and cardiac output falls below sea level values. The reduction in stroke volume is thought to be secondary to decreased ventricular filling. Exercise markedly reduces maximum cardiac output, an effect more pronounced in visitors than in natives. A 32% decrease in coronary blood flow without any evidence of myocardial ischemia has been observed after10 days at 3100 m (10,200 ft) (Grover et al., 1976). The study by Bernheim et al. (2007) reported that increased pulmonary arterial pressure in association with exercise and altitude hypoxia did not cause left ventricular (LV) diastolic dysfunction. The authors concluded that "ventricular interaction seems not to be of hemodynamic relevance in this setting." Significant increase in right ventricular (RV) wall thickness and decreased ejection fraction are observed on magnetic resonance imaging (MRI) scans in children with HA pulmonary hypertension (Ge et al., 2009). With increasing hypoxia, maximum heart rate decreased by 1 beat min^{-1} for every 130 m (about 430 ft) above 3100 m (10,200 ft). The decreased cardiac output, stroke volume, and exercise capacity noted at HA might be because of decreased preload that follows a reduction in plasma volume associated with arrival at HA. In general, systemic blood pressure is slightly lower at HA than it is at sea level. This difference is thought to be secondary to the vasodilatory effects of hypoxia on the systemic vascular smooth muscle. The incidence of hypertension at HA has been reported to be less than the frequency of occurrence at sea level (Mishra et al., 2015; Penaloza et al., 1963; Sawka et al., 1996; Singh and Chohan, 1972; San et al., 2013).

Past studies have suggested that exposure to HA induces a hyper-coagulation state in humans. Increased fibrinogen levels and a decreased clot lysis time were noted in 38 soldiers living at HA for 2 years, as compared with control subjects at sea level. Soldiers with clinical evidence of pulmonary arterial hypertension had somewhat low levels of fibrinogen, high levels of platelet factor III, and increased platelet adhesiveness. This evidence suggests that conversion to fibrin, and possibly platelet deposition, were occurring in these subjects (Singh and Chohan, 1972). The Operation Everest II project performed in a hypobaric chamber showed no changes in coagulation factors (Anholm et al., 1992; Basu et al., 2007; Bernheim et al., 2007; MacDougall et al., 1991; Rose et al., 1988).

In general, most visitors to HA notice initial weight loss, possibly because of reduced dietary intake, enhanced water loss, and loss of stored body fat. Anorexia is a common complaint of visitors to even moderate altitude. At HA, appetite and caloric intake decrease dramatically in unacclimatized persons, who generally find fat distasteful and prefer sweets. Fluid losses result from the insensible water losses associated with hyperventilation, low humidity, and diuresis induced by hypoxia and the cold environment (Huang et al., 1984; Tharion et al., 2005).

The retina of the eye has a great requirement for oxygen, making vision the first sense that is altered by the lack of oxygen. This phenomenon is demonstrated by diminished night vision even at altitudes below 3000 m (about 9600 ft). At 3048 m (10,000 ft), people require more time to learn a new task than they do at low elevations. At 6100 m (20,000 ft), impairments in sensory, perceptual, and motor performance have been demonstrated. In acute hypoxia, reduction of arterial oxygen saturation to 85% decreases a person's capacity for mental concentration and abolishes fine motor coordination. Reduction of saturation to 75% leads to faulty judgment and impaired muscular function (Hackett and Rennie, 1979).

Pulmonary arterial pressure is inversely dependent on a person's age and on the environment. At sea level, it rapidly decreases from the systemic level of the fetus to near-adult levels in the first hours or days after birth.

14.5.1 Load Carriage Strategies for High Altitudes

Load carriage forms an integral part of a soldier's daily schedule, whether he is posted in plains or at HA. Recently, an initiative to standardize the load carriage at HA for Indian populations has been started by scientists at Ergonomics Laboratory at Defense Institute of Physiology and Allied Sciences (DIPAS), Delhi, India. Since inception, DIPAS has conducted research for simultaneously reducing the subject-oriented workload and improving the object-related performance of soldiers under extreme environmental conditions. The institute has worked extensively on evaluating load carriage ensembles (LCe), starting from school bags of young children to the combat load of soldiers at HA, standardization of load carriage for sea level, desert, jungle, and different altitudes. The institute has been designing appropriate ergonomic backpacks for carrying loads at HA and standardizing the loads to be carried at HA.

Existing LCe and magnitude of load carried by Indian soldiers at HA still are similar to that at low level, even though the physiological, biomechanical, and cognitive debility at HA are established facts. Under combat situations, a normal soldier quite often is blinded by smoke or deafened by noise, his mobility is impaired by terrain, his dexterity is impaired by protective clothing, and he is cognitively impaired by high stress. Prior incorporation of ergonomic design principles is vital, because such efforts would improve sustainability in extreme situations encountered both in peacetime and wartime, allowing scientists to formulate evidence-based countermeasures to better equip personnel at high and extreme high altitudes. Load carriage criterion is now more important with introduction of newer equipment, arms, and ammunition intended for increased lethality and sustainability of the soldier. Battlefield mobility and maneuverability are important issues that should never be compromised in any sort of conflict or combat situations. They could be optimally achieved with decreased magnitude of load and adequate and even distribution of load with more freedom of mobility of limbs.

African women often carry head-supported loads of up to 60% of their body weight far more economically than army recruits carry equivalent loads in backpacks. Nepalese porters routinely carry head-supported loads equal to 100%–200% of their body weight for many days up and down steep mountain footpaths at HA even more economically than African women. Female Nepalese porters are known to carry loads that are heavier by 10% of their body weight than the maximum loads carried by the African women, with 25% smaller metabolic cost (Bastien et al., 2005). In a study by Majumdar et al. (2004) 10 infantry soldiers underwent five different load carriage operations at two different speeds—3.5 and 4.5 km h^{-1}—and four gradients (0%, 5%, 10%, and 15%) of treadmill walk to evaluate changes in physiological parameters. The same subjects underwent 15 variations of load carriage operation with two walking speeds on level ground, and six variations of load carriage in five gradients (0%, 5%, 10%, 15%, and 20%) of treadmill walk at 2.5 km h^{-1} speed for biomechanical stress evaluation. They were subjected to 22 variations of load carriage operation on level ground to evaluate the kinetic (ground reaction force) response during load carriage. Physiological and biomechanical stresses were found to increase with the increase in load, gradient, and speed of walk during load carriage operation (Majumdar et al., 2004). Based on this study, it was recommended that the optimal load for carrying combat items by an infantry soldier is 21.3 kg up to 4.5 km h^{-1} walking speed on level ground (Pal et al., 2009). The maximum load could be a backpack load of 10.7 kg up to 10% inclination at 3.5 km h^{-1} walking speed. Soldiers could carry a combat load up to 21.3 kg at 15% inclination if they were allowed to walk at their self-selected comfortable cadence. The kinematics and kinetics of low magnitude load carriage on level ground were reported by Majumdar et al. (2010, 2013) respectively, for Indian soldiers. Chatterjee et al. (2017) reported physiological responses of load carriage on medically fit Indian Infantry soldiers at Delhi (215 m) and at two HA locations, Leh (3505 m) and Tangtse (4300 m) after complete acclimatization. Volunteers carried incremental loads up to 30 kg at two walking speeds (2.5 and 3.5 km h^{-1}, respectively) on treadmills under controlled laboratory conditions at baseline and two altitudes. Results showed that there were significant reductions in VO$_2$max and maximum heart rate (HRmax) with rising altitude. Physiological responses increased with increment in load magnitude, altitude and speed. Based on the load carriage performance of the participants at two altitudes, few recommendations were made for load carriage at HA. At the height of 3500 m, maximum load of 32% of body weight (BW) could be allowed for longer durations (8 h workday) with necessary rest, water, and food. Maximum load magnitude of 45% of BW was recommended at this height for load carriage operations of only 2 h duration. At 4300 m, during level walking, load carriage operations of 32% of BW at a speed of $3.5 \, \text{km h}^{-1}$ was recommended for 2 h. At this altitude, 45% of BW could be permitted for carriage for 8 h at a slower speed. Lower loads (below 32% of BW) were suggested as ideal for carriage for long durations at both altitudes. To sustain the physiological stress and continue the load carriage activity at HA regions, it was suggested that the soldiers walk slowly (2.5–$3.5 \, \text{km h}^{-1}$) as recommended earlier by Kinoshita (1985). However, such recommendations are for reference only and cannot be applied directly while climbing steep mountain gradients for long duration under such ambient environment conditions in which one simultaneously faces fluid loss, exhaustion, probable injuries, and other stresses.

Load carriage ensembles and magnitude of load carried have become increasingly important with the introduction of newer equipment, arms, and ammunition intended for increased lethality and sustainability of the soldiers. At HA, the physical movements and associated demands involved in load carriage vary according to the nature of task undertaken and the factors influencing an individual's capacity to perform these activities. These possible determinants of an individual's load carriage ability might include age, anthropometric attributes, anaerobic and aerobic power, muscle strength, body composition, gender, and subjective response to different HA factors, such as hypoxia and cold. Other relevant factors might be dimensions and placement of loads, biomechanical factors, nature of terrain and gradient, and clothing. The energy cost of walking with loads have been found to primarily depend on walking speed, body weight, and load weight together with the terrain factors of surface type and gradients. The prediction of energy expenditure influenced by these variables might provide valuable information for assessing severity of proposed task of load carriage. The biomechanical and physiological stresses act synergistically on the soldiers, resulting in increased energy expenditure and injury risk and decrease in safety and efficiency. These stresses might further hamper their navigational ability, skilled performance, maneuverability, marksmanship, and combat performance. Standardizing the loads for carriage under extreme altitude conditions is of urgent need for optimized performance of soldiers deployed under such adverse conditions around the globe.

A study by Cymerman et al. (1981) investigated the applicability of a prediction equation for energy expenditure during load carriage at HA. The equation was validated previously at sea level. Oxygen uptake (VO_2) was determined in five young men at 4300 m while they walked with backpack loads of 0, 15, and 30 kg at treadmill grades of 0%, 8%, and 16% at $1.12 \, \text{m s}^{-1}$ for 10 min. Maximal oxygen uptake, determined on the cycle ergometer, was $42.2 \pm 2.3 \, \text{mL min}^{-1} \, \text{kg}^{-1}$ at sea level and $35.6 \pm 1.7 \, \text{mL min}^{-1} \, \text{kg}^{-1}$ at altitude. There were no significant differences in daily VO_2 at any specific exercise intensity on days 1, 5, and 9 of exposure, nor were there any differences in endurance times at the two most difficult exercise intensities. Endurance time for 15 and 30 kg loads at 16% grade were 7.3 and 4.2 min, respectively. Measured energy expenditure was compared with that predicted by the formula of Pandolf et al. (1977) and found to be significantly different (Cymerman et al., 1981). The differences could be attributed to measurements at metabolic rates exceeding 730 W or 2.1 L min^{-1} VO$_2$. These data indicated that the prediction equation could be used at altitude for exercise intensities not exceeding this upper limit. The observed deviations from predicted values at the high exercise intensities could be attributed to the occurrence of appreciable oxygen deficits and the inability to achieve steady-state conditions.

14.5.2   Mechanisms or Interventions Applied in Improving Performance at High Altitudes

Acclimatization to environmental hypoxia initiates a series of metabolic and musculo-cardio-respiratory adaptations that influence oxygen transport and use. Although it is clear that having adequate acclimatization or being born and brought up at altitude is necessary to achieve optimal physical performance at altitude, scientific evidence to support the potentiating effects after return to sea level, however, currently is unclear. Despite this, elite athletes continue to spend considerable time and resources training at altitude, misled by subjective coaching opinions and the inconclusive findings of a large number of uncontrolled studies. Scientific investigations have focused on the optimization of the theoretically beneficial aspects of altitude acclimatization, which include increases in blood hemoglobin concentration, elevated buffering capacity, and improvement in the structural and biochemical properties of skeletal muscle. However, not all aspects of altitude acclimatization are beneficial: Cardiac output and blood flow to skeletal muscles decrease, and hypoxia is responsible for a suppressed immune function and increased tissue damage mediated by oxidative stress. Future research needs to focus on these less beneficial aspects of altitude training, the implications of which pose a threat to both the fitness and the health of the elite competitor.

For athletes, HA produces two contradictory effects on performance. For explosive events (sprints up to 400 m, long jump, triple jump), the reduction in atmospheric pressure means there is less resistance from the atmosphere and the athlete's performance generally would be better at HA. For endurance events (races of 800 m or more), the predominant effect is the reduction in oxygen, which generally reduces the athlete's performance at HA. Sports organizations acknowledge the effects of altitude on performance. The International Association of Athletics Federations (IAAF), Monaco, Sweden, has ruled that performances achieved at an altitude greater than 1000 m will be approved for record purposes, but carry the notation of "A" to denote they were set at altitude. The 1968 Summer Olympics was held at altitude in Mexico City and resulted in the best athletes in the world setting records for most short sprint and jump events. Other records were also set at altitude in anticipation of those Olympics. Bob Beamon's record in the long jump held for almost 23 years and has been beaten only once without altitude or wind assistance. Many of the other records set at Mexico City later were surpassed by marks set at altitude.

Athletes could take advantage of altitude acclimatization to increase their performance upon a return to sea level, however, this might not always be the case. Any positive acclimatization effect might be negated by a detraining effect as the athletes usually are not able to exercise with as much intensity at HAs as compared to sea level. This resulted in development of the altitude training modality known as "live-high, train-low," whereby the athlete spends many hours per day resting and sleeping at a high altitude, but performs a significant portion, or all, of his or her training at a lower altitude. A series of studies conducted in Utah in the late 1990s by researchers Ben Levine, Jim Stray-Gundersen, and others, showed significant performance gains in athletes who followed such a protocol for several weeks. Other studies have shown performance gains from performing some exercising sessions at HA, yet living at sea level. The performance-enhancing effect of altitude training could be because of increased RBC count, more efficient training, or changes in muscle physiology. An increase in altitude leads to a proportional fall in the barometric pressure and a decrease in atmospheric oxygen pressure, thereby producing hypobaric effects that affect all the body organs, systems, and functions to different degrees. Chronically reduced pO_2 causes individuals to adapt and adjust to physiological stress. These adaptations are modulated by many factors, including the degree of hypoxia related to altitude, time of exposure, exercise intensity, and individual conditions. Exposure to HA elicits a response that contributes to several adjustments and adaptations being an environmental stressor. These adaptations include increase in hemoglobin concentration, ventilation, capillary density, and tissue myoglobin concentration, which, in turn, influence exercise capacity and endurance performance. Training methods, such as live high, train low and train high, live low, have been used to understand the changes in the physical condition in athletes and how the physiological adaptations to hypoxia could enhance performance at sea level. Studies need to focus on how physiological adaptations to hypoxic environments influence performance, and which protocols are used most frequently to train at HA. While a number of published studies exist to guide endurance in athletes with the best practices regarding implementation of altitude training, an unanswered question is the proper timing of return to sea level before major competitions. Evidence suggests that the de-acclimatization response of hematological, ventilatory, and biomechanical factors with return to sea level likely interact to determine the best timing for competitive performance (Bigard et al., 1991; De Paula and Niebauner, 2012; Levine and Stray-Gundersen, 1997, 2005; Norris et al., 2012; Stray-Gunderson et al., 2001).

The most fundamental aspect of occupational performance at altitude is maintenance of oxygen supply to the tissues. Humans and other animals have defenses against a low oxygen state (hypoxia). Important mechanisms include an increase in breathing (ventilation), which begins when the oxygen pressure in the arterial blood (PaO_2) decreases (hypoxemia) and is present for all altitudes above sea level. It increases with altitude and is our most effective defense against low oxygen in the environment. The process whereby breathing increases at HA is called ventilatory acclimatization. An unacclimatized person is unlikely to survive above an altitude of 20,000 ft, whereas acclimatized persons have been able to climb to the summit of Mount Everest (29,029 ft; 8848 m) without artificial sources of oxygen (Bartsch and Gibbs, 2007; Barnheim et al., 2007; Basu et al., 2007; Douglas et al., 1913; Furian et al., 2015; Grocott et al., 2009; Grover et al., 1966; Kojonazarov et al., 2007; Mirrakhimov and Strohl, 2016; Sawka et al., 1996).

As mentioned earlier in this chapter, frostbite at high altitudes is one of the most important causes of morbidity. The primary method to prevent frostbite is to keep oneself warm by wearing adequate clothes, preferably in layers, trapping the air and reducing the risk. People also should completely avoid the consumption of tobacco and alcohol. People should not unnecessarily expose themselves to the cold and the windy climate, but, if necessary, they should wear proper clothing and use a waterproof moisturizer (Taylor, 2011).

Oxygen is required for the sustained production of energy, and when oxygen supply is reduced (hypoxia), tissue functions become oppressed. Among all the organs, the brain is the most sensitive to lack of oxygen, and the central nervous system is important in the regulation of breathing. When we breathe a low-oxygen mixture, the initial response is an increase in ventilation, but after 10 min or so the increase is blunted to some extent. While the cause for this blunting is not known, it could be depression of some central neural function related to the ventilation pathway and has been called hypoxic ventilatory depression. Such depression has been observed shortly after ascent to HA. The depression is transient, lasting only a few hours, possibly because there is some tissue adaptation within the central nervous system.

Nevertheless, some increase in ventilation usually begins immediately while ascending to HA, although more time is required before maximum ventilation is achieved. On arrival at altitude, increased carotid body activity attempts to increase ventilation, and thereby raises the arterial oxygen pressure back to the sea level value. An increase in breathing causes an increased excretion of carbon dioxide (CO_2) in the exhaled air. When CO_2 is in body tissues, it creates an acid aqueous solution. When it is lost in exhaled air, the body fluids, including blood, become more alkaline, thus altering the acid-base balance in the body. The dilemma here is that ventilation is regulated not only to keep oxygen pressure constant, but also for acid-base balance. Carbon dioxide regulates breathing in the opposite direction from oxygen. Thus, when the CO_2 pressure (i.e., the degree of acidity somewhere within the respiratory center) rises, ventilation also rises, and when it falls, ventilation also falls. On arrival at HA, any increase in ventilation caused by the low oxygen environment leads to a fall in CO_2 pressure, causing alkalosis and acting to oppose the increased ventilation. Therefore, the dilemma on arrival is that the body cannot maintain constancy in both oxygen pressure and acid-base balance. Human beings require many hours and even days to regain proper balance (Cruz et al., 1980; Mishra et al., 2015; Taylor, 2011).

Another protective mechanism induces the kidneys to increase alkaline bicarbonate excretion in the urine, which compensates for the respiratory loss of acidity, thus helping to restore the body's acid-base balance toward sea level values. The renal excretion of bicarbonate is a relatively slow process. For example, acclimatization might require seven to 10 days while ascending from sea level to 4300 m (14,110 ft). This action of the kidneys, which reduces the alkaline inhibition of ventilation, once was thought to be the major reason for the slow increase in ventilation following ascent, but more recent research assigns a dominant role to a progressive increase in the sensitivity of the hypoxic-sensing ability of the carotid bodies during the early hours to days following ascent to altitude. This is the interval of ventilatory acclimatization, in which ventilation rises in response to low arterial oxygen pressure even though the CO_2 pressure is falling. The time required for acclimatization increases with increasing altitude, consistent with the concept that greater increase in ventilation and acid-base adjustments require longer intervals for renal compensation to occur. A sea level native might require three to 5 days to acclimatize at 3000 m, but it might require 6 weeks or more for complete acclimatization at altitudes above 6000–8000 m. When the altitude-acclimatized person returns to sea level, the process reverses. That is, arterial oxygen pressure now rises to the sea level value and ventilation falls. Now there is less CO_2 exhaled, and CO_2 pressure rises in the blood and in the respiratory center. The acid-base balance shifts toward acidic pH, and the kidneys must retain bicarbonate to restore balance. Although the time required for the loss of acclimatization is not well understood, it seems to require approximately the same amount of time as the acclimatization process itself. If so, then a return from altitude hypothetically gives a mirror image of altitude ascent, with one important exception: Arterial oxygen pressure immediately becomes normal on descent.

Individual susceptibility varies with regard to the time required for and magnitude of the ventilatory acclimatization to a given altitude. One very important reason is the large variation between individuals in their ventilatory response to hypoxia. For example, at sea level, if one holds CO_2 pressure constant so that it does not confound the ventilatory response to low oxygen, some people exhibit little or no increase in ventilation, while others present a very large (up to five-fold) increase. The ventilatory response to breathing low-oxygen mixtures seems to be an inherent characteristic of an individual, because family members behave more alike than unrelated individuals. People who have poor ventilatory responses to low oxygen at sea level also seem to have smaller ventilatory responses over time at HA. There might be other factors causing interindividual variability in acclimatization, such as variability in the magnitude of ventilatory depression, in the function of the respiratory center, in sensitivity to acid-base changes, and in renal handling of bicarbonate, but these have not been evaluated (Cruz et al., 1980; Douglas et al., 1913; Mishra et al., 2015, Moore, 2001; Moore et al., 1998; Zubieta-Castillo et al., 2007).

Poor sleep quality, particularly before ventilatory acclimatization, is a factor that impairs occupational efficiency. Many things interfere with the act of breathing, including emotions, physical activity, eating, and the degree of wakefulness. Ventilation decreases during sleep, as does the capacity for breathing to be stimulated by low oxygen or high CO_2, resulting in a decrease in both respiratory rate and depth of breathing. At HA, where there are few oxygen molecules in the air, the amount of oxygen stored in the lung alveoli between breaths is less. Therefore, if breathing ceases for a few seconds (called apnea, which is a common event at HA), the arterial oxygen pressure falls more rapidly than at sea level. (Anholm et al., 1992; De Paula and Niebauer, 2012; San et al., 2013).

Some work situations, particularly in the Andes, require a worker to spend several days at altitudes above 3000–4000 m, and then to spend several days at home, at sea level. The particular work schedules (days to be spent at altitude and at sea level) usually are determined by the economics of the workplace more than by health considerations. However, a factor to be considered in the economics is the interval required both for acclimatization and loss of acclimatization to the altitude in question. Particular attention should be placed on the worker's sense of well-being and job performance on arrival and the first day or two thereafter; regarding fatigue, time required to perform routine and nonroutine functions, and errors made. Strategies also should be considered to minimize the time required for acclimatization at altitude and to improve functions during the waking hours (Gore and Hopkins, 2005).

14.6   CONCLUSIONS

Ergonomics or human factors is the scientific study of people at work, aiming to reduce stress, eliminate injuries and work-related maladies, resulting in enhanced performance and safety. This might be achieved by designing tasks, workspaces, and equipment that are compatible with physical capabilities of operators. These adjustments will ensure the mental and physical wellness of individuals over longer duration of occupational exposure and increase their sustainability under adverse environmental conditions. Around the globe, variability within a population and across different populations are important phenomena that influence human-machine compatibility issues for different workforces operating under different working environments. Extensive variability and limitations in terms of body dimensions, strength, physical reachability, and cognitive abilities pose bottlenecks in the design of any equipment, workstation, or facility for the target population, apart from extremes of environmental vagaries, such as those experienced in HA and extreme HA.

This chapter discusses how application of different acclimatization strategies found in native highlanders might be used effectively for acclimatization of lowlanders to the environmental adversities of HA and enhance performance through physical activities. Different ergonomic interventions should be applied to enhance performance at HA and prevent occurrences of any overuse of physical impairment. Ergonomics principles might be employed successfully to identify best-suited physical activities to minimize occupational risk potentials of HA extreme climatic conditions and improve performance through physical activity. This chapter also discusses the strategies that might be applied toward habitability and acclimatization at HA and extreme HA for improved occupational health, safety, and performance.

Acknowledgments

The author is extremely grateful to the editors of this book for invitation to write this chapter. The author also is grateful to Director, DIPAS, Delhi, for permission to submit this chapter to the editors for publication.

References

Anholm, J.D., Powles, A.C., Houston, C.S., Sutton, J.R., Bonnet, M.H., Cymerman, A., 1992. Operation Everest II: arterial oxygen saturation and sleep at extreme simulated altitude. Am. Rev. Respir. Dis. 145 (4), 817–826.

Bärtsch, P., Gibbs, J.S.R., 2007. Effect of altitude on the heart and the lungs. Circulation 116 (19), 2191–2202.

Bastien, G.J., Schepens, B.W., Patrick, A., Heglund, N.C., 2005. Energetics of load carrying in Nepalese porters. Science 308 (5729), 1755–1756.

Basu, M., Malhotra, A.S., Pal, K., Prasad, R., Kumar, R., Prasad, B.A.K., Sawhney, R.C., 2007. Erythropoietin levels in lowlanders and high-altitude natives at 3450 m. Aviat. Space Environ. Med. 78 (10), 963–967.

Bernheim, A.M., Kiencke, S., Fischler, M., Dorschner, L., Debrunner, J., Mairbäurl, H., Maggiorini, M., Brunner-La Rocca, H.P., 2007. Acute changes in pulmonary artery pressures due to exercise and exposure to high altitude do not cause left. Ventricular diastolic dysfunction. Chest 132 (2), 380–387.

Bigard, A.X., Brunet, A., Guezennec, C.Y., Monod, H., 1991. Skeletal muscle changes after endurance training at high altitude. J. Appl. Physiol. 71 (6), 2114–2121.

Bock, J., Hultgren, H.N., 1986. Emergency maneuver in high-altitude pulmonary edema. JAMA 255 (23), 3245–3246.

Chatterjee, T., Bhattacharyya, D., Pramanik, A., Pal, M., Majumdar, D., Majumdar, D., 2017. Soldiers' load carriage performance in high mountains: a physiological study. Milit. Med. Res. 1 (4), 1–9.

Cruz, J.C., Reeves, J.T., Grover, R.F., Maker, J.T., McCullough, R.E., Cymerman, A., Denniston, J.C., 1980. Ventilatory acclimatization to high altitude is prevented by CO_2 breathing. Respiration 39 (3), 121–130.

Cymerman, A., Pandolf, K.B., Young, A.J., Maher, J.T., 1981. Energy expenditure during load carriage at high altitude. J. Appl. Physiol. 51 (1), 14–18.

De Paula, P., Niebauer, J., 2012. Effects of high altitude training on exercise capacity: fact or myth. Sleep Breath. 16 (1), 233–239.

Douglas, C.G., Haldane, J.S., Henderson, Y., Schneider, E.C., Webb, G.B., Richards, J., 1913. Physiological observations made on Pike's peak, Colorado, with special reference to adaptation to low barometric pressures. Philo. Trans. Royal Soc. London Series B 203 (294–302), 185–318.

Furian, M., Latshang, T.D., Aeschbacher, S.S., Ulrich, S., Sooronbaev, T., Mirrakhimov, E.M., Aldashev, A., Bloch, K.E., 2015. Cerebral oxygenation in highlanders with and without high-altitude pulmonary hypertension. Exp. Physiol. 100 (8), 905–914.

Ge, R.L., Ru-yan, M., Hai-hua, B., Xi-peng, Z., Hai-ning, Q., 2009. Changes of cardiac structure and function in pediatric patients with high altitude pulmonary hypertension in Tibet. High Alt. Med. Biol. 10 (3), 247–252.

Gore, C.J., Hopkins, W.G., 2005. Counterpoint: positive effects of intermittent hypoxia (live high, train low) on exercise performance are not mediated primarily by augmented red cell volume. J. Appl. Physiol. 99 (5), 2055–2057.

Grocott, M.P., Martin, D.S., Levett, D.Z., McMorrow, R., Windsor, J., Montgomery, H.E., 2009. Arterial blood gases and oxygen content in climbers on Mount Everest. N. Engl. J. Med. 360 (2), 140–149.

Grover, R.F., Lufschanowski, R., Alexander, J.K., 1976. Alterations in the coronary circulation of man following ascent to 3,100 m altitude. J. Appl. Physiol. 41 (6), 832–838.

Grover, R.F., Vogel, J.H., Voigt, G.C., Blount Jr., S.G., 1966. Reversal of high altitude pulmonary hypertension. Am. J. Cardiol. 18 (6), 928–932.

Hackett, P.H., Rennie, D., 1979. Rales, peripheral edema, retinal hemorrhage, and acute mountain sickness. Am. J. Med. 67 (2), 214–218.

Hashmi, M.A., Rashid, M., Haleem, A., Bokhari, S.A., Hussain, T., 1998. Frostbite: epidemiology at high altitude in the Karakoram mountains. Ann. R. Coll. Surg. Engl. 80 (2), 91–95.

Hornbein, T.F., Townes, B.D., Schoene, R.B., Sutton, J.R., Houston, C.S., 1989. The cost to the central nervous system of climbing to extremely high altitude. N. Engl. J. Med. 321 (25), 1714–1719.

Huang, S.Y., Alexander, J.K., Grover, R.F., Maher, J.T., McCullough, R.E., McCullough, R.G., Moore, L.G., Weil, J.V., Sampson, J.B., Reeves, J.T., 1984. Increased metabolism contributes to increased resting ventilation at high altitude. Respir. Physiol. 57 (3), 377–385.

Kinoshita, H., 1985. Effects of different loads and carrying systems on selected biomechanical parameters describing walking gait. Ergonomics 28, 1347–1362.

Kojonazarov, B.K., Imanov, B.Z., Amatov, T.A., Mirrakhimov, M.M., Naeije, R., Wilkins, M.R., Aldashev, A.A., 2007. Noninvasive and invasive evaluation of pulmonary arterial pressure in highlanders. Eur. Respir. J. 29 (2), 352–356.

Leshem, E., Pandey, P., Shlim, D.R., Hiramatsu, K., Sidi, Y., Schwartz, E., 2008. Clinical features of patients with severe altitude illness in Nepal. J. Travel Med. 15 (5), 315–322.

Levine, B.D., Stray-Gundersen, J., 1997. Living high, training low: effect of moderate-altitude acclimatization with low-altitude training on performance. J. Appl. Physiol. 83 (1), 102–112.

Levine, B.D., Stray-Gundersen, J., 2005. Point: Positive effects of intermittent hypoxia (live high, train low) on exercise performance are mediated primarily by augmented red cell volume. J. Appl. Physiol. 99 (5), 2053–2055.

Lewis, T., 1930. Observations upon the reactions of the vessels of the human skin to cold. Heart 15, 177–208.

Little, M.A., Hanna, J.M., 1977. The responses of high-altitude populations to cold and other stresses. In: Baker, P.T. (Ed.), The Biology of High-Altitude Peoples, International Biological Programme, 14th ed. Cambridge University Press, Cambridge, pp. 251–298.

MacDougall, J.D., Green, H.J., Sutton, J.R., Coates, G., Cymerman, A., Young, P., Houston, C.S., 1991. Operation Everest II: structural adaptations in skeletal muscle in response to extreme simulated altitude. Acta Physiol. 142 (3), 421–427.

Majumdar, D., Pal, M.S., Majumdar, D., 2010. Effects of military load carriage on kinematics of gait. Ergonomics 53 (6), 782–791. ·

Majumdar, D., Pal, M.S., Majumdar, D., Kumar, R. Banerjee, P. K., 2004. Biomechanical and physiological studies on the optimization of load carriage in Indian Army Personnel. Report No. DIPAS/06/2004.

Majumdar, D., Pal, M.S., Pramanik, A., Majumdar, D., 2013. Kinetic changes in gait during low magnitude military load carriage. Ergonomics 56 (12), 1917–1927.

Milner, A.D., 1998. Effects of 15% oxygen on breathing patterns and oxygenation in infants: infants are probably safe in aircraft. BMJ 316 (7135), 873–880.

Mirrakhimov, A.E., Strohl, K.P., 2016. High-altitude pulmonary hypertension: an update on disease pathogenesis and management. Open Cardiov. Med. J. 10 (1), 19–27.

Mishra, A., Mohammad, G., Norboo, T., Newman, J.H., Pasha, M.Q., 2015. Lungs at high-altitude: genomic insights into hypoxic responses. J. Appl. Physiol. 119 (1), 1–15.

Moore, L.G., 2001. Human genetic adaptation to high altitude. High Alt. Med. Biol. 2 (2), 257–279.

Moore, L.G., Niermeyer, S., Zamudio, S., 1998. Human adaptation to high altitude: regional and life-cycle perspectives. Am. J. Phys. Anthropol. 107 (S27), 25–64.

Norris, J.N., Viirre, E., Aralis, H., Sracic, M.K., Thomas, D., Gertsch, J.H., 2012. High altitude headache and acute mountain sickness at moderate elevations in a military population during battalion-level training exercises. Mil. Med. 177 (8), 917–923.

Pal, M.S., Majumdar, D., Bhattacharyya, M., Kumar, R., Majumdar, D., 2009. Optimum load for carriage by soldiers at two walking speeds on level ground. Int. J. Ind. Ergon. 39 (1), 68–72.

Pandolf, K.B., Givoni, B., Goldman, R.F., 1977. Predicting energy expenditure with loads while standing or walking very slowly. J. Appl. Physiol. 43 (4), 577–581.

Parmeggiani, L., 1983. Encyclopaedia of Occupational Health and Safety, third ed. International Labour Organisation, Geneva.

Penaloza, D., Banchero, N., Sime, F., Gamboa, R., 1963. The heart in chronic hypoxia. Biochem. Clin. 2, 283.

Ponsot, E., Dufour, S.P., Zoll, J., Doutrelau, S., N'Guessan, B., Geny, B., Hoppeler, H., Lampert, E., Mettauer, B., Ventura-Clapier, R., Richard, R., 2006. Exercise training in normobaric hypoxia in endurance runners. II. Improvement of mitochondrial properties in skeletal muscle. J. Appl. Physiol. 100 (4), 1249–1257.

Purkayastha, S.S., Selvamurthy, W., 2000. Soldier at high altitude: problem and preventive measures. Defense Sci. J. 50 (2), 183–198.

Ramaswamy, S.S., Majumdar, N.C., 1955. Energy cost of walking in two different types of ground. Indian J. Physiol. Allied Sci. 9, 113–118.

Ramaswamy, S.S., Dua, G.L., Raizada, V.K., Dimri, D.P., Vishwanathan, V.R., Madhavaiah, J., Srivastava, T.N., 1963. Study of load carriage at high altitude part I: relative effects of the magnitude of load carried and the steepness of the terrain on the optimum speed of march. Proc. Nat. Inst. Sci. 30A (5), 567–575.

Reynolds, R.D., 1996. Effects of cold and altitude on vitamin and mineral requirements. In: Marriott, B.M., Carlson, S.J. (Eds.), Nutritional Needs in Cold and High-altitude Environments: Applications for Military Personnel in Field Operations. National Academies Press, pp. 214–244. Committee on Military Nutrition Research, Institute of Medicine, Washington, DC, USA.

Rose, M.S., Houston, C.S., Fulco, C.S., Coates, G., Sutton, J.R., Cymerman, A., 1988. Operation everest. II: nutrition and body composition. J. Appl. Physiol. 65 (6), 2545–2551.

San, T., Polat, S., Cingi, C., Eskiizmir, G., Oghan, F., Cakir, B., 2013. Effects of high altitude on sleep and respiratory system and their adaptations. Sci. World J. 2013, 241569.

Sawka, M.N., Young, A.J., Rock, P.B., Lyons, T.P., Boushel, R., Freund, B.J., Muza, S.R., Cymerman, A., Dennis, R.C., Pandolf, K.B., Valeri, C.R., 1996. Altitude acclimatization and blood volume: effects of exogenous erythrocyte volume expansion. J. Appl. Physiol. 81 (2), 636–642.

Singh, I., Chohan, I.S., 1972. Blood coagulation changes at high altitude predisposing to pulmonary hypertension. Br. Heart J. 34 (6), 611.

Stray-Gundersen, J., Chapman, R.F., Levine, B.D., 2001. "Living high, training low" altitude training improves sea level performance in male and female elite runners. J. Appl. Physiol. 91 (3), 1113–1120.

Taylor, A.T., 2011. High-altitude illnesses: Physiology, risk factors, prevention, and treatment. Ramb. Maimon. Med. J. 2 (1), e0022.

Tharion, W.J., Lieberman, H.R., Montain, S.J., Young, A.J., Baker-Fulco, C.J., DeLany, J.P., Hoyt, R.W., 2005. Energy requirements of military personnel. Appetite 44 (1), 47–65.

West, J.B., 2016. Barcroft's bold assertion: all dwellers at high altitudes are persons of impaired physical and mental powers. J. Physiol. 594 (5), 1127–1134.

Wyss-Dunant, E., 1953. Acclimatization. In: Kurz, M. (Ed.), The Mountain World. Allen and Unwin, London, pp. 110–117.

Young, A.J., Reeves, J.T., 2002. Human adaptation to high terrestrial altitude. Med. Aspec. Harsh Environ. 2, 644–688.

Young, A.J., Evans, W.J., Fisher, E.C., Sharp, R.L., Costill, D.L., Maher, J.T., 1984. Skeletal muscle metabolism of sea level natives following short-term high-altitude residence. Eur. J. Appl. Physiol. Occup. Physiol. 52 (4), 463–466.

Zubieta-Castillo, G., Zubieta-Calleja, G.R., Zubieta-Calleja, N., Zubieta-Calleja, N., 2007. Facts that prove that adaptation to life at extreme altitude (8848 m) is possible. In: Lukyanova, L., Takeda, N., Singal, P.K. (Eds.), Adaptation Biology and Medicine: Health Potentials. In: 5, Narosa Publishing House, New Delhi, pp. 347–355.

CHAPTER 15

Yogic Practices for High-Altitude Ailments

Preenon Majumdar*, Dhurjati Majumdar†

*Kalinga Institute of Medical Sciences, Bhubaneshwar, India, †Defence Research and Development Organization, New Delhi, India

Abbreviations

ACTH	adrenocorticotrophic hormone
AMS	acute mountain sickness
BCE	before common era
BDNF	brain-derived neurotrophic factor
BFP	body fat percentage
DBP	diastolic blood pressure
$EtCO_2$	end tidal carbon dioxide
f_B	breathing frequency
FEV_1	forced expiratory volume in one second
FVC	forced vital capacity
HA	high altitudes
HACE	high-altitude cerebral edema
HAPE	high-altitude pulmonary edema
HR	heart rate
HVR	hypoxic ventilator response
MAP	mean arterial blood pressure
MVV	maximum voluntary ventilation
PEFR	peak expiratory flow rate
RR	respiratory rate
SaO(2)	arterial oxygen saturation
SBP	systolic blood pressure
SN	surya namaskar
SpO_2	peripheral capillary oxygen saturation
VO_2	volume of oxygen consumption
VO_2max	maximum aerobic capacity

Management of High Altitude Pathophysiology. https://doi.org/10.1016/B978-0-12-813999-8.00015-X

15.1 INTRODUCTION

Health protagonists refer to yoga as the unification of mind and body through physical activity with breath-control management. The disharmony between our thought processes and physical activity can lead to various illnesses, particularly psychomotor ones that eventually lead to deteriorating effects on physiological functions of the body. Research about yoga has proved beneficial effects on these functions, as well as environmental stressful conditions especially those found at high altitudes. Slow deep-breathing exercises as recommended in yoga reduces systemic and pulmonary blood pressure at high altitudes, however, it does not change pulmonary gas diffusion. Yoga practices or interventions also are effective against being overweight, and having hypertension, high glucose levels, and high cholesterol. Yoga can act as panacea for many, if not all, health hazards, but unlike other practices or therapies it is an institution in itself. It requires rigorous discipline, thorough practice and dedication. Although there are no observed or reported side-effects of yoga, the only drawback is that sometimes the beneficial effects on physiology occurs slowly, which demands the patience of a researcher and the concerned subject. Climbers are trained for yogic practices at a lower altitude for mountainous conditions, whereas real time training is a prerequisite. This chapter presents the concepts of yoga practices, their benefits and credibility to help people perform better at high altitudes.

15.2 YOGA AND ITS EVOLUTION

Yoga has invaded every aspect of our culture and life. The word "yoga" derives from the Sanskrit word "yuj" meaning "to yoke, or to unite." This leads to the popular translation of yoga as "union." In general, yoga is perceived as a set of principles and practices designed to promote health and well-being through the integration of mind, body, and breath. Depictions of yoga-like poses that have been found in archeological digs in India indicate the beginning of yoga as early as 3000 BCE. Today's yoga can trace its roots through four periods known as the Vedic period, the preclassical period, the classical period, and the postclassical period. Each period marks a meaningful shift or transition in the focus and evolution of yoga as we now know it (Feuerstein, 2001).

During the Vedic period, roughly 2000–600 BCE, yogic practices centered on surpassing the limitations of the mind, the strong dualist notion of reuniting the physical world with the spiritual world and studying the Vedas, the elaborate spiritual hymns and rituals of the time, which paid primary attention to the nature of reality and the roots of existence (Feuerstein, 2001).

The preclassical period, ranging from 600 to 200 BCE, is marked by the Upanishads, roughly 200 texts that exert heavy influence on later Hindu philosophy, and by the Bhagavad Gita (500–200 BCE), which is considered to be among the yoga's oldest written texts. The Gita's 700 verses detail the nature of reality, and the various "yogas" ("karma" or action, "bhakti" or devotion, and "jnana" or wisdom) that can be used to acquire transcendent understanding and liberation. In these texts, the foundations of meditation and the concept of a path to enlightenment began to emerge strongly (Feuerstein, 2001).

The classical period is characterized by Patanjali's Yoga Sutras (200 BCE), which attempted to define and standardize yoga through "eight limbs" of study and practice. Patanjali's eight limbs, (the Yamas, Niyamas, Asana, Pranayama, Pratyahara, Dharana, Dhayana, and Samadhi) (Fig. 15.1). The underlying philosophy developed in the Yoga Sutras came to be called raja yoga, from which almost all of modern yoga stems (Feuerstein, 2001). These eight aspects of yoga encompass what yoga gurus consider the five layers of the human being: the anatomical, physiological, mental, intellectual, and spiritual bodies (Stone, 2009).

The postclassical period following Patanjali is marked by a proliferation of texts, philosophies, and styles that extends from roughly the 1st century until the late 19th century. During this period, an interest in the study and purification of the body, known as hatha yoga, emerged. The primary and most comprehensive treatise of hatha yoga comes from Yogi Swatmarama's Hatha Yoga Pradipika, written around the 15th century. The advent of hatha yoga marks the first serious consideration of the physical exercises that have become today's

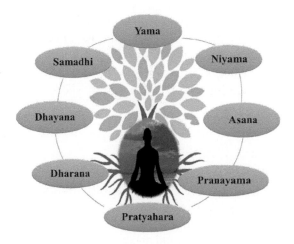

FIG. 15.1
Patanjali's eight limbs.

yoga poses, many of which can be traced back to the Pradipika and other concurrent texts (Feuerstein, 2001).

Worldwide, people now are inclined to practice asana (array of physical exercises and postures), pranayama (regulation of breath and specific breath exercises), and meditation to purify and rebuild the physical body, mainly with an attempt to achieve a higher level of consciousness.

Asana consists of the physical poses and exercises traditionally used as a means to condition and prepare the body for deep and extended periods of meditation. Today, it is the most popular form of yoga practice. Typical poses range from simple standing, seated, and supine positions to complex and challenging articulations of the body. Pranayama uses breath exercises to regulate the flow of energy (*prana*) through subtle energy channels (*nadis*) within the body.

Modern meditation practices blend elements described in Patanjali's *Dhayana*, and in *Dharana* (concentration) and *Pratyahara* (sense withdrawal). Traditional meditation practices often incorporate exercises involving the placement of attention on a *dristi* (focal point), which might include the breath and a chanted or thought mantra.

15.3 BENEFITS OF YOGA (ASANA, PRANAYAMA, AND MEDITATION)

During the last decade, yoga and meditation practices have become popular mainstream activities. Yoga is now a multibillion-dollar industry. Researchers and practitioners believe that yogic practices appeal to the public because they provide physical and mental health benefits and create a sense of community. These practices are being used worldwide in every sphere of life, including schools, colleges, industry, government offices, corporate sectors, hospitals, military, rehabilitation and therapies, sports, diseases, and prisons.

Yoga and meditation are not mutually exclusive because yoga traditionally incorporates meditation and because both practices aim to focus the mind and use the breath as a tool.

A number of reports have suggested various beneficial effects of yoga in different age groups, with improvement in physiological, physical, and psychological functions (Field et al., 2010). Mind-body practices that include relaxation have been used traditionally by people across cultures to improve health and serve as a path of spiritual awakening (Wolsko et al., 2004). Regular yoga practitioners have been found to have lowered resting heart rate (Bharshankar et al., 2003), blood pressure (Bharshankar et al., 2003), breath rate (Gopal et al., 1973), and metabolic rate (Chaya et al., 2006; Chaya and Nagendra, 2008).

Yoga also has been found to improve many features of metabolic syndromes, such as obesity (Balk, 2011; Seo et al., 2012), hyperlipidaemia (Pal et al., 2011; Telles et al., 2010a, b; Vyas et al., 2008), hyperglycemia (Mohta et al., 2009; Sahay, 2007; Telles et al., 2010a, b), and hypertension (Bhavanani and Madanmohan, 2012; Okonta, 2012; Selvamurthy et al., 1988).

The ability of yoga to induce relaxation and relieve stress has been widely reported (Parshad, 2004; Telles et al., 2008). Yoga is reported to relieve workplace stress (Hartfiel et al., 2011), examination stress (Gopal et al., 2011), and stress-induced inflammation (DiNardo, 2009). Yoga also has been found to improve many clinical conditions, such as anxiety (Agte and Chiplonkar, 2008; Gupta et al., 2006; Javnbakht et al., 2009), depression (Vedamurthachar et al., 2006), negative mood states (Yoshihara et al., 2011; Wood, 1993), and posttraumatic stress disorder symptoms in war veterans (Carter and Byrne, 2004; Rosenthal et al., 2011), tsunami survivors (Telles et al., 2007), hurricane refugees (Descilo et al., 2010; Gerberg and Brown, 2005), and flood survivors (Telles et al., 2010a, b). Li and Goldsmith (2012) recommended yoga as potential adjunct to pharmacologic therapy for patients with stress and anxiety.

The effect of hatha yoga on the improvement of cardiopulmonary status has been established. Joshi et al. (1992) showed that in a group of 75 males and females (mean age 18.5 years), after 6 weeks of practice there was significant increase in forced vital capacity (FVC), forced expiratory volume in 1 s (FEV_1), peak expiratory flow rate (PEFR), maximum voluntary ventilation (MVV), breath holding time, and significant decrease in breathing frequency (f_B). Raju et al. (1997) and Ray et al. (2001) reported significant improvement in overall cardiovascular endurance of young subjects who were given varying periods of yoga training compared to others who performed other types of exercises. Selvamurthy et al. (1988) showed that selected yogic practice could decrease serum epinephrine and norepinephrine levels in essential hypertensive people (Selvamurthy et al., 1988). Yoga has been shown to improve glutathione and redox status (Pal et al., 2013; Sinha et al., 2007). Pal et al. (2014) reported that a significant decrease of cortisol and adrenocorticotrophic hormone (ACTH) and increases in serotonin, dopamine, and brain-derived neurotrophic factor (BDNF) were noted following yogic practices. They stated that yogic practices might help prevent age-related degeneration by changing cardiometabolic risk factors, autonomic functions, and BDNF in healthy males. In 2017, Mandal et al. showed that yoga improves postural balance.

Several studies, especially from India, have explored the beneficial effects of yoga for high altitudes (HA) acclimatization and maintaining physical and mental fitness in order to keep Indian soldiers battle ready (Halder et al., 2012; Harinath et al., 2004; Joseph et al., 1981; Nayar et al., 1975; Ray et al., 2001; Rao, 1968; Selvamurthy et al., 1988; Sinha et al., 2014).

15.4 HIGH-ALTITUDE AILMENTS AND YOGA

About 2.5% of the world's population live at HA regions. Living at HA is challenging because of various physical and environmental hazards. >100,000 Indian soldiers posted in the country's northern and eastern mountainous areas are subjected to the rigors of altitude calamities daily. HA are characterized by low oxygen pressure resulting in hypoxia or oxygen deficiency in the body, extreme cold, high velocity winds, low humidity, and increased ultraviolet radiation. These environmental adversities result in a number of HA maladies affecting a person's physiological, biomechanical, and cognitive abilities.

Different adaptive mechanisms come into play depending on whether a person's stay at HA is of short or long duration. Health hazards associated with HA include acute mountain sickness (AMS), high-altitude pulmonary edema (HAPE), high-altitude cerebral edema (HACE), cold injury, dehydration, dryness, snow blindness, and sunburn, resulting in significant reduction in physical and mental performance. It has long been accepted that HA maladies would occur whenever man inhibits these regions and that prevention of these ailments might not be feasible.

The term altitude refers the vertical height above sea level. Practically, HA does not have any precise definition. An elevation of 2700 m (9000 ft) and above is considered as HA, because above this point most people develop signs and symptoms associated with low partial pressure of oxygen and hypoxia.

The most common ailment at HA is AMS, which develops within a few days of exposure to HA. The most common symptoms are headache, nausea, anorexia, insomnia, lassitude, vomiting, and dizziness. AMS is determined by the altitude reached, rate of ascent, physical activity, and degree of preacclimatization (Roach et al., 1998). Pathophysiological features associated with AMS include hypoventilation, impaired gas exchange, fluid retention, redistribution of blood flow, and increased sympathetic drive (Hackett et al., 1998). The symptoms and extent of AMS vary from person to person, with the hypoxic ventilator response (HVR) of the individual playing a vital role. Individual differences in HVR at sea level influence the variable ventilatory response during acclimatization to HA (Reeves et al., 1993) up to 42%. Moore et al., 1986, demonstrated a clear relationship between low ventilatory response to hypoxia and AMS. Rathat et al. (1992) reported that 80% of AMS-susceptible subjects could be predicted by the ventilatory and cardiac responses to hypoxia during exercise. Another study suggested that hypoventilation could be important in the pathophysiology of AMS (Burtscher et al., 2004).

As a physiological response, ventilation increases upon exposure to HA-hypoxic environment, leading to raised levels of alveolar, arterial, and tissue oxygenation. Lower levels of peripheral capillary oxygen saturation (SpO_2)

were reported in individuals susceptible to AMS than those who were not (Burtscher et al., 2004).

Acute exposure to HA results in reduction of physical work capacity and exercise performance. Chatterjee et al. (2017) reported a drop of about 16.5% and 29.2% of maximum aerobic capacity (VO_2 max) compared to sea level of a group of Indian soldiers after 1 month of their stay at 3505 and 4300 m, respectively. Altitude, however does not appear to affect every person equally. A decrease in SpO_2 with increasing hypoxia accounts for about 86% of the variation in the decrement in physical performance in trained individuals (Ferretti et al., 1997). Individuals with high hypoxic ventilatory response (HVR) presumably have higher alveolar and arterial oxygen pressure, which in turn improves the SpO_2. Therefore, a more vigorous ventilatory response could help people at HA to avoid AMS and might help improve their performance at moderate and extreme altitudes. Himashree et al. (2016) conducted a study on >200 fully acclimatized soldiers in the Indian army posted to HA regions at Karu, Leh, India, at an altitude of 3445 m. The soldiers were divided into two groups of equal size. The control group carried out the routine activities for physical training in the Indian army. The intervention group practiced a comprehensive yoga package, including physical asanas, pranayama, and meditation, and did not perform routine physical training. Both groups were monitored during their activities. A comprehensive range of anthropometrical, physiological, biochemical, and psychological parameters were measured: height and weight, body fat percentage (BFP), heart rate (HR), respiratory rate (RR), systolic and diastolic blood pressures, peripheral saturation of oxygen, (end tidal carbon dioxide; $EtCO_2$), chest expansion, pulmonary function, physical work capacity (VO_2max), hematological variables, lipid profile, serum urea, creatinine, liver enzymes, blood glucose, and anxiety. Measurements were made at baseline and postintervention. The yoga group showed a significant improvement in health indices and performance as compared to the control group. They had lower weights, BFPs, RRs, systolic and diastolic blood pressures, and anxiety scores. They also had a significantly higher $EtCO_2$, forced vital capacity, forced expiratory volume in the first second (FEV_1), and VO_2max. The yoga group showed a significant reduction in serum cholesterol, low-density lipoprotein, triglycerides, and blood urea as compared to both their preyoga levels and to the exercise group. The study concluded that practice of yoga facilitates improvements in health and performance at HA and is superior to routine training with physical exercises (Himashree et al., 2016).

A study in 2007 showed that well-performed yoga induces long-term changes in respiratory function and control. Researchers monitored the ventilatory, cardiovascular, and hematological parameters in 12 Caucasian yoga trainees and 12 control sea level residents, at baseline and after 2-week exposure to HA (Pyramid Laboratory, Nepal; 5050 m), 38 active lifestyle high-altitude natives

(Sherpas) and 13 contemplative lifestyle HA natives with practice of yoga-like respiratory exercises (Buddhist monks) studied at 5050 m. At baseline, HVR, red blood cell count, and hematocrit were lower in Caucasian yoga trainees than in controls. After 14 days at altitude, yoga trainees showed similar oxygen saturation, blood pressure, RR interval compared to controls, but lower HVR (-0.44 ± 0.08 vs -0.98 ± 0.21 L/min/m/%SaO(2); $P < .05$), minute ventilation (8.3 ± 0.9 vs 10.8 ± 1.6 L/min, $P < .05$), breathing rate (indicating higher ventilatory efficiency), and lower red blood cell count, hemoglobin, hematocrit, albumin, erythropoietin, and soluble transferrin receptors. Hypoxic ventilatory response in monks was lower than in Sherpas (-0.23 ± 0.05 vs -0.63 ± 0.09 L/min/m/%SaO(2); $P < .05$) and the values were similar to baseline data of Caucasian yoga trainees and Caucasian controls, respectively. Red blood cell counts and hematocrits were lower in monks as compared to Sherpas. In conclusion, Caucasian subjects practicing yoga maintain a satisfactory oxygen transport at high altitudes, with minimal increase in ventilation and with reduced hematological changes, resembling Himalayan natives. Respiratory adaptations induced by the practice of yoga might represent an efficient strategy to cope up with altitude-induced hypoxia (Bernardi et al., 2007).

Harinath et al. (2004) studied the effect of practicing selected yogic asanas (postures) for 45 min and pranayama for 15 min during the morning on a group of 15 healthy (25–35 years) volunteers for 3 months. The control group of the same number and age performed body flexibility exercises for 40 min and slow running for 20 min during morning hours for 3 months. They observed that yogic practices for 3 months resulted in an improvement in cardiorespiratory performance and psychologic profile. The plasma melatonin also showed an increase after 3 months of yogic practices. The systolic blood pressure, diastolic blood pressure, mean arterial pressure, and orthostatic tolerance had no significant correlation with plasma melatonin. However, the maximum night time melatonin levels in yoga group showed a significant correlation ($r = 0.71$, $P < .05$) with well-being score (Harinath et al., 2004).

Sinha et al. (2004) showed that surya namaskar (SN), a group of yogic exercise consisting of 12 postures could be a substitute of aerobic exercise (Fig. 15.2). SN seemed to be ideal as it involves both static stretching and slow dynamic component of exercise with optimal stress on the cardiorespiratory system, reinforcing the beneficial effect of yoga on cardiovascular function.

A study in 2013 demonstrated that the increase in systolic blood pressure (SBP), diastolic blood pressure (DBP), and mean arterial blood pressure (MAP) were greatest with standing postures as compared to yogic inversion and floor postures (Miles et al., 2013). Sinha et al. (2014) however, showed that the pressor response in Indian army trainees was attenuated during SN practice compared to yoga-trained proficient and semiproficient individuals,

FIG. 15.2

The set of 12 postures in surya namaskar (SN).

possibly because of attainment of better autonomic balance by the trainees. Army trainees in this study also were involved in other outdoor aerobic exercises that have the potential to reduce blood pressure. Yogic practice is a mixture of aerobic and anaerobic activities. The researchers concluded that the greater reduction of sympathetic reactivity during SN in army trainees could be because of their well-maintained aerobic fitness level.

It is established that most HA ailments occur because of sympathetic overactivity; regular practice of yoga, pranayama, and meditation stimulate the autonomic balance more toward parasympathetic activity, which helps maintain the hemodynamic balance. Hatha yoga clearly has benefits for cardiopulmonary endurance through body and breath control, including relaxation techniques. These benefits manifest clinically as improved lung capacity, increased oxygen delivery, decreased VO_2 and respiratory rate, and decreased resting heart rate, resulting in overall improved exercise capacity at sea level and HA (Raub, 2002).

The intense stretching and muscle conditioning associated with attaining and holding yoga postures increases skeletal muscle oxidative capacity and decreases glycogen use, possibly caused by increased vascularization, increased intramuscular oxygen and glycogen stores, increased oxidative enzymes, or by increased numbers of mitochondria (Shephard and Astrand, 2000).

The slow increase in lung capacity (FEV_1, FVC) associated with well-practiced yoga breathing increases oxygen delivery to highly metabolic tissues (muscle). The slow breathing rates associated with yoga breathing have been shown to substantially reduce chemo-reflex response to hypoxia (Röggla et al., 2001).

15.5 CONCLUSIONS

Several research studies have established the beneficial effects of yoga, breathing exercises, and meditation on the physiological, metabolic, and psychological parameters at sea level, hypoxic, and simulated altitude conditions. Adequate yoga practices before ascending to HA is highly beneficial in HA acclimatization. The number of studies about these beneficial effects in actual HA environments are few, however, and this area needs extensive investigation.

References

Agte, V.V., Chiplonkar, S.A., 2008. Sudarshan kriya yoga for improving antioxidant status and reducing anxiety in adults. Altern. Complem. Ther. 14 (2), 96–100.

Balk, J.L., 2011. Yoga for weight loss. Altern. Med. Alert 14 (5), 49–53.

Bernardi, L., Passino, C., Spadacini, G., Bonfichi, M., Arcaini, L., Malcovati, L., Bandinelli, G., Schneider, A., Keyl, C., Feil, P., Greene, R.E., 2007. Reduced hypoxic ventilatory response with preserved blood oxygenation in yoga trainees and Himalayan Buddhist monks at altitude: evidence of a different adaptive strategy? Eur. J. Appl. Physiol. 99 (5), 511–518.

Bharshankar, J.R., Bharshankar, R.N., Deshpande, V.N., Kaore, S.B., Gosavi, G.B., 2003. Effect of yoga on cardiovascular system in subjects above 40 years. Indian J. Physiol. Pharmacol. 47 (2), 202–206.

Bhavanani, A.B., Madanmohan, Z.S., 2012. Immediate effect of chandra nadi pranayama (left unilateral forced nostril breathing) on cardiovascular parameters in hypertensive patients. Int. J. Yoga 5 (2), 108–111.

Burtscher, M., Flatz, M., Faulhaber, M., 2004. Prediction of susceptibility to acute mountain sickness by SaO2 values during short-term exposure to hypoxia. High Alt. Med. Biol. 5 (3), 335–340.

Carter, J., Byrne, G., 2004. A Two-Year Study of the Use of Yoga in a Series of Pilot Studies as an Adjunct to Ordinary Psychiatric Treatment in a Group of Vietnam War Veterans Suffering From Post-traumatic Stress Disorder. Online document at: www.Therapywithyoga.com. Accessed 27 November 2018.

Chatterjee, T., Bhattacharyya, D., Pramanik, A., Pal, M., Majumdar, D., Majumdar, D., 2017. Soldiers' load carriage performance in high mountains: a physiological study. Mil. Med. Res. 4 (1), 6.

Chaya, M., Nagendra, H., 2008. Long-term effect of yogic practices on diurnal metabolic rates of healthy subjects. Int. J. Yoga 1 (1), 27–32.

Chaya, M.S., Kurpad, A.V., Nagendra, H.R., Nagarathna, R., 2006. The effect of long-term combined yoga practice on the basal metabolic rate of healthy adults. BMC Complement. Altern. Med. 6 (1), 1–6.

Descilo, T., Vedamurtachar, A., Gerbarg, P.L., Nagaraja, D., Gangadhar, B.N., Damodaran, B., Adelson, B., Braslow, L.H., Marcus, S., Brown, R.P., 2010. Effects of a yoga breath intervention alone and in combination with an exposure therapy for post-traumatic stress disorder and depression in survivors of the 2004 Southeast Asia tsunami. Acta Psychiatr. Scand. 121 (4), 289–300.

DiNardo, M.M., 2009. Mind-body therapies in diabetes management. Diabetes Spectr. 22 (1), 30–34.

Ferretti, G., Moia, C., Thomet, J.M., Kayser, B., 1997. The decrease of maximal oxygen consumption during hypoxia in man: a mirror image of the oxygen equilibrium curve. J. Physiol. 498 (1), 231–237.

Feuerstein, G., 2001. The Yoga Tradition: Its History, Literature, Philosophy, and Practice. Hohm Press, Prescott, AZ.

Field, T., Diego, M., Hernandez-Reif, M., 2010. Tai chi/yoga effects on anxiety, heart rate, EEG, and math computations. Complement. Ther. Clin. Pract. 16 (4), 235–238.

Gerbarg, P.L., Brown, R.P., 2005. Yoga: A breath of relief for hurricane Katrina refugees. Curr. Psychiatr. Ther. 4 (10), 55–67.

Gopal, K.S., Bhatnagar, O.P., Subramanian, N., Nishith, S.D., 1973. Effect of yogasanas and pranayamas on blood pressure, pulse rate, and some respiratory functions. Indian J. Physiol. Pharmacol. 17 (3), 273–276.

Gopal, A., Mondal, S., Gandhi, A., Arora, S., Bhattacharjee, J., 2011. Effect of integrated yoga practices on immune responses in examination stress: a preliminary study. Int. J. Yoga 4 (1), 26–32.

Gupta, N., Khera, S., Vempati, R.P., Sharma, R., Bijlani, R.L., 2006. Effect of yoga-based lifestyle intervention on state and trait anxiety. Indian J. Physiol. Pharmacol. 50 (1), 41–47.

Hackett, P.H., Yarnell, P.R., Hill, R., Reynard, K., Heit, J., McCormick, J., 1998. High-altitude cerebral edema evaluated with magnetic resonance imaging: clinical correlation and pathophysiology. JAMA 280 (22), 1920–1925.

Halder, K., Chatterjee, A., Kain, T.C., Pal, R., Tomer, O.S., Saha, M., 2012. Improvement in ventilatory function through yogic practices. Al Ameen J. Med. Sci. 5 (2), 197–202.

Harinath, K., Malhotra, A.S., Pal, K., Prasad, R., Kumar, R., Kain, T.C., Rai, L., Sawhney, R.C., 2004. Effects of hatha yoga and omkar meditation on cardiorespiratory performance, psychologic profile, and melatonin secretion. J. Altern. Complem. Med. 10 (2), 261–268.

Hartfiel, N., Havenhand, J., Khalsa, S.B., Clarke, G., Krayer, A., 2011. The effectiveness of yoga for the improvement of well-being and resilience to stress in the workplace. Scand. J. Work Environ. Health 37 (1), 70–76.

Himashree, G., Mohan, L., Singh, Y., 2016. Yoga practice improves physiological and biochemical status at high altitudes: a prospective case-control study. Altern. Ther. Health Med. 22 (5), 53–59.

Javnbakht, M., Kenari, R.H., Ghasemi, M., 2009. Effects of yoga on depression and anxiety of women. Complement. Ther. Clin. Pract. 15 (2), 102–104.

Joseph, S., Sridharan, K., Patil, S.K., Kumaria, M.L., Selvamurthy, W., Joseph, N.T., Nayar, H.S., 1981. Study of some physiological and biochemical parameters in subjects undergoing yogic training. Indian J. Med. Res. 74 (1), 120–124.

Joshi, L.N., Joshi, V.D., Gokhale, L.V., 1992. Effect of short-term pranayam practice on breathing rate and ventilatory functions of lung. Indian J. Physiol. Pharmacol. 36 (2), 105–108.

Li, A.W., Goldsmith, C.A.W., 2012. The effects of yoga on anxiety and stress. Altern. Med. Rev. 17 (1), 21–36.

Miles, S.C., Chun-Chung, C., Hsin-Fu, L., Hunter, S.D., Dhindsa, M., Nualnim, N., Tanaka, H., 2013. Arterial blood pressure and cardiovascular responses to yoga practice. Altern. Ther. Health Med. 19 (1), 38–45.

Mohta, N., Agrawal, R.P., Kochar, D.K., Kothari, R.P., Sharma, A., 2009. Influence of yogic treatment on quality of life outcomes, glycemic control, and risk factors in diabetes mellitus: randomized controlled trial. Exp. J. Sci. Heal. 5 (3), 147.

Moore, L.G., Harrison, G.L., McCullough, R.E., McCullough, R.G., Micco, A.J., Tucker, A., Weil, J.V., Reeves, J.T., 1986. Low acute hypoxic ventilatory response and hypoxic depression in acute altitude sickness. J. Appl. Physiol. 60 (4), 1407–1412.

Nayar, H.S., Mathur, R.M., Kumar, R.S., 1975. Effects of yogic exercises on human physical efficiency. Indian J. Med. Res. 63 (10), 1369–1376.

Okonta, N.R., 2012. Does yoga therapy reduce blood pressure in patients with hypertension? An integrative review. Holist. Nurs. Pract. 26 (3), 137–141.

Pal, A., Srivastava, N., Tiwari, S., Verma, N.S., Narain, V.S., Agrawal, G.G., Natu, S.M., Kumar, K., 2011. Effect of yogic practices on lipid profile and body fat composition in patients of coronary artery disease. Complement. Ther. Med. 19 (3), 122–127.

Pal, R., Singh, S.N., Halder, K., Tomer, O.S., Mishra, A.B., Saha, M., 2013. Effect of yogic practices on age-related changes in oxygen metabolism and antioxidant-redox status. J. Exp. Integr. Med. 3 (4), 305–312.

Pal, R., Singh, S.N., Chatterjee, A., Saha, M., 2014. Age-related changes in cardiovascular system, autonomic functions, and levels of BDNF of healthy active males: Role of yogic practice. Age 36 (4), 9683.

Parshad, O., 2004. Role of yoga in stress management. West Indian Med. J. 53 (3), 191–194.

Raju, P.S., Prasad, K.V.V., Venkata, R.Y., Murthy, K.J.R., Reddy, M.V., 1997. Influence of intensive yoga training on physiological changes in six adult women: a case report. J. Altern. Complem. Med. 3 (3), 291–295.

Rao, S., 1968. Oxygen consumption during yoga-type breathing at altitudes of 520 m and 3800 m. Indian J. Med. Res. 56 (5), 701–705.

Rathat, C., Richalet, J.P., Herry, J.P., Larmignat, P., 1992. Detection of high-risk subjects for high-altitude diseases. Int. J. Sports Med. 13 (S 1), S76–S78.

Raub, J.A., 2002. Psychophysiologic effects of hatha yoga on musculoskeletal and cardiopulmonary function: a literature review. J. Altern. Complem. Med. 8 (6), 797–812.

Ray, U.S., Sinha, B., Tomer, O.S., Pathak, A., Dasgupta, T., Selvamurthy, W., 2001. Aerobic capacity and perceived exertion after practice of hatha yogic exercises. Indian J. Med. Res. 114, 215–221.

Reeves, J.T., McCullough, R.E., Moore, L.G., Cymerman, A., Weil, J.V., 1993. Sea-level PCO2 relates to ventilatory acclimatization at 4300 m. J. Appl. Physiol. 75 (3), 1117–1122.

Roach, R.C., Greene, E.R., Schoene, R.B., Hackett, P.H., 1998. Arterial oxygen saturation for prediction of acute mountain sickness. Aviat. Space Environ. Med. 69 (12), 1182–1185.

Röggla, G., Kapiotis, S., Röggla, H., 2001. Yoga and chemoreflex sensitivity. Lancet 357 (9258), 807.

Rosenthal, J.Z., Grosswald, S., Ross, R., Rosenthal, N., 2011. Effects of transcendental meditation in veterans of operation enduring freedom and operation Iraqi freedom with post-traumatic stress disorder: a pilot study. Mil. Med. 176 (6), 626–630.

Sahay, B.K., 2007. Role of yoga in diabetes. J. Assoc. Physicians India 55, 121–126.

Selvamurthy, W., Ray, U.S., Hegde, K.S., Sharma, R.P., 1988. Physiological responses to cold (10°C) in men after six months' practice of yoga exercises. Int. J. Biometeorol. 32 (3), 188–193.

Seo, D.Y., Lee, S., Figueroa, A., Kim, H.K., Baek, Y.H., Kwak, Y.S., Kim, N., Choi, T.H., Rhee, B.D., Ko, K.S., Park, B.J., 2012. Yoga training improves metabolic parameters in obese boys. Korean J. Physiol. Pharmacol. 16 (3), 175–180.

Shephard, R.J., Astrand, P.O. (Eds.), 2000. Endurance in Sport. second ed. Blackwell Science Ltd., London.

Sinha, B., Ray, U.S., Pathak, A., Selvamurthy, W., 2004. Energy cost and cardiorespiratory changes during the practice of surya namaskar. Indian J. Physiol. Pharmacol. 48 (2), 184–190.

Sinha, S., Singh, S.N., Monga, Y.P., Ray, U.S., 2007. Improvement of glutathione and total antioxidant status with yoga. J. Altern. Complem. Med. 13 (10), 1085–1090.

Sinha, B., Sinha, T.D., Pathak, A., Tomer, O.S., 2014. Effects of yoga training on blood pressure response during surya namaskar following 11 months of yoga practice in army men and yoga-trained individuals. Int. J. Clin. Exp. Physiol. 1 (1), 51–56.

Stone, M., 2009. Yoga for a World out of Balance: Teachings on Ethics and Social Action. Shambhala Publications, Boston, MA.

Telles, S., Naveen, K.V., Dash, M., 2007. Yoga reduces symptoms of distress in tsunami survivors in the Andaman Islands. Evid. Based Complement. Alternat. Med. 4 (4), 503–509.

Telles, S., Patra, S., Montesoo, S., Naveen, K.V., 2008. Effect of yoga on somatic indicators of stress in healthy volunteers. J. Indian Psychol. 26 (1–2), 52–57.

Telles, S., Naveen, V.K., Balkrishna, A., 2010a. Serum leptin, cholesterol, and blood glucose levels in diabetics following a yoga and diet change program. Med. Sci. Monit. 16 (3), LE4–LE5.

Telles, S., Singh, N., Joshi, M., Balkrishna, A., 2010b. Post-traumatic stress symptoms and heart rate variability in Bihar flood survivors following yoga: a randomized controlled study. BMC Psychiatry 10 (1), 18.

Vedamurthachar, A., Janakiramaiah, N., Hegde, J.M., Shetty, T.K., Subbakrishna, D.K., Sureshbabu, S.V., Gangadhar, B.N., 2006. Antidepressant efficacy and hormonal effects of sudarshana kriya yoga (SKY) in alcohol-dependent individuals. J. Affect. Disord. 94 (1), 249–253.

Vyas, R., Raval, K.V., Dikshit, N., 2008. Effect of raja yoga meditation on the lipid profile of postmenopausal women. Indian J. Physiol. Pharmacol. 52 (4), 420–424.

Wolsko, P.M., Eisenberg, D.M., Davis, R.B., Phillips, R.S., 2004. Use of mind–body medical therapies. J. Gen. Intern. Med. 19 (1), 43–50.

Wood, C., 1993. Mood change and perceptions of vitality: a comparison of the effects of relaxation, visualization, and yoga. J. R. Soc. Med. 86 (5), 254–258.

Yoshihara, K., Hiramoto, T., Sudo, N., Kubo, C., 2011. Profile of mood states and stress-related biochemical indices in long-term yoga practitioners. BioPsychoSocial Med. 5 (1), 1–8.

Further Reading

Bhavanani, A.B., Sanjay, Z., Madanmohan, 2011. Immediate effect of sukha pranayama on cardiovascular variables in patients of hypertension. Int. J. Yoga Ther. 21 (1), 73–76.

Mondal, K., Majumdar, D., Pramanik, A., Chatterjee, S., Darmora, M., Majumdar, D., 2017. Application of yoga as an effective tool for improving postural balance in healthy young Indian adults. Int. J. Chinese Med. 1 (2), 62–69.

Index

Note: Page numbers followed by "*f*" indicate figures and "*t*" indicate tables.